Biodynamics

Biodynamics

Why the Wirewalker Doesn't Fall

Bruce J. West, Ph.D.
Lori A. Griffin, Ph.D.

WILEY-LISS

A JOHN WILEY & SONS, INC., PUBLICATION

Library of Congress Cataloging-in-Publication Data:

West, Bruce J.
   Biodynamics : why the wirewalker doesn't fall / Bruce J. West, Lori A.
Griffin.
     p. cm.
   ISBN 0-471-34619-5 (cloth)
   1. Biophysics.   2. Dynamics.   3. Biomechanics.   I. Griffin, Lori.   II.
Title.
   QH506.W475 2004
   571.4—dc21   2003009325

Printed in the United States of America.

10 9 8 7 6 5 4 3 2 1

Contents

Part III Fields—How Complex Systems Move

Part IV Data Analysis—
What We Can Learn from Measurements

Preface

In this monograph, we attempt to communicate to the reader a sense of the excitement and mystery associated with the processes of learning and scientific discovery. We have taken what is often a dry and dusty discipline and washed away the stultifying influence of scholars, to produce a monograph that can be read with pleasure by any student of science. Our purpose is to present a holistic picture of human movement and locomotion; one that pulls together the equally arcane disciplines of mechanics, thermodynamics, information propagation, and the statistics of time series. The final result is not your usual textbook, although there are diagrams, problems, and worked-out examples. Instead the book has many characteristics usually associated with a research monograph: speculations, arguments that are meant to provoke response and stimulate thinking, and connections made to modern research discoveries that are usually reserved for advanced texts. All this is done with a minimum of mathematical formalism.

All too often it appears that one is forced to choose between two kinds of texts which discuss the human body and locomotion. On one side are those books that present a turgid collection of discrete and continuous equations, bracketing definitions and proofs, on page after lifeless page. These are usually written with a view towards getting the mathematics right and an almost complete disregard for language and the spirit of science discovery. On the other side are those well-written texts that present superficial popularizations of the science and which dilute any discussion of fundamental mechanisms and avoid mathematics altogether. These latter books leave the reader with little formalism on which to build. Neither of these two approaches communicates well to the modern student who wants substance, but who may not have the traditional mathematical background to take in undigested information. *Wirewalker* is aimed toward that curious student, who has heard about the recent discoveries in biology and medicine, but never thought they could penetrate the mathematics barrier.

What makes the present manuscript different from the handful of somewhat similar books, is that we adopt a holistic approach to science, using scaling methods that are simply explained and which reduce the mathematics to its simplest possible form. This approach allows us to discuss the fundamental mechanisms that control human movement and locomotion on land and in water, without losing the reader in a labyrinth of mathematical detail. To maximize communication and minimize anxiety, we found it useful to adopt an informal, almost conversational style in the manuscript, without sacrificing the clarity necessary to explain, understand, and use such concepts as fractals, chaos and randomness involved in kinesis. Therefore our book has the information content of a textbook, but without the pedantic trappings. *Wirewalker is* intended for the intellectually curious person who wants to understand how the fast and slow movements of a wire walker, balanced high above the sawdust, are related to human movement and locomotion in general, but who lacks the mathematical background to solve dynamical equations, or who, with the passage of time, has forgotten how to solve such equations.

Is the beating of the human heart really regular, or is it chaotic? Is health quantifiable? Why do the elderly fall for no apparent reason? These and many other questions are addressed here from the perspective that the human body is a complex system, interacting in a dynamical fashion with an intricate and often unknowable environment. Simple walking, breathing, and the beating of the heart, are examples of the kinds of entangled fractal networks encountered in the human body. We discuss ways of extracting and processing information from monitors of these systems, using nonlinear dynamical techniques, and explore valuable implications of these diagnostics for students and practitioners of medicine, biology, and biophysics. And yes, we do supply answers to the questions posed above as well as to other related queries.

A number of people have assisted us in arriving at the present format of *Wirewalker.* Eric Eason and Denise Camera helped us to: Explain things in an understandable manner, develop problem sets, and identify potential sources of embarrassment that we should and did eliminate from the book. Other individuals such as Heather Fredrick, Ryan Murray, Patrick Ryan, and Joseph Tranquillo read early versions of the manuscript and suggested interesting questions for the student as well as editorial changes. Rebecca Dotson, Christopher Wilkerson and Jennifer West provided guidance and instruction on the computer graphics. We thank each and every one of them for their contributions, both large and small, and in particular we thank Dr. Richard Moon for sharing the data on which Chapter 18 is based.

Finally, Bruce West thanks his wife Sharon West, for her tolerance and good humor, as well as for the artwork used throughout the book. Lori Griffin thanks Giulio Bloise for sharing an adventure to North Carolina in order for her to participate in this project.

BRUCE J. WEST
LORI A. GRIFFIN

Durham, North Carolina
August 2003

Biodynamics

PART I

INTRODUCTION TO SCIENCE

What we have attempted to do in the first part of the book is provide a perspective for the student that is different from what is traditionally found in science courses. The perspective we promote is that no knowledge is irrelevant in the study of science, and that science is concerned with learning how to think about the world in a way that will facilitate our understanding of the world. Science is a way of knowing but, more than that, science is a way of testing what we think we know.

You never enter the same river twice, said an Eastern philosopher, emphasizing that the river is not a thing, but a continually changing dynamical structure that is never the same from one entry to another. All dynamical systems share this property of uniqueness in time. Humans never do anything in exactly the same way twice, no matter how repetitious an activity; the no-hitter, the four minute mile, or the flawless dive are all the result of continual practice. Most of all, athletes train for consistency of outcome, but no matter how hard they train, there is always variability in their performance. Their goal is to achieve an average level of play that far exceeds that of the competition, and to reduce the variation around this level of play.

The contrast between average behavior and the variation around that average is magnified in the professional athlete. When Michael Jordan did not make more than 30 points in a game, the fans were disappointed. Jordan's average level of play was so far above that of anyone else on the court that spectators expected him to do better than average all the time. But this, of course, is a contradiction, since an average implies that sometimes you do better and sometimes you do worse. In the same way, not every stock suggested by your broker makes money, not every painting by Picasso is a masterpiece, and not every movie made by Orson Wells wrenches the heart. In some of these situations, we think that we might understand the average, but we almost never understand the fluctuations. In part, we do not understand the fluctuations because there have been very few systematic studies of variability. It is

Biodynamics: Why the Wirewalker Doesn't Fall. By Bruce J. West and Lori A. Griffin
ISBN 0-471-34619-5 © 2004 John Wiley & Sons, Inc.

the average behavior that is the more important, isn't it? The answer to this question is an emphatic "no."

In the engineering sciences, the dual character of time series measurements is identified with signal and noise. The signal is taken as the smooth, continuous, predictable, large-scale motion in a process. The notions of signal and predictability go together, in that signal implies information, and very often the mechanical interpretation of information has to do with our ability to associate that information with mechanistic, predictable processes in the system. Noise, on the other hand, is typically discontinuous, small-scale, erratic motion that is seen to disrupt the signal. The noise is assumed, by its nature, to contain no information about the system, but rather to be a manifestation of the influence of the unknown and uncontrollable environment on the system's dynamics. Noise is considered undesirable and is filtered out of time series whenever possible.

The signal-plus-noise paradigm constructed by engineers and used by scientists to model human locomotion is replaced in this book with the paradigm of the high wirewalker. In the circus, high above the crowd and without a net, the tightrope walker (wirewalker) carries out smooth average motions plus rapid erratic changes of position, just as in the signal-plus-noise paradigm. However, in the case of the wirewalker, the rapid changes in position are part of the walker's dynamical balance. Far from being noise, these apparently erratic changes in position serve to keep the wirewalker's center of gravity above the wire, so he/she does not fall. Thus, both aspects of the time series for the wirewalker's position constitute the signal (both the smooth long-time part and the discontinuous short-time part), and contain information about the wirewalker's dynamics. If we are to understand how the wirewalker retains his or her balance on the wire, we must analyze the fluctuations, that is, the wirewalker's fine tuning to losses of balance.

This, then, is our approach to understanding human locomotion. First we develop the physical laws and principles necessary to provide a theoretical context for the large-scale, smooth aspects of locomotion. Then, rather than attributing the observed fluctuations around this smooth, or average, behavior to noise, we analyze these fluctuations for information content, using ideas from dimensional analysis, nonlinear dynamics, and scaling. We find that these fluctuations do, like the movements of the wirewalker, maintain the delicate balance necessary for a physiological system to retain function and contribute to the phenomenon of biodynamics.

CHAPTER 1

Ways of Knowing

Objectives for Chapter One

- Understand the difference between the mathematical model of a phenomenon and the properties of reality that the mathematics are attempting to model.
- See that science is not absolutely certain with regard to the outcome of individual experiments. Science is statistical in nature and allows for the quantification of errors.
- Microscopic and macroscopic processes are not independent of one another and, consequently, the seen and the unseen influence each other all the time.
- It is not always possible to predict the outcome of experiments, even when we know the equations that determine the motion of a system. Deterministic equations do not always predetermine the result.
- The human body is a complex, open, dynamical system that should not be described in terms of averages. Traditional averages like heart rate, breathing rate, and metabolic rate are less important than the fluctuations around these average values.

Introduction

Overview

In this first chapter we introduce the image of a wirewalker to sharpen the dynamical concepts we explain and use in the remainder of the book. The goal is to make clear that one does not need a great deal of mathematical formalism to understand the exciting new concepts in science as they apply to human locomotion. We show that the relationship between our everyday understanding of heat and the scientist's

Biodynamics: Why the Wirewalker Doesn't Fall. By Bruce J. West and Lori A. Griffin
ISBN 0-471-34619-5 © 2004 John Wiley & Sons, Inc.

understanding of random atomic motion are closely related, in order to emphasize this point. This demonstration also highlights the fact that all of science is statistical in nature, which means that the predictions that scientists so loudly proclaim can only be made about average, not instantaneous, properties of systems. The variability of phenomena is unpredictable and yet it is this variability that we eventually show to be the ultimate indicator of the health of a system.

The Physicist and the Poet

The difference between an average value and fluctuations suggests an important distinction between ways of obtaining knowledge and what we consider knowledge to be. The humanist and the scientist appear to view the world in incompatible ways, but we argue that the methods used in each area are not so different. The traditional idea that science is certain, with which the humanist has been repeatedly clubbed, is shown to be false using the dynamical concept of chaos. We argue that both the humanist and scientist are working to penetrate the mist in which the complexity of the world is shrouded, and each hopes to provide a coherent picture of who and what we are. Herein, we use the tools of science to understand why the wirewalker, moving on a slender thread, high above the crowd, does not fall. If the explanation as to why the wirewalker does not fall should also provide a metaphor for a greater understanding of our world, so much the better.

1.1 HUMAN LOCOMOTION IN PERSPECTIVE

Biodynamics Is a Synthesis of Many Sciences

Biodynamics brings together elements from the physical and life sciences, incorporating aspects of physics, chemistry, physiology, and mathematics. As pointed out by Winter [1] regarding biomechanics, we can find this blending of disciplines in the writings of Leonardo da Vinci and Galileo Galilei, and subsequently in the studies of Lagrange, Bernoulli, Euler, and Young.* These names will become familiar to the reader, if they are not already. The definition of biodynamics adopted here is broader than the mere application of the laws of mechanics and dynamics to living systems, even though such applications are included. In order for the student to gain insight into the true complexity of biodynamics, we include in our discussion other areas of study, such as thermodynamics, hydrodynamics, and statistics. Our purpose is to synthesize those aspects of the various disciplines necessary to understand the locomotion of the human body in a variety of environments.

Is Locomotion Complicated?

We do not restrict our considerations to the mechanical aspects of biodynamics. Although it is true that one can view the combination of muscles, ligaments, and the skeletal structure as if it were a mechanical system used for locomotion, there is much more to the story. Activating muscles requires energy. This energy is contained in the food we eat. The body metabolizes food and converts it to a form that

---

*See the Appendix for minibiographies of the most important scientists in the various fields under discussion.

enables muscles to carry out their various tasks. For the body to utilize energy, the heart transports the oxygen from the lungs using the blood that it pumps through the body. Herein, we lay the foundation for understanding how the various contributions to biodynamics act separately and together to keep us alive, healthy, and mobile.

Mechanics versus Dynamics

There is often some confusion between the familiar term biomechanics and the more general term biodynamics. For example, Hay [2] introduces students of sports medicine and athletic competition to what is known about human locomotion. This body of knowledge was formerly known as kinesiology, but subsequently outgrew that name from the perspective of the discipline, contributed to our understanding of human movement, and became biomechanics. Here we further extend the scope of that analysis to biodynamics. On the other hand, Winter's book [1] on biomechanics focuses on measurement and interpretation. The former category contains data acquisition, including what we can measure, what equipment we use, and what techniques have been developed to process the data. The latter category includes mechanics, the branch of physics dealing with both statics and the motion of mass as it applies to biological systems.

Who Is this Book For?

It is important to recognize that each research group using biomechanics has a different set of questions they wish to have answered:

Athletes and coaches question how to enhance performance.

Physicians and physical therapists question how to reduce pathologies and overcome the effects of trauma.

Physical scientists question how to establish the principles necessary to understand the human body from a general biophysical perspective.

The difficulty has always been that the various groups do not share a common language and consequently cannot and do not talk to one another. In this book, we address the communications problem, with a view toward making powerful research models accessible to a broad spectrum of users. In particular, we shall eventually come to the phenomena of human gait, heartbeat, and breathing, to which we apply the methodologies developed.

Knowledge Is Not Unlimited

The question of how we gain and use knowledge to understand the world and our role in it has many answers. The environmentalist might say that we are part of a vast ecological system and because we are sentient we have a responsibility to use our knowledge for maintaining rather than destroying the environment. A theologian might say that we are part of God's creation and we should use our knowledge for the glory of the creator. There are, indeed, many ways to know the world and we briefly discuss the difference between what passes for information in the sciences and in the humanities. We demonstrate that although science is experimentally based, it shares with the humanities a kind of uncertainty in what we can ultimately know about the world and about ourselves in it.

Coupling of Various Scales

One of the fundamental properties of the human body is that it requires a relatively constant temperature in order to function. We must probe rather deeply into the microscopic nature of matter in order to understand the temperature regulation process. This delving into the nature of heat brings up all kinds of issues: predictability, randomness, and complexity, to name a few. The understanding of heat and temperature compel us to examine how processes at the microscopic level are tied to what happens at the macroscopic level. A simple example of this is how the molecules of aspirin relieve a headache; events on small scales have indirect, large-scale influences.

Measures of Various Scales

Science is an experimental activity, whether it is physics, biology, or any of the other scientific disciplines. But to do an experiment, the scientist must have a clear idea of what is to be measured, and this implies having a metric of one kind or another. These metrics are expressed in terms of certain fundamental units. The fact that fundamental *units* cannot be defined, but only determined experientially through experiments, requires a certain amount of discussion. Time, length, and mass are three of the fundamental units of the physical world. We have experience with each of them, but they are still worthy of closer inspection because much of what we have learned regarding their properties is not true.

The Physics of Biomechanics

The physical basis of all of biomechanics rests on the three pillars of Newton's laws of motion. Therefore, any introduction to biodynamics would be incomplete without a discussion of how these laws fit into the scheme of locomotion. We give a truncated historical discussion of the laws to set the stage for future development and to show that much can be understood without the mathematics.

Dimensional Analysis

We are able to avoid the mathematics of Newton's laws, in large part, by introducing the method of *dimensional analysis*. One way to define this method is by recognizing that all physical observables have units, so that phenomena that are determined by various combinations of physical observables also have units (dimensions). The units (dimensions) on one side of an equation representing a physical process must equal the units (dimensions) on the other side of the equation. If there are two factors of length in the product on the left-hand side of an equation, say an area, then there must be two factors of length in the products on the right-hand side, say a length times a width. From this simple observation, we are able to construct descriptions of rather complicated events without solving Newton's equations of motion. Dimensional analysis is a new way for nonspecialists to think about phenomena, and its mastery would be time well spent.

Gravity Is the Great Unseen Force

The force of gravity is discussed in some detail, in part, because it is ubiquitous, but also because the understanding of it was the first of the great intellectual achievements of the physical sciences. Newton's law of gravity stated that the same dynamics that applies to humans on the earth's surface applies with equal validity on the moon, sun, and stars. Finally, it is gravity that, in large part, determines how and why we walk and run the way we do.

Question 1.1

Describe in your own words a phenomenon in which you might observe "signal and noise." Identify each part and explain why it might be important to be conscious of both parts, and to know which is which.

Question 1.2

Do you agree that the physicist and the poet actually use similar methods to view the world? This is not a yes–no question; you must justify your answer.

Question 1.3

What are some of the other sciences that our study of biodynamics will include? Give a brief description of the science, not just its name.

Physics Is Background

1.2 WHAT IS BIODYNAMICS?

Success has a hundred fathers, whereas failure is an orphan. In the same way, biodynamics has a hundred definitions, each one stressing a different aspect of our attempt to understand the nature of mechanical behavior of motion in biological systems. Here we take biodynamics to mean the study of physical laws, usually mechanical laws, and their application to movement of and in living organisms. In particular, we are interested in the application of biodynamical insights to the locomotor systems of the human body. However, because relative physiology often helps us to better understand how the human body works, we occasionally consider other animals as well.

Engineering Is Infrastructure

Biodynamics, as a science, is useful in the design and engineering of everything from footwear to safety belts. As an art, it is indispensable to physical therapists. Coaches find it to be extremely practical, and physicians treat it as one of the many subdisciplines that they need to master on the way to getting a medical degree. In engineering departments, biodynamics (often taught under the rubric biomechanics) is taught in terms of mechanics, with vectors, statics of rigid bodies, dynamics of rigid bodies, and strength of material, along with a variety of other disciplines. Biodynamics in engineering is an abstract discipline involving significant mathematical modeling. One might say the goal of the engineer is to construct a formal mathematical model of movement and thereby understand human locomotion.

Sport Is One Application

Kinesiology departments, where most coaches and physical therapists receive their fundamental training, focus less on the mathematics and basic physics and more on the application of the models, even though those models have been developed in other disciplines. For example, the understanding of human gait is often based on the physical properties of a pendulum, or the assumption of regularity in stride. In the latter case, the fluctuations in the stride interval are smoothed since the variability in these intervals are random and thought to be due to noise. Kinesiology focuses less on the theoretical physical mechanisms, say those producing the fluctuations in human gait, than it does on the practical implications and applications of what we know experimentally. The goal of kinesiology might be summarized as describing and predicting ways to empirically improve performance in walking, running, and jumping.

Medicine Is Another Application

Next, there is the way in which the physician understands biodynamics—the movement of the human body as the operation of a collection of physiological organs. A general practitioner might consider each organ as a subsystem in a larger complex system that interacts with many, if not all, of the other subsystems (organs). On the other hand, the specialist might see a particular organ and its interconnections as a system (for example, the cardiovascular system to the cardiologist). This person would be primarily concerned with the subsystems of this specific system, these being the heart, arteries, veins, and the sympathetic and parasympathetic nervous systems, and how these all work together to constitute a healthy cardiovascular system. How one divides the body into systems and subsystems is a matter of convenience, dependent on the questions one chooses to ask. For example, one might wish to determine if it is safe for a rehabilitated person to walk without a walker, or to determine if walking has deteriorated due to disease.

Other Applications

Of course, these separate perspectives of biodynamics are not as simplistic as indicated here. The physical therapist learns some physiology, the physician learns some engineering, and the coach learns some mathematics. The point of this discussion is that all of these various disciplines, as well as others, use and need to understand biodynamics. However, there is no text that teaches the physics needed by the therapist without the mathematics of the engineer. There is no monograph that makes clear what the implications of the mathematical models are without burdening the physician with unnecessary mathematical detail. There is no set of lecture notes that spans the full range of biodynamics in such a way that the engineers do not feel their time has been wasted in taking the course. In this book, we attempt to satisfy these various groups, with their different interests, questions, and motivations, and to do so in an interesting way.

1.2.1 No Mathematics?

The Map is Not the Territory

Of course, the reader is free to ask how we intend to teach physics and the models used in biodynamics without subjecting the reader to the mathematics of engineering courses. We start by noting that biodynamics, like other sciences, is experimental in nature, and that a mathematical model is a description of reality and is not the reality itself. This means that one can eventually resolve, by means of experiment, all disagreements as to matters of truth, independently of the logic of one's arguments. Mathematics is, after all, the logic of science.

Summary of Experimental Facts

Scientists often regard a mathematical equation as the direct summary of vast amounts of experimental data, in which case it is an empirical relation. An example of such an empirical relation is Newton's law of cooling, which states that the rate at which an object cools down is directly proportional to the difference between the temperature of the object and the temperature of the surrounding air. Thus, the harder you work out in the gym, the hotter you get, and the more rapidly you cool down when you stop exercising. That is why it is a good idea to put on a light jacket between workouts. The jacket slows the rate of cooling by not allowing the heated air immediately adjacent to the skin to escape and be replaced with cooler air.

Mathematical Models Are the Only Way

Folk wisdom provides another example of summarizing experimental information. One can bring a high fever under control by putting the suffering individual into an ice bath. The large temperature difference will quickly lower the fever by dissipating the generated heat. This reduction of the body temperature keeps the fever from doing damage to microscopic processes and objects, such as to the proteins and their interactions in the brain.

Definitions of Relationships

One can also use mathematical equations to define the relationships among generic quantities, variables that capture the essential nature of the phenomenon being measured. Concepts like force and mass do not depend on any single cluster of experiments, but are generic concepts, useful in any experiment involving physical bodies. All physical bodies have mass. All bodies influence one another through forces. Therefore, biodynamics will, at least, depend on these concepts. In any event, we shall find that mass and force have a mathematical relation that allows one concept, say mass, to be determined by experiment. We then consider mass to be fundamental, and determine the other concept (force) by the mathematical relation (thus making force a derived, or less fundamental, concept). In some sense, scientists consider the result of experiment to be more fundamental than the result of theory. An experimental quantity is seen to have more reality, or more truth, than the corresponding theoretical quantity.

Reasoning and Not Facts Lead to Errors

People have used facts to support invalid theories or to come to false conclusions, but that does not invalidate the facts. Given a certain set of facts and a properly reasoned argument, if a false conclusion is reached then the theory on which the argument is based is flawed in some way, not the facts. Facts are given by nature and reason is given by human beings, so if one encounters a contradiction, first look to the human side of the equation for a resolution. Of course, sometimes a given set of facts may fail to support a valid theory. However, the failure to support a valid theory may not be because the facts are untrue, but may be because the facts are incomplete, or inaccurately presented, or have an unacceptable amount of error.

Truth Is Observable

Therefore, what we mean by *truth* in the present context is not some lofty ideal to which we aspire, but rather the dirty reality of experimental facts. To say that something is a true fact is redundant, but to say that something is an untrue fact is a contradiction in terms. There can be no such thing as an untrue fact. All facts are, by definition, true, in the sense that they are, or have been, observed in the world. Thus, when facts conflict, look for a flaw in the experiment or for a fallacy in the presentation. Facts are facts; all else is deception.

1.2.2   Thermodynamic Phenomenology

Second Law of Thermodynamics

Let us consider one of the kinds of facts we have been discussing. It is an experimental fact that two isolated objects at different temperatures, when placed in direct physical contact with one another, will both come to the same temperature after some period of time. Further, this new temperature will lie between the two initial temperatures the objects had when they were isolated from one another. If we do this experiment once, we can report the result as a specific fact. If, however, we do this experiment a large number of times, with many different-shaped objects, with different materials, and at different temperatures, and always get the same general result, then we might say that nature tends to be this way. Or, driven by our success, we might say that this fact is not just true of the experiments we have done, but true of all experiments independently of where and when they are done. But when such a statement is consistent with all possible experiments we can conceive of, we have what we call a law of nature, which is the case of the inequality of temperatures. The law in the case of temperature differences is called the second law of thermodynamics—a rather imposing name for a relatively simple observation.

What Is Temperature?

The framework for physical arguments, such as those leading to the second law of thermodynamics, is called *phenomenology,* in which phenomena are explained by relationships that have been uncovered by means of experiments. We shall use the experiments of biodynamics to guide us in the formation of general laws, which, like the second law of thermodynamics, do not require mathematics for its understanding, that is, a mathematical formalism is not required for the understanding of phenomena at a fairly intuitive level. Of course, one might ask more penetrating questions such as: What do we mean by the temperature that is measured in these experiments? Why should the temperature be lowered in the object with the higher temperature and why should it be raised in the object at the lower temperature? The level of insight required to answer these questions requires a somewhat deeper understanding of the physical world, and it is this deeper level we explore in this book.*

The Human Body Is an Open System

We are interested in phenomenology because the human body is an open system and it is open in at least three distinct ways. The first way the body is open is thermodynamically—the body exchanges energy with its environment. We exchange energy in the form of heat, sweating in the sun or cooling in the shade. The second way the body is open is that the body is affected in an overall manner through external forces, such as gravity, acting on it. The unforced, or free, situation in any system is one in which every force in the system is internal and in balance, so there is no net movement. In the human body, there are a myriad of internal forces, including muscles pulling on bones, the contraction of the heart muscles, the motion of oxygen across the membrane of the lungs, and so on. Finally, the body is open to the flux of things through it, for example, material (food and waste) and energy (fuel). Here *flux* means the rate at which a quantity of stuff changes per unit of time, for example, the rate at which food is ingested, metabolized, transformed into waste, and ejected from the body.

Different Disciplines Dovetail

We should also emphasize that in biology it is difficult, if not impossible, to separate biodynamics from thermodynamics. This is a consequence of the fact that in order to move, the body "generates" its own energy, by metabolizing food, and releases waste (energy it cannot use) in many forms. One of the waste products is heat. Therefore, to understand the thermal regulation of the body requires an understanding of thermo-

---

*For those that cannot wait for an explanation, the Second Law of Thermodynamics is based on the interpretation of temperature in terms of the average kinetic energy of the microscopic particles that make up the material. The higher the kinetic energy, the faster the particles move. At the interface of the two bodies, particles of the two kinds are colliding with one another. The hotter particles are giving up momentum and energy to the cooler particles, thereby heating them up. It is this transfer of kinetic energy that cools down the hotter body and heats up the cooler body. When the average kinetic energy of the two bodies are the same, the same amount of energy is transferred both to and from each body, yielding no net transfer of energy. The two bodies are then said to be in equilibrium and to have the same temperature.

dynamics, which is the change in heat over time. Furthermore, the foundations of thermodynamics are the mechanical laws of motion for the microscopic particles that make up all physical objects, including the human body. Therefore the same laws that determine how we run and jump also determine why we sweat when we exercise.

Two Kinds of Theories

In physics, there are two kinds of theories: the fundamental theory that is based on one or more universal principles that generate all the predictions of interest, and the phenomenological theory that is based on one or more universal observations that generate all the predictions of interest. An example of a fundamental theory is the motion of collections of particles that can be determined from the universal principle of the conservation of energy. We shall talk a great deal about such things in mechanics, dynamics, biomechanics, and biodynamics. An example of a phenomenological theory is the communication inside the body by means of electrical impulses that can be understood from the application of the fundamental principles of electricity and chemistry, but which is not in itself fundamental. We shall also have occasion to discuss heat and thermodynamics in the latter context, but, for the moment, let us continue our discussion of just the facts.

Science and Humanities Are Separate

It is this reliance on experimental information, facts, and quantifiable uncertainty that has been used historically to separate the sciences from the humanities. Some individuals are attracted to the dependability of facts in science, whereas others find the concrete aspects of scientific investigation to be too confining and long for the freedom of artistic creativity. However, the artist, like the scientist, is constrained by his/her medium of expression.

Question 1.4

What is the definition of biodynamics? How is it different from biomechanics?

Question 1.5

Explain why mathematics is often necessary to describe physical phenomena, but very often is not used to describe biological phenomena.

Question 1.6

Why is thermodynamics necessary for understanding biological phenomena?

1.3 KINDS OF KNOWING

Two cultures

C. P. Snow, famous author and scientist, argued that we live in two separate, distinct, and nonoverlapping cultures [3]. Science and mathematics form one of these cultures and the humanities and arts form the other. He believed that the members of these two cultures did not communicate with one another because they spoke different languages. It is our belief that the differences between the sciences and the humanities are more apparent than real. For example, in literature one often proclaims one piece of work to be exceptional and another to be trite, based on the consistency of the actions of the characters, or the development of situations, or the loving way in which the author creates word pictures. A character who acts inconsistently is not believable. To resolve a difficult literary situation, the writer develops an implied psychological profile of the individual, which, although not overt in the character's actions, does provide a background explanation for the behavior. For example, the serial killer who is the youngest member of a large family may have a father who is hard-working, alcoholic, and abusive; this, although not pinpointing the explanation for the killer's behavior, provides sufficient background to develop a rationale for the behavior. People love to have things explained, or rather they love for things to have an explanation, whether they understand the connection between the event and the explanation or not.

Kinematics Is the Study of Motion

With regard to the desirability of an explanation, the science of biodynamics is the same. We attempt to describe or explain the actions and behaviors of people regardless of whether we understand the "why" or the "cause" of the action or behavior. For example, the body of knowledge called *kinematics* falls into this category. This is the study of the branch of *mechanics* dealing with the motion of systems of material particles without reference to the forces that act on the systems. We can describe the motion of the air in this room if we specify the location and the velocity of each and every one of the air particles at a given time. The future motions of particles are determined by their present positions and velocities, so that given the latter we can predict the former. In a complex system such as the body, there are so many constraints that we cannot predict the motion of individual particles. Rather, we predict the motion of aggregates of particles such as the arm or leg. How one constructs such models of large numbers of particles will be taken up in due course.

Different Levels of Knowledge

The same is true of more complex objects, say the particles that make up the runner in the 100 meter sprint. At one level, you can describe kinematically the actions of the runner and make satisfactory predictions based on his stride length if you are, say, the coach of the runner. If you are a dietician, in charge of what the runner eats, you will have to know more than what you can learn kinematically, say, what the individual's metabolic rate is. Further, if you are the team doctor, you will have to know more still. How satisfactory the level of understanding is determined by how you intend to use the knowledge.

We Experiment All the Time

We evaluate the behavior of literary characters using our own experience as the basis of comparison and later judge if the personality extensions made by the author are believable or not. Would we have been as heroic, as romantic, as tenacious, or as devious and cruel as the protagonist? Perhaps we may even adopt some of the characteristics of the protagonist in our everyday life, try them out, and see if they fit. The rationality of Conan Doyle's Sherlock Holmes, the fatalism of the characters of Albert Camus, and the mysticism of Herman Hesse are all affectations with which one has burdened one's friends at some point. This modification of one's personality to accommodate an attractive feature from something we have read, or seen in a movie, is an experiment. Not as precise or as quantitative as those done in the physical sciences, but an experiment nonetheless. One of the differences between this type of experiment and those done in physics is that it is probably not reproducible, which is to say that another person who attempts the same experiment will achieve a totally different result. There is little or no predictability of outcome in these kinds of social experiments.

Experimental Results Are Variable

There has been an avalanche of "how-to" or "self-help" books in the past 20 years or so. These books are predicated on the idea that social experiments, such as "confronting one's fears," "being assertive," "sharing from the child within," should, more or less, yield the same results for everyone. But this is often not the case. The variety of outcomes is due partly to the fact that social situations are so complicated that we fail to see those aspects that differ from case to case. In fact, it is typical that the same person attempting the same social experiment, such as asking for a date or attempting to be amusing within a group, may yield exactly opposite results on two successive tries. In part, this is the result of not being able to control the feelings and actions of others in the same way one can control the outcome of an experiment in the laboratory. It is believed that this does not happen in the physical sciences. It is thought that one obtains the same result for the same experiment over and over and over again. This assumed predictability and reproducibility of outcome may be one of the reasons that certain introverted individuals find the sciences so attractive.

1.3.1   Science Is Statistical

Uncertainty, Chance, and Error

However, the above image of the predictability of science is not accurate. What one finds in the real world is that the results of two experiments are never exactly the same. No matter how sophisticated the apparatus, no matter how well designed the experiment, no matter how good the method, there are differences in experimental results. Two hundred years ago these differences were attributed to experimental error, that is, to nonsystematic differences in the experimental outcome due to minute changes in the initial conditions over which the experimenter has no control. The paradigm for such fluctuations in experimental results was the *Law of Frequency of Errors*. This law states that the random changes in the outcomes of well-designed experiments are guided by the mean value, the most probable outcome of an experiment, and the fluctuations around this mean value. The errors or fluctuations that are large occur much less frequently than those that are small. In fact, if the fluctuations in outcome from experiment to experiment are independent of one another, and the basic character of the experiment is the same from case to case, then the errors form a bell-shaped curve, peaked at the mean value of the experimental outcome. This bell-shaped curve persists even after we have worked to suppress every imaginable source of error in the experiment. No matter how sharp the result, if we look more carefully we find this law to be lurking in the shadows. Thus, the bad news is that no set of experiments yield the same results. The good news is that the changes from experiment to experiment, although random, follow a law, a law that we reexamine from time to time.

What is Predictability?

So what is it that scientists mean when they say the result of an experiment is *predictable*? They are referring to certain average properties of the phenomena. When we say our heart rate is 60 beats per minute, or our body temperature is 98.6° Fahrenheit, or any other number we associate with the human body, we are talking about average values. These averages are the established norms that are the result of hundreds and sometimes thousands of years of data that we have accumulated and averaged. One level of understanding is to predict these average values and how they change under different conditions of the system and of the environment. A deeper level of understanding is achieved if we can predict the distribution of fluctuations in addition to the average values, and how this distribution changes with changes in the system and the environment. This relatively harmless-looking statement is at the heart of everything we have to say. We maintain that there is as much, and often more, information contained in the fluctuations of measurements, which is usually ignored, than there is in average values. Further, if we are sufficiently clever we can tease this information out of the data and use it to understand such complex phenomena as human locomotion.

Nothing Is Exactly Reproducible

One may object to such sweeping generalizations and point out that social scientists do, in fact, make predictions about social experiments; and to the extent that the experiments are reproducible, they may be said to be scientific. Of course, this is not the case with complex phenomena, whether social or physical, about which we do not obtain, nor do we expect to obtain, the same result with each experiment. Rather, we expect that the result will vary from experiment to experiment. The resulting change may not be dramatic, but it is usually sufficient to show that there is no one clear answer, and so an ensemble of experiments must be performed. In these cases, it is the measured quantity, averaged over the ensemble of experiments, that is reproducible. We shall have more than ample time to stress the scientific study of such phenomena, but, for the moment, let us retain our focus on the comparison between the liberal arts and biodynamics.

What Does "Scientific" Mean?

It may be of some value to emphasize what is required for an experiment to be considered *scientific*. First, the experiment must involve a phenomenon that is *regular,* which is to say, there is a pattern that can be identified uniquely with the phenomenon. Second, the phenomenon must be one that can be *regulated;* that is, the scientist must have certain parameters that are under his/her control. A change in value of a parameter will then yield a discernible change in the experimental output. Traditionally, this dependence has been thought to be linear, so that the change in output is proportional to the change in the parameter value. Finally, the experiment must be *repeatable,* if not exactly, then at least on average. It is this *approximate* repeatability that allows for the generation of an ensemble of experiments.

Some Things Are Observed but Not Controlled

Of course, there are other kinds of scientific activity, such as the observation of waves on the ocean surface or the behavior of large groups of people. It is not possible to do a laboratory experiment on ocean waves, but it is possible to observe the properties of such waves and to then compare those properties with waves that can be generated in a wind tank in a laboratory. When the observed properties of the uncontrolled ocean waves can be explained in terms of the regulated waves in the wind tank, we think we understand the former in terms of the latter. This is typically how we scale from what we learn in the laboratory to what we observe, but cannot regulate, in the real world. The laboratory is our microcosm of reality, much as a good novel provides insight into the human condition.

1.3.2 Conflict Resolution

Experiment Is the Great Arbitrator

A literary controversy may be based on two completely opposite interpretations of a character's motivation, and conducting an experiment may do nothing to resolve the conflict. However, a controversy in biodynamics may always be resolved by experiment because every situation is unique and, in principle, can only be produced by a "single and unique set of circumstances." Assuming, of course, that the results are in fact distinguishable, they are assumed to be outside the possible statistical fluctuations of the experiment.

Make a Guess and Test It

The experimental approach to understanding the world is strictly scientific. It makes Nature the ultimate arbitrator of all disputes and there is no Court of Appeal. We formulate a hypothesis that is useful in describing a given phenomenon and then design an experiment to test the truth of the hypothesis. In the jargon of experimental design, we attempt to falsify the hypothesis, since an experiment cannot prove a hypothesis true. It can either yield results consistent with the hypothesis or yield a result that contradicts the hypothesis. This is an important point, so let us go over it again. An experiment cannot prove a hypothesis true with absolute certainty, but it can prove a hypothesis false with absolute certainty. To see this, let us make up a hypothesis: *The hearts of small animals beat faster than do the hearts of large animals.*

Ordering Experimental Results

The hypothetical experiment we will design measures the heart rates of all species of animals with sufficient sampling to determine the average heart rate and average body weight of the adults of each species. We then independently order the heart rates and body weights of each species. If the ordering in rates and the ordering in body weights is the same, then we can say that the hypothesis is true for all the types of animals measured. However, we cannot assert that the hypothesis is true for types of animals whose heart rates and body weights we have not measured. Therefore, the hypothesis would be considered conditionally true. On the other hand, if one single species of animal contradicted the presumed order, the hypothesis would be absolutely false.

Hypotheses Are Sometimes Nonsense

This experimental testing of hypotheses sounds very professional and even scientific. The difficulty with the reasoning is that we do not make isolated hypotheses in science. There is generally a body of theory and a usually much larger body of experimental information that must be considered in addition to the proposed experiment. Therefore, the proposed experiment is often only one small piece of a much larger puzzle. Further, the likelihood of a positive experimental result, if the hypothesis is not true, is often so small that we do in fact take a positive experimental result to be evidence for the truth of the hypothesis. In our example, as we investigate many more kinds of animals, all of which exhibit the ordering behavior, we come to the conclusion that our statement is not just a hypothesis, but is an empirical rule. Then if we encounter a type of animal that does not obey the rule, we try to understand what is different about this kind of animal, and why it does not satisfy the established regularity.

Two Kinds of Rules

Empirical rules can be of two kinds: qualitative and quantitative. A qualitative rule would be of the kind given by the ordering relation of heart rate and body weight in the example. A quantitative rule would be one that not only provides the ordering, but would indicate how far apart two kinds of animals are. If the heart rate of the lighter animal were twice that of the heavier animal, then the heavier one might be, say, one hundred times heavier. We just made these numbers up, but you get the idea. In the qualitative case, we can only know the ordering, whereas in the quantitative case we have a quantitative measure that also tells us how far apart things are. Historically, science, and especially physical science, has endorsed the quantitative over the qualitative, but we shall find that both quantitative and qualitative have their uses.

Averages Are Unique

It is apparently the uniqueness and quantitative character of a physical process that makes the physical sciences different from the humanities. However, when the outcome of a physics experiment changes from one experiment to another, we have a set of numbers that fluctuate around some average value. It is the average value to which the empirical rules and, ultimately, the laws of physics apply. We mentioned that these fluctuations, though random, also satisfy certain laws such as the *Law of Large Numbers*. This is what the law of frequency of errors has evolved into over the years. It is the existence of such laws, applicable to even random fluctuations, that makes the evolution of physical systems unique.

Uncertainty Is Consistent with Determinism

It is the apparent lack of such laws that makes the behavior of an individual so arbitrary in society. Put more positively, it is this apparent lack of predictability of the behavior of the individual that we call free will. Herein, we argue that the existence of natural laws is not as restrictive as we once believed and, in particular, deterministic behavior is not inconsistent with free will. We emphasize that we are not discussing quantum mechanics here, but continue at the level of Newton's laws. Further, predictability is not as pervasive in physics as we were once taught, and spontaneity in the humanities is completely consistent with scientific determinism. This observation is also consistent with Newton's laws.

Absolute Predictability Is a Myth

The great astronomer and mathematician Marquis de Laplace so firmly believed in deterministic forces that he argued that if he knew the locations and velocities of all the particles in the universe at a given instant, he would be able to predict all the future motion of the universe. The fact that we cannot do this he attributed to our inferior brains and asserted that God could certainly do this by using Newton's equations of motion. This arrogance was a clear articulation of a belief in the determinism and predictability of the physical world, and that the existence of uncertainty and statistics was the result of incomplete knowledge. The position of Laplace is in fundamental conflict with the humanist concept of free will. His view is also in conflict with much of the understanding of mechanics that has developed in the last 30 years and which goes under the heading of nonlinear dynamics and chaos theory. We shall address these ideas in due course.

1.3.3   History and Heat

History is Worthy but It Is Not Science

We do not pursue a systematic historical perspective here, in part because of the limitations of time and space, but, more importantly, because the history of science and the philosophy of science are not in themselves science. They are interesting and worthy pursuits, but they are not our concern here. We are primarily concerned with two things: the study of how biodynamical and biothermal phenomena behave as they do, independently of how we want them to behave; and how we conceptualize or model that behavior. It is important to stress that the physical sciences, unlike literature, philosophy, the other humanities, and much of the social sciences, are not concerned with why things behave as they do, since, ultimately, we do not know why, we only know how.

Heat and Motion

We introduced the notion of *biothermal processes* above. For the limited purposes here, we take biothermal to mean the transfer of heat from the body to the environment and back again to the body. There are a number of different perspectives that one can adopt in studying biothermal processes. Here we take the limited perspective of thermodynamics and attempt to relate the transfer of heat to the motion of microscopic entities, whether they are cells, molecules, or atoms. This approach is actually less mathematical than using the transport equations found in most engineering texts on the subject and actually began in the distant past with the philosophers of Greece.

The Motion of Atoms Is Deterministic

Some philosophical arguments are as germane today as when they were first articulated. One of these arguments has persisted for 2500 years and has to do with the source of randomness and uncertainty in the natural sciences and in our lives. The Greek philosopher Democritus introduced the idea of the atom into the natural sciences. He proposed that all matter is made of tiny particles flying around in mostly empty space. In addition, he also argued that these particles travel along paths that are deterministic in character, which is to say that nothing happens by chance. Randomness is, therefore, a consequence of the immense number of particles contained in every cubic centimeter of matter and our ignorance about where they are and how fast they are moving. Given an intellect of sufficient magnitude, one that could know the positions and velocities of all these particles, the complete future would be apparent. For mere mortals like us, such complete information is not available and if it were available we would not know what to do with it. Thus, randomness is associated with this state of incomplete knowledge. Our uncertainty about the present state of the world, or some small part of it, is projected forward in time into a multiplicity of possible futures.

The Motion of Atoms is Random

On the other hand, another Greek philosopher, Epicurus, believed that *randomness* was not a result of the number of particles, but was rather a consequence of the uncaused behavior of particle motion. He believed that the elimination of uncaused events was contrary to the notion of free will. It may be difficult for us to understand the idea of an uncaused event, given the indoctrination the children of the industrial revolution have had from early childhood. We have heard repeatedly that there is a cause for everything. Sometimes, the situation is so complicated that the cause is obscured, but it is nonetheless there somewhere. This was not the situation in ancient Greece, where the idea that a spontaneous event, an event completely disconnected from the past, was as likely to occur as one generated by the past. Such spontaneous events would, of course, be random and unpredictable. Thus, Epicurus accepted atomic theory, but he also believed in spontaneous events.

So, Who Was Right?

We are still uncertain as to the fundamental source of randomness in physical phenomena. It has not been established if randomness is a consequence of the large number of particles, or if it is the result of the nature of the interactions among the particles, or perhaps even a combination of these two quite different effects. It is worth mentioning that resolving such philosophical speculations could determine our understanding of free will and the knowability of the future, and would be considered by some to be a worthy pursuit.

Randomness and Uncertainty

In the above discussion we introduced one of the more subtle concepts entering into the physical sciences, that is, the existence and role of randomness [4, 5]. In the present context the idea of randomness may be related to the difference between an "act" and an "event." Following Turner [6], we distinguish between the two by noting that an event has a symmetry in time, in that there is no difference between knowing an event can happen and knowing that an event did happen, so no additional information is gained by the occurrence of an event. On the other hand (here we replace Turner's word "act" with the less value-laden word "action"), an action has an asymmetry in time, in that what is known about a process is fundamentally different before and after an action. An event may be predicted by the situation preceding it; an action may not be so predicted. However, even though we cannot predict the onset of an action, in retrospect we may say that an action is consistent or understandable given the preexisting situation.

Self-Organization Is Ubiquitous

Turner also argued that this distinction between event and action allows for the notion of "freedom" to be reintroduced into a deterministic universe, and for a clear separation to be made between what a thing is (*ontology*) and how it is known (*epistemology*). Note that we use the term "action" to include the "nonconscious" process of self-organization made by the formation of such things as stable neural connections within the brain, the patterns of branches within the mammalian lung, and oscillating biological activities in the human body, as well as the development of myths, religions, and political organizations in society.

1.3.4 Predictability and Chaos

Two Sources of Randomness

From one perspective, the unpredictability of free actions has to do with the large number of elements in the system, so many in fact, that the behavior of the system ceases to be predictable [7]. On the other hand, we now know that having only a few dynamical elements in a system does not ensure predictability or knowability. It has been demonstrated that the irregular time series observed in such disciplines as biology, kinesiology, physics, and physiology, are at least in part due to nonlinear dynamical interactions and the phenomenon of chaos [8]. Technically, *chaos* means that the behavior of a nonlinear system depends sensitively on where a system starts its motion. This is often called "a sensitive dependence on initial conditions of the solutions to a set of nonlinear, deterministic, dynamical equations." Practically, chaos means that the solutions to such equations look erratic and may pass all the traditional tests for randomness even though they are deterministic. In Figure 1.1 we compare a random time series with a chaotic time series.

Typical Time Series Are Erratic

First of all, a *time series* is a set of values of some selected variable at different points in time, say the discrete moves of the price of a particular stock on the New

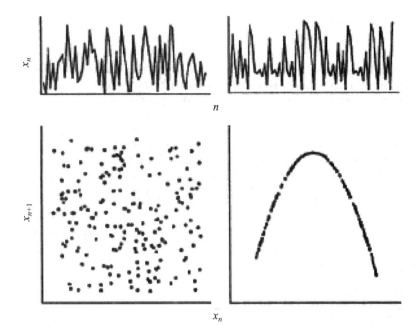

Figure 1.1. Top: On the left is a random time series obtained using a random number generator on a computer. On the right is a chaotic time series generated using the iteration equation (1.1). Bottom: The values of the data points are plotted versus one another, that is, the X_{n+1} point is graphed versus the X_n point. The difference between the truly random, on the left, and the chaotic data points, on the right, is clear. (Taken from Ref. [5] with permission.)

York Stock Exchange. The variable may also vary continuously in time, like the angular position of the right leg as a person walks at a normal speed on level ground. We do not consider such real data as yet but, instead, we generate a set of discrete data points using a random-number generator on a computer. For the purist, it should be pointed out that one cannot generate a random number on a computer, only a pseudorandom number, since the computer can only operate deterministically. The set of N points so generated is denoted X_n, where $n = 1, 2, \ldots, N$ in Figure 1.1. In the upper-left-hand corner of the figure, the data points are connected by straight-line segments, which aid the eye in visualizing the random process. In order to generate these points, a mean and variance are specified in the appropriate computer program. This random time series does not look very different from the one depicted in the upper-right-hand corner of the figure. However, this latter time series is generated in a very different way.

Nonlinear Maps Generate Random Sequences

The data points on the right in the figure are generated using a nonlinear deterministic equation. The dynamical equation is the recursion relation (called a map since it maps a given point into one that was not previously known), often called the logistic equation:

$$X_{n+1} = 3.95X_n(1 - X_n) \tag{1.1}$$

An initial value is chosen, $n = 0$, $0 < X_0 < 1$, and the equation is then iterated by inserting that value into the right-hand side of the equation and generating the next term in the sequence, which is then inserted into the right-hand side and so on. The iteration is done on a computer N times to give the sequence $X_0, X_1, \ldots, X_N$. This sequence of values is plotted versus iteration number (time) and then connected by straight-line segments in the upper-right-hand time series. Thus, the output of this simple nonlinear, deterministic, dynamical equation mimics random behavior in time. As discussed by Bassingthwaighte et al. [5], the two time series, depicted in Figure 1.1, by construction, have the same statistical properties, approximately the same mean, and approximately the same variance. Both time series look random, but we now know that not everything that fluctuates erratically in time is random. The time series on the left is what is traditionally called random since it was generated using a computer to select "random" numbers and therefore to generate noise. *The time series on the right, on the other hand, is not random at all in the sense of noise; it is completely deterministic, since it was generated by the deterministic map [Equation (1.1)]. However, this time series would pass all the traditional tests for a statistical process.*

Statistical Independence Means It Is Really Random

The term chaos is perhaps a poor word to describe the above phenomenon, since in common usage chaotic means disordered or unstructured. Here chaos does not mean disordered at all; the meaning is closer to that of constrained randomness, or randomness confined to an interval. What distinguishes the chaotic time series from the random time series is depicted in the two lower plots in Figure 1.1. The data on the lower left shows what one might expect as the relation between two successive random data points, that is, no relation at all. The value of X_{n+1} is plotted on the vertical axis and X_n is plotted of the horizontal axis, so each point in the plane has the coordinates (X_n, X_{n+1}). If the horizontal and vertical coordinates are random numbers, independent of one another, there will be a set of random points on the plane with no apparent pattern, and that is what we observe.

Chaos Is a Different Kind of Randomness

The set of points on the lower right in Figure 1.1 is quite different. In this figure, there is a smooth deterministic relation between two successive values of the variable X. How can such a smooth relation lead to such erratic behavior in the graph of the variable versus the iteration number depicted in the upper right of Figure 1.1? The reason for the erratic time series is, in part, that the order in which the various values of the variable are generated by the parabolic function $X_n(1 - X_n)$ is not apparent from the function. The order in which the values are generated is only seen in the "time series" and it is the order of appearance of the values of X that look random.

Random Time Series

As was pointed out, chaos is the result of the system's sensitivity to initial conditions. Thus, no matter how small a change one makes in the initial conditions, eventually one obtains a totally different looking time series in that the order in which the values are generated are completely different in the two generated sequences. It is in this sense that both of the time series shown are random. We are interested in random time series because biodynamical phenomena are typically recorded as time series of values, whether voltage in electroencephalograms, concentrations in O_2 intake, or body temperature, and time series are always erratic. The challenge is to properly interpret the source of the fluctuations in biodynamical time series. Are the fluctuations due to noise which is the coupling of the system to the uncontrolled environment, or are they due to chaos, an intrinsic dynamical property of the system?

1.3.5 Complexity

An Incomplete List of Complexity Characteristics

It is useful to list the properties associated with the complexity of a system, because we are seeking quantitative measures of such systems. We note that in everyday usage, phenomena with complicated and intricate features, having both the characteristics of randomness and order, are called complex. Further, there is no consensus among scientists, poets, and scholars as to what constitutes a good quantitative measure of complexity. Therefore, any list of traits of complexity is arbitrary and idiosyncratic, but, given that disclaimer, the following traits are part of any detailed characterization of complexity:

1. A complex system contains many elements, each one representing a dynamical variable.
2. A complex system typically contains a large number of relations among its elements, for example, a food web, a communications network, etc. These relations usually constitute the number of independent dynamical equations that determine the evolution of the system.
3. The relations among the elements are generally nonlinear in nature, often being of a threshold or saturation character, or, more simply, of a coupled, deterministic, nonlinear dynamical form.
4. The relations among the elements of the system are constrained by the environment and often take the form of being externally driven or having a time-dependent coupling. This coupling is a way for the system to probe the environment and adapt its evolution for maximal survival.
5. A complex system is typically a composite of order and randomness, but with neither being dominant.
6. Complex systems often exhibit scaling behavior over a wide range of time and/or length scales, indicating that no one or few scales are able to characterize the evolution of the system.

Parameters Provide Measures of Properties

The above list includes the most common properties selected to characterize complex systems, and in a set of dynamical equations, these properties can often be theoretically kept under control by one or more parameters. The values of these parameters can sometimes be taken as measures of the complexity of the system. This way of proceeding is, however, model-dependent and does not allow comparisons among the complexities of distinctly different phenomena, or, more precisely, among distinctly different models of phenomena.

Simplicity Can Generate Complexity

Therefore, if we think of random time series as complex, then the output of a chaotic generator is complex. Nevertheless, something as simple as a one-dimensional quadratic map can generate a chaotic sequence, such as the logistic map in Equation (1.1). Thus, using the traditional definition of complexity, it would appear that chaos implies the generation of complexity from simplicity. This is part of Poincaré's legacy of paradox. Another part of that legacy is the fact that chaos is a generic property of nonlinear dynamical systems, which is to say, chaos is ubiquitous; all systems change over time, and because they are nonlinear, they manifest chaotic and therefore random behavior [9].

Chaos Versus Noise

A *nonlinear system* with only a few dynamical variables can generate random patterns and therefore has chaotic solutions. So we encounter the same restrictions on our ability to know and understand a system when there are only a few dynamical elements as when there are a great many dynamical elements, but for very different reasons. Let us refer to the uncertainty arising in phenomena having a large number of degrees of freedom as *noise,* the unpredictable influence of the environment on the system of interest. Here, the environment is assumed to have an infinite number of elements, all of which we do not know, but they are coupled to the system of interest and perturb it in a random, that is, unknown, way [4]. This is very different from the dynamically generated uncertainty produced by the internal structure of the system. Thus, noise can only occur in an open system, whereas chaos can arise in a completely isolated system since it is a consequence of the internal dynamics.

Chaos and Nonlinear Dynamics

Nonlinear dynamics is the generic term used for the mathematics of phenomena that change over time, but in such a way that the output, or the rate of change of the output, is not proportional to the input. In general, chaos is a consequence of the nonlinear, deterministic interactions in an isolated dynamical system, resulting in erratic behavior having limited predictability. This is a difficult concept to digest in a single sentence, so if you have not run across it elsewhere do not worry. We shall come back to chaos and its properties again and again, so by the time you have finished reading this book, either chaos will be an old friend or your nemesis. In any event, chaotic behavior will no longer be a stranger. *Chaos is an implicit property of a complex system, whereas noise is a property of the environment in contact with the system of interest.* Chaos can, therefore, be controlled and predicted over short time intervals, whereas noise can neither be predicted nor controlled, except perhaps through the way the environment interacts with the system.

Question 1.8

What does a scientist mean by predictability, variability, and uncertainty? Discuss these concepts using biomedical examples.

Question 1.9

What is the difference between chaos and noise? Give an example of each.

Question 1.10

What is the value of experiment in the search for knowledge? What role do experiments play in science?

The World is a Matter of Perspective

There are two worlds: the humanist's world with the solid table upon which these pages rest, which the poet describes in terms of the lineage of the oak, and the scientific world with an ephemeral desk consisting of swarming microscopic particles in the vast emptiness that, like a collection of bees, push on the underside of this paper to hold it aloft. In the first world, we can feel the desk and our hand, see the words on the page, everything making contact with everything else. In the second world, there is no direct contact. The force I perceive due to the particles in my hand interact remotely with the particles in the table and I am taught to interpret this as direct contact. The pages of this book are revealed to your brain by the light that is remotely reflected from the page and which travels the intervening distance to where the photons are absorbed by the cells (rods and cones) in your eyes. It is not a matter of which of the two worlds is more real—the poet's mind or the scientist's brain—for they are the same world. It is a matter of which is the more useful. The world is as it is, and both the scientist and the humanist attempt to make sense out of it. Herein, we endeavor to make the perspective of the scientist a useful one for you to adopt. We assume that you already value the view of the poet.

Methods and Meaning

We suspect that an approach that synthesizes methods and meaning will help individuals to be active participants in a technological society. Until now, it has been left to the student to accomplish the task of synthesis alone. It is probably fair to say that most students have not been able to successfully synthesize physics and literature, or physics and psychology, or physics and anything else for that matter. Physics and mathematics, with their unbending laws and apparently unflinching certainty, seem to stand in stark contrast to other disciplines, beckoning, like the sirens of *Ulysses,* for the others to follow. The rocks are strewn with broken theories from the social sciences and the wreckage from the humanities lies in the depths below.

Dimensional Analysis and Scaling

If we adopt the traditional method of presentation, then, even leaving out many of the technical details, the abundance of physics, engineering, and mathematics topics, in terms of variety, would dominate the presentation, leaving little or no time for synthesis with anatomy and physiology. We, therefore, choose a completely different approach, one that allows us to obtain the appropriate description of the phenomena without using the traditional calculus, and, in fact, using only a little algebra. This technique is dimensional analysis and *scaling,* and its approach enables us to determine the states and dynamics of a system simply from knowledge of the units involved in the phenomenon and a few basic concepts such as energy, force, and momentum.

The Body Is Self-Regulating

The human body is a complex dynamical system that is in continual contact with the environment. If the temperature of the environment is higher than that of the body, there will be absorption of heat by the body. If the temperature of the environment is lower than that of the body, there will be a flux of heat out of the body. In attempting to regulate body temperature, when heat is absorbed, the body attempts to generate less heat internally. In the corresponding way, when the flux of heat is outward, the body generates more heat internally to make up for it. Other examples come to mind, having to do with how the body is stressed and when the demand for oxygen is increased, as during exercise, when we breath more deeply and the heart beats faster. Of course, meeting your future spouse for the first time might provoke such a reaction as well.

CHAPTER 2

Ways of Thinking

Objectives of Chapter Two

- Appreciate that there is more than one way to think about and understand any given phenomenon.
- Extend science beyond quantitative aspects of a phenomenon and incorporate the qualitative aspects as well.
- Learn how to use the dimensions of length, time, and mass to determine how a process changes over time without knowing the equations of motion.
- Experience how fractals provide a new way to grasp the meaning of complicated events.
- Recognize and highlight the physical basis of locomotion.

Introduction

A New Way of Thinking

In this chapter, we introduce the reader to a new way of thinking about the physical world, one that is often taught to engineers, rarely to physicists, and never to medical students. This way of thinking is determined by an understanding of length, time, and mass, which is to say, an understanding of the dimensions of things. Students are taught Newton's laws of motion in most colleges, as well as how the equations resulting from those laws describe the motion of physical objects. Although these laws are valid, we see no reason why students should have to learn the technical mathematical methods for solving the subsidiary equations if there exist more direct methods for modeling the dynamics of physical and biological phenomena. Dimensional analysis is one method by which we can directly use what we know

Biodynamics: Why the Wirewalker Doesn't Fall. By Bruce J. West and Lori A. Griffin
ISBN 0-471-34619-5 © 2004 John Wiley & Sons, Inc.

about a process to determine its dynamics, without constructing or solving the equations of motion.

Dimensional Analysis Rather than Dynamics

Dimensional analysis relies on the role measurement plays in the sciences, particularly in physics and biology. Historically, measurement has been the basis of the importance of quantification in the physical and life sciences, from the experiments of Galileo and Newton, along with those of Harvey, up to the present day. The purpose of this chapter is to show the student how they can describe the dynamics of physical bodies without solving or, in fact, without necessarily knowing, the equations governing that motion. In addition to dynamics, the geometry of fractal objects is revealed, since fractal geometry seems to be nature's way of constructing physiological systems. Some algebra is required, but that is as bad as it gets.

2.1   WHAT ARE DIMENSIONS?

Fundamental Dimensions

The fundamental dimensions in mechanics are length, time, and mass. It may appear that the number of distinct kinds of dimensions are as arbitrary as the unit of the dimension itself. However, this is not the case. In fact, if one starts from the laws of motion in mechanics, it is possible to build up the dimensions used in the less basic areas of physics. In 1832, Gauss proposed the dimensions of length, mass, and time as being absolute and the basis for the development of the science of mechanics as formulated by Sir Isaac Newton in 1687. These dimensions were thought to be irreducible in that they cannot be derived from one another, nor can they be resolved from anything more fundamental; they can only be directly experienced. This is another way of saying that these units can only be operationally defined through experiment. Therefore, other dimensions used in mechanics are called derived dimensions.

Derived Dimensions Are Not Fundamental

In contrast to fundamental dimensions, there are derived dimensions, for example, speed, the distance traveled divided by the amount of time required to cover that distance. Speed uses two of the basic dimensions: length and time. On the other hand, the dimensions of force, momentum, and energy require the use of all three fundamental dimensions, as we shall find out. Thus, the sciences of mechanics and dynamics can be completely developed in terms of these three basic dimensions, whereas the study of electricity and magnetism requires the introduction of additional dimensions, for example, the concept of electrical charge. Furthermore, the science of thermodynamics requires the introduction of the notion of temperature, also through experiment. We shall address these new, nonmechanical dimensions, in due course.

2.1.1   Measurement and Lengths

Measurement Is Assigning Numbers to Properties

One thing that makes biodynamics different from the liberal arts is that it is quantitative and exact, at least in principle. This property is a consequence of measurement. The idea of measurement is so common today that we forget that it appears only when there is a high degree of civilization. The unit of length, that is, the material standard of length, may be moved from place to place without change. In particular, it may be moved end-to-end in a sequence to generate a longer, equally subdivided scale. Today, we define measurement as the assignment of numbers to represent properties, but there are at least two distinct ways of doing this. One is direct, as in finding the distance one has traveled. The other is indirect, being the result of inference, as in the determination of the distance between atoms in a piece of metal. If the numbers we assign to the property are to have any meaning, we must determine a ratio by comparing one magnitude with another of the same kind.

Errors Are Associated with Theory

It is also useful to point out that errors in measurement accumulate. Therefore, if we make errors in the measurement of distance as well as errors in the measurement of time, then these errors are compounded in the determination of derived quantities such as speed. This is something one should bear in mind when reading about data sets and in thinking about the origins of all that variability in the data. In addition, the notion of error depends on there being a theoretical quantity that is being measured; for example, the length of the table or the temperature of the room, or, more abstractly, a swimmer's ideal speed.

Distance Is Not Displacement

Next, consider the related but distinct concepts of distance (length) and displacement (also a length). Distance is given by the number of times one places a reference length, a fancy term for ruler, end-to-end to measure the distance between two points. For instance, measuring how far a javelin is thrown or a shot put is launched. Displacement also involves the distance between two points, but here the direction is also important. For example, if there is a cup sitting six inches from the edge of a table, and we move the cup seven inches, then it is important if we displace the cup towards the edge or away from the edge. Consider a runner coming down the side of a mountain along a winding path. The distance the runner travels in the race is the actual length of the path down the mountain. The displacement of the runner is the line-of-sight distance from the location of the runner to the starting point of the race on the side of the mountain.

Measurement Is Not Always Rational

It is possible to construct a reference interval for length; we only need a ruler or a stick. However, implicit in the discussion of measurement are the notions of commensurability and counting, that is, laying a ruler end-to-end to measure a length. However, the *Pythagorean Theorem* shows that one can construct a line that is *incommensurate* and therefore not measurable. For example, a triangle with two legs of unit length (the size of the ruler), has a hypotenuse whose length is $\sqrt{2}$, and therefore cannot be measured with the ruler. In general, the determination of a length is a matter of counting the number of times we place our ruler end-to-end in order to cover the distance of interest. The object being measured need not be a simple geometrical object. It could also be the path along which a person walks, for example. We discuss how such paths are defined subsequently. We also point out that we implicitly assume that the object we are measuring is sufficiently smooth so that if we make the ruler sufficiently small, then the length of the object (curve) being measured is independent of the ruler size.

Units Are the Fundamental Measures

Let us go back to the measurement of distance using the ruler. Suppose a certain magnitude, such as the foot, is named the unit of measure. We know that when we measure anything, the ruler does not fit an even number of times. One always finds a little left over, so we must approximate the fraction of a foot remaining when the ruler does not fit an integer number of times. Thus, we introduce a smaller unit of measure such as the inch. So the runner may have traveled 5297 feet and 7 inches down the side of the mountain. We can regroup these units and say the runner traveled 1 mile, 3 yards, 3 feet, and 7 inches. This is how we determine the concept of distance. We abstract from the known and familiar phenomena to the unknown, which we try to cast in terms of the known. Of course, the choice of unit is arbitrary and one could just as easily have chosen the centimeter rather than the inch as the unit of measurement. In either case, the procedure is to express the measurement as a number followed by the appropriate unit name used in the measurement. The dimension emerges from our experience, but the number associated with that dimension, the unit of measure, is selected for convenience.

2.1.2 Measurement and Time

Time is Fleeting

Unlike an interval of length, an interval of time may be used only once, and then it is gone. The measurement of time therefore requires a process that is periodic. A periodic process is one that generates a uniform sequence of time intervals with which to measure a longer time interval. A swinging pendulum does this. It takes the fixed amount of time time for the pendulum to swing out from a point in space and come back to the same point again. Length and time are therefore fundamentally different in quality. Length can be captured and held in one's hand as a standard; time can only be realized in terms of a dynamical process, to be generated when needed, but otherwise absent, except perhaps in one's mind.

Some History of Time

Consider for the moment the concept of time as a unit of measure. What we say about time usually falls into two categories. The first category is philosophy, in which we attempt to determine the intrinsic properties of time. The second category is more pragmatic and concerns the technical procedures for how to measure time, regardless of what we ultimately decide it to be. We measure a life span as three score and ten years and no philosophical argument can alter that. Prior to Aristotle, time was considered to be a mystical entity in which the past and future could be intertwined. After Aristotle, time was thought to be a linear sequential process that unfolds in one direction. Notice that without this physical concept, separated and distinct from mythology, it is not possible to formulate the notion of history. Herodotus, in the fifth century BC, became civilization's first historian. He could chronicle events in a sequential order because proper time was linear. Not only does the development of history require the adherence to a linear time, so also do the rules of rational thought.

Time Is Partly Subjective

On the other hand, the notion that time can only be directly experienced is not entirely accurate, since our experiences do not come to us in a vacuum, but are selected and interpreted by us in a cultural context. Yet, the cultural context is similarly determined by our general concepts. We shall find that the physical notion of time does not always correspond to physiological time. The shrew, with a metabolic rate hundreds of times higher than that of the elephant, lives, in the opinion of some, a full and complete life in a much shorter span of physical time. But there may be a sense in which the physiological lifetimes of the shrew and the elephant are equivalent, since the product of their metabolic rate and life span are approximately equal.

Time Is Partly Objective

For the purposes here, we accept the Aristotelian notion of linear time, at least temporarily, and adopt an arbitrary unit to delineate it. This unit is the second. The second is defined as the 1/86,400 part of a mean solar day, the mean or average being taken over a twelve-month period. Note that we refer the unit of time to a fraction of a periodic process, the rotating of the earth on its axis. The mean is taken to eliminate the seasonal variability of the length of a day, and so the average is taken over a second periodic process, the length of time required for the earth to orbit the sun.

A Second Is Cosmic

The standard unit of time is the second, but we know that some processes occur over time scales much shorter than a second, whereas others occur on time scales much longer than a second. A reference time may be created by constructing a physical device that does nothing but repeat a given operation in a specific time interval over and over again. For example, a pendulum of a given length may swing back and forth 3600 times per hour and 86,400 times per day. The "period" of such a pendulum is one second. This device could then be used to determine if the length of a day were constant by counting the number of "oscillations" of the pendulum from the passage of the sun directly overhead on two successive days at various seasons of the year.

Analogues of Mechanical Clocks

It is now possible to proceed by analogy and construct an electronic circuit that has the same properties as a physical pendulum. Such a circuit would have an output that varies periodically such that the constant time interval between successive maxima of the voltage can be made quite small. Periods of one millionth of a second (10^{-6} seconds), and smaller are readily obtained with such electronic oscillators. This oscillator goes through a million cycles in each second. Oscillators that achieve periods of a million million cycles in one second are quite common. Thus, the second is the universally accepted unit of time, with 60 seconds in a minute, 60 minutes in an hour for increasing time, and 0.001 (10^{-3}) seconds in a millisecond and 0.000001 (10^{-6}) seconds in a microsecond for decreasing intervals of time. For most of the physiological applications of interest in this book, the millisecond is probably the smallest time interval that will have any significance.

Frequency Is Cycles per Second

Instruments used to measure short time intervals express their response time in terms of periods or cycles per second, called Hertz. A 100 Hertz (Hz) device, for example, can track a signal that goes through one hundred cycles in one second. A high-frequency device may operate at, say 10 MHz [ten megaHertz (mega = 10^6)] or ten million cycles per second. This would be beyond the resolution required for most biodynamical experiments, unless perhaps one is tracing a particular dynamical process back to the level of the vibrations of individual molecules. This is not so unthinkable when we consider the influence that medications can have on human behavior, and drugs certainly operate at the molecular level. So perhaps we should not dismiss processes that operate on time intervals smaller than milliseconds.

2.1.3 Measurements Are Not Always Simple

Geometry Is the Basis of Measurement

Let us again examine our first basic unit, that of distance or displacement. Here again, the way in which we experience space dates back to the Greeks, and to one Greek in particular—Euclid—and his geometry, the bane of most high school students. In the third century BC, Archimedes, using Euclid's geometry, presented the axiom that the shortest distance between two points is a straight line. Implicit in this self-evident statement is that Euclid's space is uniform, continuous, and homogeneous. This means that space is a vast, formless emptiness in which we put things, and one region is quite like another, except for the things in it. We can, therefore, move objects around in space without changing either the objects or the space, simply because space cannot interact with the mass of objects. We find that this picture of space, one that was shared by Newton, was changed by Einstein in 1905. He showed that space and time were inextricably linked together and that the universe is one of four dimensions, not three. Some 10 years later, Einstein also showed that space possesses properties tied to the material bodies contained within it. In particular, space bends in the vicinity of material bodies; the more material (greater mass), the greater the bending, and the shortest distance between two points is not always a straight line, but is a geodesic in a curved space.

Fractal Geometry Replaces Euclidian Geometry

In traditional measurement theory, we assume that there is a length associated with an object that is independent of how we choose to measure it. This is objective reality. Or is it? We shall find that this rather obvious notion of length is often wrong and there are many physical and biological phenomena for which this traditional concept of length is not appropriate. Such phenomena are called *fractals* and we shall have occasion to thoroughly discuss them somewhat later. For the moment, we shall denote as fractal a curve or object whose length is determined by the length of the ruler used to measure it. In fact, we shall find that given two points a finite distance apart (say you can put the index finger of your left hand on one point and the index finger of your right hand on the other point), you can connect your fingers with a fractal curve that has an infinite length. Further, if the curve is closed, although it is infinitely long, it will enclose a finite area. A fractal curve is sketched in Figure 2.1.

Fractal Curves Are Strange

In Figure 2.1, we depict a curve with a substantial number of reversals in direction, reversals that occur on many different length scales. In the top curve, the length of the ruler is four and the length of the curve is six units long, that is, we set the ruler end-to-end six times to determine the length of the curve. However, we see that a

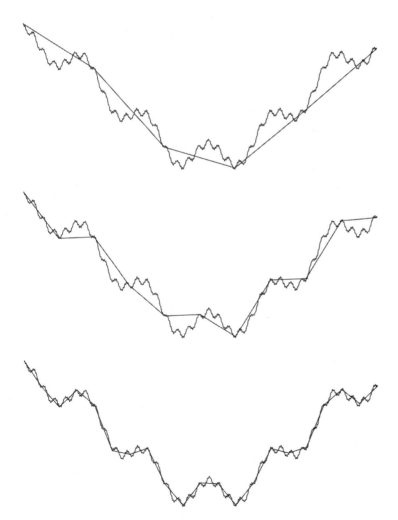

Figure 2.1. The length of the crinkled curve is seen to be a function of the length of the ruler used to measure it. The shorter the ruler, the longer is the apparent length of the curve. Note that the relation between the size of the ruler and the length of the curve is not linear.

great deal of detail is being left out of the determination of the length. Therefore, the size of the ruler is too great and we must reduce it to get a more accurate determination of the curve's length. In the bottom curve, the length of the ruler is one and the measured length of the curve is twenty-four of these new units long, as compared to twelve of these new units obtained in the measurement of the curve's length just above. We continue the process of reducing the size of the ruler and measuring the length of the curve, with the expectation that when the ruler becomes sufficiently small, the curve will appear smooth.

A Fractal Curve is Infinitely Long

The fractal idea is that a fractal curve will never become smooth but will instead continue to become longer and longer as the ruler with which we measure it becomes shorter and shorter. The traditional notion is that eventually we will reach a scale size over which the curve is smooth and any ruler that size, or smaller, will give the "true" length of the curve, one that is independent of the size of the ruler. A fractal curve violates this intuition, so that no matter how small we make the ruler, the length of the curve will be longer. This is the mathematical notion of a fractal and is a consequence of the idea of *self-similarity*. We shall discuss the fractal concept more fully at the appropriate time.

Scales Within Scales Within Scales

Of course, the world is never defined exactly by mathematical equations or concepts. So with regard to physical fractals, one cannot continue to reduce the size of a ruler indefinitely without encountering new phenomena. Consider the scaling downward from the human body. Humans are of the order of a meter in size and our appendages, like our hands, are on the order of 10^{-1} meters (10 centimeters). The "cells" observed in our skin are on the order of 10^{-3} m, that is, a millimeter or so. The grooves of a long-playing record, zooplankton, and protozoa are 10^{-4} m in size. At the next scale down, 10^{-5} m, the cells that form tissue are fully distinct. In the human body there are a hundred times more cells than the known number of stars in the galaxy. The smallest living cells are only about 10^{-4} m in diameter, whereas few of the simpler cells grow larger than 10^{-5} m. Every completed cell has a nucleus packed with enormously long molecules called DNA. The long, twisted molecular ladder of the double helix has only a few turns on the scale of 10^{-8} m. The separate amino acids become distinguishable at the 10^{-9} m, as do the hydrogen bonds.

New Phenomena Always Occur

On the scale of 10^{-10} m the laws of quantum mechanics were first identified through the formation of atoms. All atoms obey a single structural principle; they consist of a minute center surrounded by a diffuse cloud of electrons that are attracted to the center. Electrical forces govern the motion of the electrons, but quantum laws restrict the effect of the electrical force, making the interaction inherently probabilistic. This distribution of electrons describes the stable charge density intertwining atoms and not the orbits of individual electrons, since in the microscopic domain, there are no orbits in the sense that we view them in celestial mechanics; each electron is distributed throughout space. These charge distributions are due to the outermost electrons and are shared by atoms that are bonded together. The atom's innermost electrons form a spherical shell of diameter 10^{-11} m surrounding the atomic nucleus.

Mass Is Mostly Not There

It is remarkable how small the *atomic nucleus* is compared with the atom itself, that is, with the *electron cloud*. The scale 10^{-12} m must be realized before the atomic nucleus begins to appear, but 10^{-15} m is attained before the individual *nucleons* comprising the *atomic kernel* can be discerned. The nucleus contains the major part of the *atomic mass*, even though it is the electrons that fill up space. These are, of course, the fundamental building blocks of matter and lead us to our third basic physical unit, *mass*.

2.1.4   Measurement and Mass

Mass Is Not Weight

Mass is by far the most subtle of the three basic units. The term mass, in common use, denotes the quantity of matter possessed by an object. In the physical sciences, however, mass has a more precise meaning. Mass is the resistance of a body to changes in velocity. Like time and distance, mass is an experimental and experiential quantity that requires experiment to refine its meaning. Galileo determined that objects moving with a constant speed tended, in the absence of retarding forces such as friction, to continue to move in the same way forever. This property of moving objects to continue moving in an unchanging way is called inertia. Mass is a measure of this inertia. Note that this definition has nothing to do with the weight of an object; you will recall that weight is a force. Note also that mass is a concept that one can never "prove" in a mathematical sense, but rather constitutes a summary of vast amounts of physical experiments. Since we cannot produce an equation to elucidate mass, nor can we construct a logical argument to make its meaning clearer, we do the only thing left and suggest an experiment to demonstrate its properties.

Experiment Determines Mass

Consider the following experiment. Suspend two metal balls having the same diameter and made of the same material from a support by slender wires, as depicted in Figure 2.2. Displace the left ball, M_L, a given distance along the armature from its equilibrium position of rest in the vertical position (this also corresponds to a height s above its resting position) and release it. Upon impact of the moving ball with the stationary ball, M_R, M_L will come to rest and M_R will be kicked out a maximum distance equal to the initial displacement of M_L (or very nearly so) and will also achieve a height s. The fact that those two maximum distances, or heights, are the same leads one to conclude that the masses of the two balls are equal: $M_L = M_R$. If M_R were made of a different material from M_L and we measured its maximum displacement to be twice the initial displacement of M_L, or twice the height, we could conclude that the mass of M_R is half that of M_L. It is only through experiments such

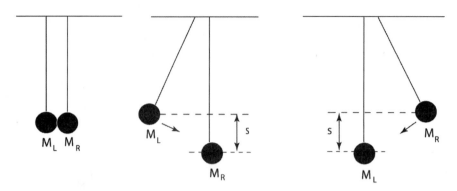

Figure 2.2. Two balls suspended by strings, labeled M_L and M_R, are depicted. Initially, they are in the equilibrium positions shown on the far left. The ball M_L is moved a distance to the left and then released, while M_R remains at the equilibrium. After M_L is released, it swings to the right and strikes M_R. M_L then comes to rest at its equilibrium position and M_R swings out to the maximum displacement shown.

as this one that the concept of mass acquires meaning, and it was through such experiments that Galileo first defined mass and inertia.

Mass Measures Inertia

From this experiment, we see that if two identical spherical bodies undergo a direct collision, the change in velocities of the two spheres are equal in magnitude but opposite in direction. On the other hand, if the bodies are made of the same material, but the radius of the first is λ times the radius of the second, r, then the magnitude of the velocity changes are in the proportion λ^{-3}, since the volume of a sphere is a constant times the cube of the radius, that is, $(4/3)\pi r^3$. From this result, it appears that the change in velocity depends on the relative volumes of the spheres, when they are made of the same material. If the materials of the two spheres are different, one being of density ρ_L and the other of density ρ_R, then the proportion in the velocity changes is $\rho_L/[\lambda^3\rho_R]$. Note that this ratio is the same as the ratio of the masses of the two objects, M_L/M_R, since $M = (4/3)\pi r^3\rho$ is the volume times the density. Density, the mass per unit volume, is a physical concept derived from the concepts of mass and volume, the latter being derived from the concept of length.

Measurement Units and Logic Work Together

Euclidean space, Aristotelian time, and Galilean mass form the basis of classical physics and therefore biodynamics. The first two concepts, involving space and time, have been around for nearly two and one half millennia, but a quantitative science could not be made from natural philosophy until the third quantity, mass, was identified and introduced into the conceptual scheme some 300 years ago. This is the linear worldview adopted by essentially everyone in civilized society, and which was included in the logic of Aristotle. Once an abstract, linear, and continuous space and time had been developed, Aristotle turned his attention to the rules of rational thought—the rules of logic. These rules were codified by the Greek philosophers into a standardized system that survives today. Not only does it survive, but a facility with Aristotle's logic is a prerequisite for success in virtually every professional, technological, and literate position in civilized society. Many forms of aberration are tolerated by society, but not irrationality.

List of Useful Quantities

In Table 2.1, we list a number of physical quantities that we use in this book, starting with the undefined measures of mass, length, and time. A fourth fundamental quantity, useful in thermodynamics as well as in physiology [10], is temperature. We shall discuss the concept of temperature in a bit more detail after we have discussed some additional thermodynamic concepts. Each of the quantities in the table shall be discussed in detail at a later time; we record them here for future reference.

A Map Is Important in Unfamiliar Territory

A person can always collaborate with someone having the mathematical expertise they lack if they learn something of the other's language. It is therefore important to learn what the equations are, but not as an exercise in rote learning. The purpose of learning the equations is to understand how mathematics and the phenomena being described are related to one another. The map (mathematics) is not the reality (phenomenon), but if you have ever gone on a trip, far from the places that are familiar to you, you realize the importance of having a map with you. Even a crude and inaccurate map is often better than relying on the testimony of the local inhabitants. The locals might not intentionally mislead you, but perhaps they do not travel much themselves.

Question 2.1

What are the fundamental limitations on what we can know about mass, distance and time? How do these limitations determine what we can ultimately know about the world? (Answer in 300 words or less.)

Question 2.2

What is the difference between psychological and physical time? If you were walking along a smooth, level road at noon and then again at midnight, what

Table 2.1. A list of the four basic physical quantities—mass, length, time, and temperature—and one derived quantity—mass density. The notation for the dimension of each quantity along with a typical unit is also listed.

Quantity	Description	Dimension	Units
Mass	Measure of inertia	$[M]$	kilogram
Length	Measure of displacement	$[L]$	meter
Time	Measure of duration	$[T]$	second
Mass density	Mass/volume	$[\rho] = [ML^{-3}]$	kg/m^3
Temperature	Measure of heat	$[\Theta] = [°K]$	degrees Kelvin

would be the difference in your perception of the passage of time? (Answer in 300 words or less.)

2.2 METHOD OF DIMENSIONS

Various Systems of Units

In physics, there are three absolute systems of units. These systems depend on the chosen units. The first is the centimeter-gram-second (CGS) system, which uses the centimeter for length, the gram for mass, and the second for time. The second system is the English system, which uses the yard for length, the slug for mass, and again the second for time. Finally, there is the meter-kilogram-second (MKS) system, where the length is in meters (100 centimeters), the mass is in kilograms (1000 grams), and time is in seconds. With these various systems of units, it is necessary to be able to convert from one system to another. We shall not use the English system, in part because the pressure to form puns using the English unit of mass would be overpowering.

2.2.1 Conversion between Units

A Specific Example of Measurement

Let us consider an example. Suppose the displacement of the weight at the end of the arm of a pendulum is measured as 10 cm and we want to convert this displacement to inches. This is an empirical question; a question that can only be answered by experiment or by looking up the answer in a book. However, looking up the answer in a book only means that you use the result of someone else's experiment. If we measure the displacement with an American ruler, we find that the displacement is approximately 3.94 in. This is not an equality; that is, 10 cm does not equal 3.94 in exactly, but the true value of the displacement is well approximated by the measured value. By true value we mean that if a measuring instrument more sensitive than a ruler and an eye were used to measure the displacement of the pendulum, the same first three numbers would be obtained—3, 9, and 4—but more decimal places would also be determined; for example, the measured value would be 3.945 in. Now that we see how we convert from one unit of measure to another in a specific case, let us examine how conversion works in general.

Relations between Inches and Centimeters

If the property of a given process is measured and assigned the magnitude A in the units $[A]$ that result in its numerical value being equal to a, we can write

$$\frac{A}{[A]} = a \tag{2.1}$$

Note that the quantity A has both magnitude a and units $[A]$. Let us suppose that A is the displacement of a pendulum from its vertical resting position, as in the example above. In that example we had $a = 10$ and $[A]$ denotes the units of centimeters. In words we read Equation (2.1) as: "The displacement of the pendulum, A, is 10 centimeters from its equilibrium position." Now we measure the same displacement using a ruler with different units, $[B]$, and obtain the value b, so that

$$\frac{B}{[B]} = b \tag{2.2}$$

Note that since we are measuring the same displacement in both cases, displacement must be the same for both, that is, $A = B$. Thus, we may write

$$A = a[A] = B = b[B] \tag{2.3}$$

where we take cognizance of the convention of writing the number followed by the unit. We may therefore transform from one system of units to the other using the ratios of the measurements, so that the ratio of the units becomes

$$\frac{[A]}{[B]} = \frac{b}{a} \tag{2.4}$$

The bad news is that there is some algebra in our discussion. The good news is that it does not get much worse than Equation (2.4).

Conversion Factor

From the present measurement, we interpret the bracketed ratio in Equation (2.4) to be

$$\frac{[B]}{[A]} = \frac{a}{b} = \frac{\text{inches}}{\text{centimeters}} \approx 2.54 \tag{2.5}$$

so there are approximately 2.54 centimeters in one inch. The precision of the conversion is determined by how many decimal places of accuracy we obtain in Equation (2.5). Thus, knowing this conversion factor (or one obtained from a more careful measurement) we would not need to measure the displacement again. We can use the right-hand side of Equation (2.5) to obtain

$$\frac{10}{b} \approx 2.54$$

so that $b \approx 3.94$. Therefore, the displacement of the pendulum is 10 centimeters or 3.94 inches.

Pounds and Grams

As another example, suppose that $[A] =$ pounds, and the magnitude of the weight measured is $a = 2.2$. How many kilograms is this? Here again, we need the ratio of the basic units in the two systems of measurement. This time the units are those of force, so that we again determine from experiment that Equation (2.4) yields

$$\frac{\text{pounds}}{\text{kilograms}} \approx 0.454$$

Now the ratio of the magnitudes is

$$\frac{b}{2.2} \approx 0.454$$

so that the measured force in the metric system is $b \approx 1$. Thus, the measured weight of 2.2 pounds is 1 kilogram. Note that in different conventions, the kilogram can be a measure of mass as we have used it here, but it can also be a measure of force, in which case it corresponds to 9.8 Newtons. Therefore, we have that 2.2 lb equals 9.8 N, or 1 N is about 0.2 lb, approximately the weight of an apple. The average person often thinks of mass and weight as being interchangeable. To add to the confusion, it is customary in the metric system to specify the quantity of matter in units of mass (grams or kilograms); however, in the United States, quantities of matter are given in units of weight (pounds) and the slug is never used in polite society. We shall comment on this further when discussing certain biological models in which the discussion is in terms of mass, but the data is in terms of weight. This is quite important in comparative physiology, because most of the relations across species involve the mass (weight) of the animal.

2.2.2   Using a Ruler

The Length of an Object

In Equation (2.4), we see that the numerical value of a physical quantity varies inversely with the unit used to measure it. Expressed in terms of length measurements, the number of times, N, we lay a ruler of length η end-to-end to cover a given distance varies inversely with the length of the ruler:

$$N \propto \frac{1}{\eta} \tag{2.6}$$

In Equation (2.6), the proportionality factor is determined by the units of the ruler being used. Thus, if we decrease the unit of measurement from a meter to a centimeter, we increase the number of times we place the ruler down by a factor of one hundred.

Derived Units Are Also Useful

Most quantities in biodynamics depend on derived rather than basic units. In other words, most measurements are indirect rather than direct. For example, a velocity requires the separate measurement of a distance and a time. Even the apparently direct measurement of speed by the speedometer in a car is not as direct as it would appear. The speedometer actually measures the number of revolutions per second of a tire. The distance traveled in a second is the number of the revolutions per second multiplied by the circumference of the tire. Thus, if you change the size of the tire, the speedometer will misread the speed of the car because the circumference of the tire is different from that for which the speedometer was calibrated. Therefore, derived units not only depend on multiples or fractions of the fundamental units, but also on powers and combinations of powers of the units of length, mass, and time.

For example, if l is a typical length of an object, then its area is given by l^2, and its volume is given by l^3. The area has a dimension of two in length, so we may denote its units by $[L^2]$, and the volume has a dimension of three in length, so its units are denoted by $[L^3]$. Here, the square brackets signify "the dimension of" the quantity being considered. In the same way, we can express the speed in terms of the dimensions of its units as

$$[S] = [LT^{-1}] = \frac{[L]}{[T]} \tag{2.7}$$

We read this expression as "the dimension of speed, S, is the dimension of length divided by the dimension of time." Here, the term dimension is used synonymously with the term unit.

Number and Nature

In the above notation, we can rewrite the relation between the number of times a ruler is laid end-to-end to determine the length of a curve and the length of the ruler in Equation (2.6) as

$$N = \frac{[L]}{\eta} \tag{2.8}$$

Here, the units of the ruler are distance and, therefore, multiplying by the units $[L]$ cancels out the units of the ruler in the denominator, η leaving a dimensionless number, N. We determined previously that this simple relation, which is so intuitively obvious, is not applicable to a wide range of natural phenomena; phenomena that are so irregular that we can never find a ruler such that Equation (2.8) is satisfied.

Phenomenology and Epistemology

Now, we come to the central question: How do we use the above dimensions to understand physical phenomena? We answer this question by introducing the method of dimensions in a sequence of steps. First, we present the logic of the procedure and then apply the method to a number of physical phenomena; see for example, the book by Huntley [11]. The examples are developed to highlight certain physical principles, but they also reveal the power of this way of thinking. This new way of thinking is very useful in biodynamics.

2.2.3 A Way of Thinking

General Dimensions for Measurable Quantities

Suppose that a measured quantity of interest, A, has the dimensions p, q, and r relative to the units of length, mass, and time, respectively. We can then write for the units of $[A]$ in the above notation

$$[A] = [L^p M^q T^r] = [L^p]\,[M^q]\,[T^r] \tag{2.9}$$

where the units are seen to factor. Consider some example cases. If A is a velocity, $A = V$. Then since $[V] = [L]/[T]$ in Equation (2.9), there is one factor of length, so $p = 1$; no factor of mass, so $q = 0$; and an inverse factor of time, so $r = -1$. If A is a momentum $A = P$. Then since $[P] = [M]\,[V]$ there is one factor of length, so $p = 1$; one factor of mass, so $q = 1$; and an inverse factor of time, so $r = -1$. If A is an acceleration $A = a$. Then since $[a] = [L]/[T^2]$ there is one factor of length, so $p = 1$; no factor of mass, so $q = 0$; and two inverse factors of time, so $r = -2$. We maintain that any quantity in classical mechanics and biomechanics can be written in the form of Equation (2.9). If you do not remember the units, refer to Table 2.1 and Table 2.2.

Intuition Over Integration

Do not be overly concerned if concepts listed in Tables 2.1 and 2.2 are not familiar to you. If you did know all these quantities and how to properly interpret them, you would be well ahead of the game. Of course, there is a substantial difference between knowing the definition of a force and being able to determine the motion of a body under a given set of forces. In this book, we think it is less important to have facility with the mathematical manipulations of equations chosen to model biodynamical phenomena than it is to be able to identify how physical and biological mechanisms are related to one another. An appreciation of biodynamical phenomena comes through experience; the identification of the physical and biological mechanisms within these phenomena comes only through thinking about experience.

A Force Example

Let us use these abstract equations to solve some "practical" problems, perhaps going a little more slowly to savor each of the details in the example. You may find that problems called practical by a physicist may not be viewed that way universally. In any case, let us begin by returning to the laws of mechanics. The definition of the unit of force, which is derived from Newton's second law of motion, can be determined by examining the units of the equation

$$\text{force} = \text{mass} \times \text{acceleration} \tag{2.10}$$

In this way, the dimensional formula for force is given by

$$[F] = [M]\,[LT^{-2}] \tag{2.11}$$

Table 2.2. A list of the five derived physical quantities is given, along with the notation for the dimension of each quantity and a typical unit for each

Quantity	Description	Dimension	Units
Velocity	Displacement/time	$[V] = [LT^{-1}]$	m/sec
Acceleration	Displacement/time^2	$[a] = [LT^{-2}]$	m/sec^2
Momentum	Mass × velocity	$[p] = [MLT^{-1}]$	kgm/sec
Force	Momentum/time	$[F] = [MLT^{-2}]$	Newton
Pressure	Force/area	$[P] = [ML^{-1}T^{-2}]$	poise

since the acceleration is the ratio of the number representing the velocity of a mass to the time, and the velocity is the ratio of the distance traveled by a mass to the time taken to traverse this distance, that is, a number of length units divided by the square of a number of time units (see Table 2.2). The acceleration is the change in velocity of a body over a given interval of time divided by that small time interval.

Adding Exponents

The form of Equation (2.9) has a number of implications for derived units. The first is that if three physical quantities A, B, and C are given by the product relation $C = AB$, then the numerical value of the physical quantity c is given by the product of the numerical values of the physical quantities $a \times b$. Second, the dimension of C is equal to the product of the dimensions of A and B,

$$[C] = [A]\,[B] \tag{2.12}$$

Of course, both A and B have the form of Equation (2.9), so that the dimensional equations have a product form. Suppose that $[A]$ is a force and $[B]$ is a displacement, then the dimensional equation for the product of force and displacement is

$$[C] = [MLT^{-2}]\,[L] \tag{2.13}$$

We can simplify this equation, using the rule for the addition of the exponents of products: $[L]^{\alpha}[L]^{\beta} = [L]^{\alpha+\beta}$, to obtain

$$[C] = [M]\,[L]^2\,[T]^{-2} \tag{2.14}$$

The interpretation for the quantity $[C]$ can be made from the observation that $[L]/[T]$ has the dimension of a velocity $[V]$, so that Equation (2.14) becomes

$$[C] = [M]\,[V]^2 \tag{2.15}$$

which has the dimensions of an energy. We have therefore "derived" an important result in mechanics using only dimensional analysis—force times displacement is energy. This conclusion will require a little adjustment before it can be applied, but it is basically correct, as we shall find in the next chapter.

The Fundamental Rule is Dimensional Homogeneity

The fundamental rule of dimensional analysis is that the dimension of each unit on both sides of an equation must be the same. This is referred to as *dimensional homogeneity*. Do not be intimidated by the formalism or the names of things. The relations simply mean that if there are two factors of mass on one side of the equation, then there must be a corresponding two factors of mass on the other side of the equation. The same equality of factors applies for length and time or any other unit being used in the analysis. We note that the dimensional formulas for a physical quantity do not contain numerical factors or coefficients. A physical quantity is not fully represented quantitatively by its dimensional formula. This is a weakness of the method of dimensional analysis, but we shall find that this lack of an overall coefficient is a small price to pay in order to avoid the differential and integral calculus.

The Measurement of Gravity

In the experiments conducted by Galileo in which he established the law of inertia, he also determined that the acceleration of a particle induced by the force of gravity is a constant. Galileo's interest was in falling bodies, but since he did not possess a sufficiently accurate clock, he used inclined planes to slow down the vertical descent of particles, that is, rolling balls. He determined that a ball rolling down an inclined plane, on which friction had been minimized, increased its speed by equal amounts in successive seconds. When the ball rolls in a straight line we write this observation as

$$\text{acceleration} = \frac{\text{change of velocity in a time interval}}{\text{time interval}} \tag{2.16}$$

so that if the inclined plane produced an acceleration of 3 meters/second/second (m/sec^2) then at the end of the first second the speed of the ball is 3 m/sec, at the end of the second second the speed is 6 m/sec, then 9 m/sec, and so on. When the acceleration is constant, the instantaneous speed is given by the magnitude of the acceleration multiplied by the time interval.

How Far Will a Moving Body Go?

As our first example of the application of the method of dimensional analysis, let us determine the distance traversed by a ball falling from rest under the influence of the uniform acceleration of gravity. Such a problem should be of interest to anyone planning to leap from a bridge with a bungee cord tied to his or her ankles. Curiosity would dictate that we would want to know how fast we are plummeting to earth before we reach the end of our cord.

The Acceleration of Gravity

First of all, let us introduce the letter g to denote the constant acceleration of masses toward the earth near the earth's surface. There is some small variation in this number g as one travels over the earth's surface, because the earth is not spherically symmetric. However, leaving this detail aside, for most purposes of interest to us, the value g is traditionally assigned as 981 cm/sec^2 in CGS units, 9.81 m/sec^2 in MKS units, and 32 ft/sec^2 in English units. We denote the distance through which the bungee cord jumper falls as s and express this as a function of the important physical quantities in the problem. One such important quantity is the constant acceleration of gravity g and another is certainly the time. But are there any others? Shouldn't the mass of the falling object (person) be included in the function? Don't more massive objects fall more quickly than less massive objects? This would certainly be important to a high jumper, who would want to trim down as much as possible. Let us see if this is true or not.

The Distance Fallen in a Given Time

We express the distance through which the person falls, before springing back, by the equation

$$S = Cg^\alpha t^\beta m^\gamma \tag{2.17}$$

where α, β, and γ are unknown parameters to be determined and C is an overall constant that cannot be determined using this method. Since g is an acceleration, it has the dimensions $[LT^{-2}]$, so we may use Equation (2.17) to write the dimensional equation

$$[L] = [LT^{-2}]^\alpha [T]^\beta [M]^\gamma \tag{2.18}$$

without any of those annoying overall constants. We now make the assumption that this equation is dimensionally homogeneous and, following our earlier analysis, we write the factors for length, time, and mass from each side of the equation:

$$[L] = [L]^\alpha$$
$$[T^0] = [T]^{-2\alpha+\beta} \tag{2.19}$$
$$[M^0] = [M]^\gamma$$

Equating Exponents

Note that any quantity to the zero power is unity, so when an equation does not contain a given dimension, the exponent of that dimension is zero. The unknown dimensions may be determined by equating exponents on the right-hand sides of Equations (2.19) with those on the left-hand side of the equations, including those with zero exponents:

$$\text{exponent of } L: 1 = \alpha$$
$$\text{exponent of } T: 0 = -2\alpha + \beta \tag{2.20}$$
$$\text{exponent of } M: 0 = \gamma$$

The solution to Equations (2.20) is $\gamma = 0$, $\alpha = 1$, and $\beta = 2$, so that in the original equation for the distance traversed by the falling person we can write

$$s = Cgt^2 \tag{2.21}$$

We can see from Equation (2.21) that the overall numerical coefficient C cannot be obtained using dimensional analysis. By other means, it is found that $C = 1/2$; for example, by doing the appropriate experiment. However, even without the coefficient we can see that distance increases as the second power of the time, the first power of the gravitational acceleration, and independently of the mass of the body. This last fact Galileo found almost impossible to get his contemporaries to accept. In his book, *Two New Sciences,* Galileo devotes many pages to convincing the reader that two objects dropped from a common point, although very different in weight will reach the ground at the same time [12]. Part of the difficulty lay in the fact that the dialogues were confined to logic and geometry; the algebra of Equation (2.21) would not have been welcome.

Question 2.3

What can we learn from dimensional analysis and scaling? Give some examples of physical and biological phenomena where these ideas might be useful. (Answer in 300 words or less.)

Question 2.4

Why does dimensional analysis work? (Answer in 300 words or less.)

2.3 SOME   What Do the Equations Mean?
INTERPRETATION
We have solved some equations in a fairly painless way, but what do they mean? The equation that expresses the distance as a quadratic function of time (2.21) means that the invisible force of gravity is continuously pulling on massive objects. One of the experiments done by Galileo involved the use of two inclined planes facing one another, as shown in Figure 2.3. A ball rolling down one plane from a given initial height will roll up to the same height on the second inclined plane, where its velocity will be zero. The ball will then reverse the direction of its roll and repeat the same process in the opposite horizontal direction. The sojourn on the first plane is downward with increasing velocity, due to the pull of gravity being in the vertical direction of motion. However, when the ball starts on the second inclined plane, the direction of motion of the ball and the direction of the pull of gravity are in opposition. Thus, although gravity does not change what it is doing, its influence on the ball on the second plane is the opposite of what it was on the first plane, retarding rather than augmenting the ball's motion. This explanation is not complete and we shall return to this problem later to complete our analysis.

Figure 2.3. Gallileo's experiment. Two wedges are shown. The ball starts from the upper right and rolls down the wedge due to the influence of gravity. Its velocity increases until it reaches its maximum value at the bottom and then it starts up the other wedge. The velocity is retarded as it goes up the other wedge because gravity is now working against the ball. If there is no friction, the ball will eventually reach the top of the second wedge and the process will start again.

A Hoary Example of Motion

Many high school students today know that sound is a wave phenomenon and its description is given in elementary physics texts by a wave equation. However, the wave equation had not been invented in 1686, when Newton considered the problem of calculating the speed of sound in air. He did know, however, that sound was a wave phenomenon because of the observations of such effects as reflection, refraction, and diffraction, phenomena that do not concern us here. Given the lack of formalism Newton reasoned in a way very similar to the dimensional analysis we use here. First of all, he noticed that sound is transmitted through air by the motion of air molecules. Therefore, the sound, which is a disturbance of the air, must depend on the density of the air, ρ. Second, force is required to move anything, including air molecules. He reasoned that a local force linearly binds a molecule to its local position, and to determine the effect of a macroscopic disturbance a large area must be considered. Therefore, the pressure, P (force/area) is a more reasonable quantity to consider than the force on a single air molecule. Thus, the speed of sound, V, should be determined by the pressure and mass density of the air. The dimension equation expressing this observation has the form

$$\text{speed of sound} = (\text{pressure})^\alpha \times (\text{mass density})^\beta \qquad (2.22)$$

or in terms of the dimensions,

$$\frac{[L]}{[T]} = \left(\frac{[M][LT^{-2}]}{[L^2]} \right)^\alpha \left(\frac{[M]}{[L^3]} \right)^\beta \qquad (2.23)$$

Equating the exponents of the various dimensions yields

$$\text{exponent of } L: 1 = -\alpha - 3\beta$$
$$\text{exponent of } T: -1 = -2\alpha \qquad (2.24)$$
$$\text{exponent of } M: 0 = \alpha + \beta$$

so that from Equation (2.24) we obtain $\alpha = 1/2$, $\beta = -\alpha = -1/2$. Substituting these values for α and β into Equation (2.24) we have $1 = -1/2 - 3(-1/2) = 1$, a *tautology*. In this way the speed of sound in air is found to be

$$V = \sqrt{\frac{P}{\rho}} \qquad (2.25)$$

the value obtained by Newton. You might take a moment and ponder the significance of the equation you have constructed, one that is not intuitively obvious.

Newton Was Wrong

Using the pressure of air at sea level and the corresponding mass density, both of which were known at the time, Newton calculated the speed of sound in air to be 945 ft/sec. This value of the speed of sound is 17% smaller than the observed value of 1142 ft/sec. For 130 years, the best minds in physics attempted to recover this

17%, using all the mathematical machinery that had been developed during that time, but to no avail. That is how the matter stood until 1816, when Laplace, a young astronomer and mathematician, pointed out that a slight correction to the pressure ought to be used. A correction that took into account the fact that a sound wave moves so fast that no heat is released with the passage of a sound wave. This is called the *adiabatic pressure* rather than the *isothermal pressure,* the value used by Newton. With this replacement of the pressure in Equation (2.25), Laplace obtained the observed speed of sound to within experimental error. Therefore, Newton was right even when he was wrong.

The Speed of Water Waves

We have presented an example of a force acting on a solid mass as well as in a gas. Now let us consider a physical observable having to do with water. How would you go about calculating the speed of a wave crest on the surface of water? If you have ever observed water waves at the beach or in a pond, it is apparent that short waves and long waves travel at different speeds. Further, the rising and falling of the water surface must work in opposition to gravity. Therefore, the speed of the wave crest should depend on these two quantities and, just to be safe, we throw in the density of water as well:

$$\text{wave speed} = (\text{wavelength})^\alpha \, (\text{accelation of gravity})^\beta \, (\text{mass density})^\gamma \quad (2.26)$$

so that in terms of dimensions,

$$\frac{[L]}{[T]} = [L]^\alpha [LT^{-2}]^\beta [ML^{-3}]^\gamma \tag{2.27}$$

Using the *dimensional homogeneity condition,* we equate the exponents on both sides of the equation:

$$\text{exponent of } L\text{: } 1 = \alpha + \beta - 3\gamma$$
$$\text{exponent of } T\text{: } -1 = -2\beta \tag{2.28}$$
$$\text{exponent of } M\text{: } 0 = \gamma$$

so that the exponent of the mass density vanishes: $\beta = 1/2$ and $\alpha = 1/2$. Thus, the speed of the wave crest, u, is given by

$$u = \sqrt{\lambda g} \tag{2.29}$$

where λ is the wavelength of the wave. These are called gravity waves because of their dependence on g and we can see that long waves travel faster than do short waves. Note also that the speed is independent of the mass density of either air or water.

Period of Water Waves

The speed of the wave crest is not the only quantity we could determine using this method. Another quantity that might be of interest is the period of the wave. We could go though the above argument, but that should not be necessary. The period of the wave should depend on the wavelength and the acceleration of gravity so that

$$\text{period of wave} = (\text{wavelength})^\alpha \times (\text{acceleration of gravity})^\beta \qquad (2.30)$$

which, in dimensional terms, yields

$$[T] = [L]^\alpha [LT^{-2}]^\beta \qquad (2.31)$$

Equating the exponents on both sides of the equation yields

$$\text{exponent of } [T]: 1 = -2\beta$$
$$\text{exponent of } [L]: 0 = \alpha + \beta \qquad (2.32)$$

so that $\beta = -1/2$ and $\alpha = 1/2$. Inserting these values of the exponents into Equation (2.30) gives us the expression for the period of a gravity wave in terms of the wavelength λ:

$$\tau = \sqrt{\lambda/g} \qquad (2.33)$$

so that long waves have longer periods in addition to moving more rapidly than shorter waves. A further conclusion we can draw, based on the fact that the frequency of a wave is one over the period, is that longer gravity waves have lower frequencies than do shorter gravity waves.

Question 2.5

Let us consider the speed of sound in water. Using scaling arguments, show how the speed of sound changes in water for a diver as she descends from 10 m below the surface to 100 m below the surface. Explain your result (answer in 300 words or less).

Question 2.6

How far can a 10 m wave travel on the ocean surface in one hour? How far can a 100 m wave travel in the same time?

This Is How Mechanics Works

2.4 **SUMMARY**

A force, defined by a mass times an acceleration, is given in a physical system and from this force the response of the system is determined. Distances traveled under constant acceleration are quadratic in time, whereas velocities under constant acceleration are linear in time. If the mass is constant and the acceleration is constant, then the force must also be constant. Therefore, two of the problems we have solved are constant force problems, and we shall have occasion in the future to discuss phenomena in which the forces involved are constant as well as those in which the force changes over time. Further, when the system is not rigid, like a volume of air, other physical quantities such as the pressure might be more appropriate for describing the dynamics of large numbers of particles rather than simple forces.

Dimensional Analysis Gives Us the Solutions to Dynamic Equations

You might be reluctant to derive equations, but only the most unreasonable of people would say that equations are not useful once we have them. Consider the equation for the speed of sound in air, given by Equation (2.25) as the square root of the ratio of the pressure to the mass density. We can use this result to draw a number of conclusions. One conclusion is that sound will travel faster in media that have a density less than that of air, given that they are at the same pressure. A second conclusion is that the speed of sound increases with increasing pressure for a constant density. Thus, the speed of sound is greater at the seashore than it is in the mountains. Of course, this conclusions assumes that the density of air does not change as we go up the mountain. This is a question that can be answered experimentally.

CHAPTER 3

Mechanics—A First Look

Objectives for Chapter Three

- Understand that mathematics is the language most natural for constructing models of scientific phenomena, but also that there are many kinds of mathematics.
- Newton's laws define what we need to know about mechanical forces in the physical world, through inertia and momentum.
- Newton's Law of Universal Gravitation indicates that the same dynamics that operate on the earth act throughout the cosmos.
- The same law that determines the period of the earth in its orbit around the sun determines the period of a pendulum.

Introduction

The Nature of the Irreducible

In this chapter, we provide a perspective on the quantitative character of the physical sciences and explore the reasons for mathematical modeling of physical and biological phenomena. It is true that mathematics is the language of science, but there are many dialects, some of which we know and some we do not. We argue that as adults we are empiricists. We know that there are many things that cannot be explained, but are simply true. We accept such fundamental events as being unexplainable and, in turn, explain everything else in terms of them. Thus, the fundamental elements of science—the measures of length, time, and mass—can be experienced but cannot be defined.

Biodynamics: Why the Wirewalker Doesn't Fall. By Bruce J. West and Lori A. Griffin
ISBN 0-471-34619-5 © 2004 John Wiley & Sons, Inc.

As the World Turns

The goal of this chapter is to use Newton's laws of motion to define mechanical force as proportional to the product of mass and acceleration in a way that reveals that this is more than a definition. Of equal importance are the sources of these forces, such as the muscles in the body and gravity. We focus on gravity as an exemplar of external forces because it is what keeps the moon orbiting the earth and what the wirewalker defies in his dance in the air across the arena. However, our emphasis is on interpretation and only incidentally on formal structure. The acrobat, like the wirewalker, in defying gravity is aware of the laws of physics without ever having stepped inside a physics lab.

3.1 SOME HISTORY

The Beginning of Mechanics

Galileo Galilei is the father of the experimental science of dynamics, the science that studies the motion of material bodies. In his investigations, Galileo actually abandoned the Greek idea of causation and sought only to correctly describe the results of his experiments [12]. For example, with regard to his experiments on falling bodies he states that "The causation of the acceleration of the motion of falling bodies is not a necessary part of the investigation."

Galileo condensed the activity of the natural philosopher into the following three tenets:

> 1) Description is the pursuit of science, not causation; 2) science should follow mathematics, that is, deductive reasoning; 3) first principles come from experiments, not the intellect.

Science Is Mathematics

In his reliance on experiments and their descriptions, Galileo relinquished the teleological reasoning of his predecessors for the precision of mathematical description. He believed that mathematics was implicit in physical phenomena; in other words, nature herself uses mathematical principles in her unfolding. According to Galileo:

> Philosophy is written in this grand book—I mean universe—which stands continuously open to our gaze, but it cannot be understood unless one first learns to comprehend the language in which it is written. It is written in the language of mathematics, and its characters are triangles, circles, and other geometrical figures, without which it is humanly impossible to understand a single word of it; without these, one is wandering about in a dark labyrinth.

Mathematics Is the Language of Science

Of course, as used here, philosophy means natural philosophy or physics. Galileo had the right idea concerning the language of mathematics in the description of physical phenomena, but he was too restrictive in his choice of mathematical dialect—the geometry of Euclid. We shall see that in the modern world science has learned to speak in many tongues. Newton embraced the philosophy of Galileo and in so doing inferred mathematical premises from experiments, rather than from physical hypotheses. In this way, Newton's three laws of motion—1) the Law of Inertia, 2) the Law of Forces and 3) the Law of Action and Reaction—enabled him to deduce all of mechanics. Newton also found it necessary to invent a new kind of mathematics, the calculus, to describe the motion of the physical bodies that fascinated him.

3.1.1 The Original Experiments

Overcoming Adversity with Experiments

It is nearly impossible today to imagine the extreme difficulty of carrying out careful experiments in Galileo's age. For example, the measurement of falling bodies was difficult because they involved continuously changing velocities, which to be measured accurately required precision clocks. Such clocks did not exist at the time. To overcome this problem, Galileo used rolling balls on inclined planes to give him experimental control over the rate at which an object descends vertically. In Figure 3.1 we see that a rolling ball "drops" one unit of distance vertically in the same time it takes to roll five units of distance horizontally. The ratio of the vertical to horizontal distance the ball moves is 1 to 5. Experiments of this kind enabled Galileo to formulate two simple principles:

> *The Principle of Inertia:* A body moving on a level surface will continue in the same direction at constant velocity unless disturbed.
>
> *The Principle of Superposition:* If a body is subject to separate influences, each producing a characteristic type of motion, it responds to each without modifying its response to the other.

Rectilinear Motion Was Studied First

We note in passing that these principles do not apply to such human activities as rolling, twisting, and diving, all of which require some kind of rotational motion. It is not that we do not think such things important, for we most surely do, but the straight-line or rectilinear motion is easier to discuss and came first historically. We shall get to the other kinds of motion in due course.

The Natural State of Matter

The principles of *inertia* and *superposition* were the outgrowth of Galileo's attempt to determine why, when a stone is thrown, the stone continues to move after the

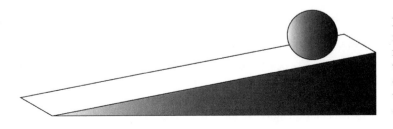

Figure 3.1. The wedge, or inclined plane, has a rise-to-run ratio of one to five. In one second, the ball "falls" a vertical distance of one unit, in the same length of time it traverses five units of distance horizontally.

thrower's hand ceases to propel it. This was the kind of question the Greeks were fond of asking. Although Galileo did not answer this question, he did make the question moot by saying that such a question is outside the realm of natural philosophy (science) and replaced the question with the principle of inertia. As explained by Mott-Smith [13], children are rationalists and believe that everything has an explanation and, therefore, continually ask the question "why." Adults are empiricists. We know that some things cannot be explained, but are simply the way they are. It is a sign of maturity that we accept certain fundamental events as being unexplainable and, in turn, explain everything in terms of them. Science can do no more. As Mott-Smith states, "Whoever would desire more, must consult the metaphysician, the philosopher, the theologian, the mystic or the charlatan."

Turning Around the Questions of Motion

Thus, instead of answering the hoary question regarding the continuing motion of the thrown stone, Galileo turned the question on its head and asked why doesn't the rock continue in its motion forever. He assumed that motion, not rest, is the natural state of matter, and that it is the resistance of the intervening medium that eventually compels the moving object to stop. But whatever the reason, all things move, and moving things do not stop of their own accord, but must be compelled to do so.

Superposition Is Another Principle

The second principle, that of superposition, is the recognition that certain phenomena, such as the motion of a body, may be separated into noninteracting components that may be understood separately, and then reassembled to understand the operation of the whole. For example, the motions of a particle in the horizontal and vertical directions are independent of one another. This leads to the surprising result that if I take a bullet and drop it from a certain height and simultaneously fire a second bullet horizontally over a flat plane, the two bullets will strike the ground at exactly the same instant. The speeding projectile falls in exactly the same way as does the bullet that has no horizontal velocity, and since it is the rate of falling that determines when the two bullets strike the ground, they impact at the same time, if not in the same way. Physically we would say that the horizontal and vertical velocities of the two bullets were produced by different mechanisms: the exploding gunpowder and the pull of gravity, respectively. How far the fired bullet travels is a matter of how long it is in flight, and the duration of its flight is determined by its height from the ground, just as in the case of the ball rolling down Galileo's inclined plane.

3.1.2 Newton's Laws of Motion

Mechanical Forces Can Be Either Fundamental or Derived

In the discussion of the law of inertia, we introduced the ideas of disturbance and influence, both being replacements for the concept of force. At this juncture, we can treat force either as a defined quantity expressed in terms of other measurable quantities, or as an experimental quantity. If force is considered to be directly experienced, then the idea of mass, introduced previously, must be modified, because force and mass are proportional to one another, the proportionality factor being the acceleration of the body. Therefore, if force is the experience, then mass must be defined in terms of force. On the other hand, if mass is the experience, then force must be defined in terms of mass. Since we have defined mass as a measure of inertia, it then follows that force is defined as the inducement of acceleration of one body by another. We denote the force acting on a mass m by the symbol F, so that if a is the acceleration of the body, we therefore have the mathematical representation for the force:

$$F = ma \tag{3.1}$$

In this notation, a bold italic letter indicates a vector (a quantity with both direction and magnitude) and a regular letter indicates a scalar (a quantity with magnitude only).

Statics Considers Forces in Balance

A force is a push or pull, and Equation (3.1) states that whenever I push or pull an object of mass m, I cause it to accelerate, that is, to change its velocity. The direction of the change in velocity (the acceleration), is the direction of the force. If I push or pull a stationary book on a table and cause it to move, then I have exerted a force on the book. If you and I stand on opposite sides of the table and we both push on the book, but it does not move, we exert equal and opposite forces on the book and it remains stationary. Equal and opposite forces acting on the same body cancel one another out, that is, they produce no net effect. Thus, we may exert a force on an object without producing a net acceleration. This is the subject of the discipline of *statics* and, as every structural engineer knows, structures are designed to have forces in balance and to avoid motion of any kind.

Action At a Distance Is Not Magic

However, not all forces require the direct physical contact we experience between the hand and the book. The bullet falling to the ground, indeed anything falling to the ground, is a consequence of the unseen force of gravity. Gravity is the invisible influence of the massive earth on any and all of the material bodies on the earth. Gravity is the force we work against in walking, running, and jumping, and understanding how our body responds to it can only add to our understanding of locomotion. Like the unseen electrical fields, which also permeate all of space, the influence of gravity is a direct result of the properties of the underlying physical field. We shall discuss these fields later.

Acceleration, Force, and Mass Are Related

Whatever their source, forces cause acceleration, that is, they induce changes in velocity. These changes can be positive or negative, increasing or decreasing, but they are always called acceleration. The word deceleration was coined to denote the process of decreasing the velocity of an object, as in reducing the speed of an automobile, and although it is useful in some contexts, such a reduction is already included in the vector definition of acceleration. Consider an apparatus with a simple hammer that we can use to strike any body with a known force. In striking two separate bodies of the same material with the same force, if the velocity induced in one body is twice the velocity induced in the other body, we say that one has twice the inertia (mass) of the other.

Newton's Three Laws in Words

Newton summarized what was known about mechanics at the time in the three laws mentioned above. In his words these laws of motion are:

> *Law I:* Every body continues in its state of rest, or of uniform motion in a right (straight) line, unless it is compelled to change that state by a force impressed on it.

> *Law II:* The change in motion [rate of change of momentum] is proportional to the motive force impressed; and is made in the direction of the right line in which that force is impressed.

> *Law III:* To every action there is always opposed an equal reaction; or, the mutual actions of two bodies are equal, and directed to contrary parts.

Inertia Is the Resistance to Changes in Motion

It is obvious that the first law is a slight modification of Galileo's principle of inertia and gives prominence to the notion of persistence. A phenomenon continues to do whatever it is doing unless an intervention occurs to change that state of behavior. That intervention, whatever it is, we define as a force. This definition is made more precise in the second law, in which both the magnitude and direction of the impressed force are related to the change in magnitude and the change in direction of the momentum. The momentum is defined by Newton as the product of the mass and the velocity of an object. The distinction between mass and weight had not been fully appreciated up until that point, and we discuss that distinction more fully subsequently. For the moment, we merely point out that the form of the force given by Equation (3.1) is only valid if the mass of an object is constant, since the momentum (mass times velocity) can change by modifying the mass of an object as well as by modifying its velocity.

Force Is Changing Momentum

Most of the situations we shall be discussing involve constant mass, so the time rate of change in the momentum will be given by the mass times the time rate of change in the velocity (acceleration), that is, Equation (3.1). At the risk of burdening the reader with additional mathematics, consider the momentum of an object to be given by p, and the change in the momentum that occurs in a time Δt to be given by Δp. The force equation replacing Equation (3.1) is then indicated by the ratio of the change in momentum to the change in time:

$$F = \frac{\Delta p}{\Delta t} \tag{3.2}$$

The relation of Equation (3.2) with Equation (3.1) is that the change in momentum can arise from a change in the velocity (acceleration) or from a change in mass.

Influence of Force on Runners

As an example of force, consider the impact of a runner's foot with the ground. The velocity of a runner is determined by how much force can be exerted against the ground as the runner pushes off with each step. The ground must absorb the impact of the runner's heel coming down, the center of gravity moving forward as the force rotates to the ball of the foot, and then the push off, keeping the center of gravity in a continuous forward motion. Now suppose that the runner has a backpack with a broken strap, so as she runs various articles fall out of the backpack. Consequently, the force with which her heel strikes the ground will be reduced over time since, with the reduction of mass, the change in momentum will also be reduced. If she runs at a constant speed, the change in force cannot be calculated using Equation (3.1) since in the example the mass is not constant. Thus, we would use Equation (3.2) instead of (3.1), since the force equation (3.2) allows for change in mass, as well as changes in velocity. Do not be overly concerned with the equations, however; you will not be required to solve such equations, only to discuss them.

Units of Force

The unit of force called the Newton (N) is defined to give one kilogram of mass an acceleration of one meter per second per second, following the practice whereby scientists enshrine the memory of certain departed colleagues by placing their name on some appropriate unit with which they had something to do. The formal definition of one meter per second per second means that an object with the speed of one meter per second is changed by this amount in each second. Therefore, a mass with a speed of one meter per second at the end of one second increases to two meters per second at the end of two seconds, increases to three meters per second at the end of three seconds, and continues in this way until acted upon by an external force.

The Third Law of Motion

How is it that we arrive at the third law? Consider two weight lifters labeled C and C' in Figure 3.2. They both lift a weight W. The first lifter exerts a vertical force F

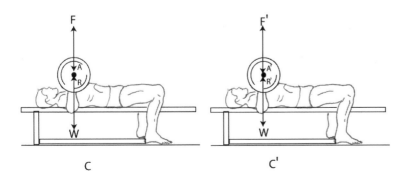

Figure 3.2. The balance of forces is proof that two forces are in fact equal to one another.

and the second a vertical force F'. If equilibrium is produced in both cases, that is, the weight is held stationary above the lifter's head, we conclude that the two forces F and F' are equal to one another, since they are both equal to the weight being lifted, W.

A Closer Look at Action and Reaction

Adapting an argument of Poincaré [14], we cannot say the weight W being lifted by the body C is directly balanced by the force F. A more thoughtful analysis reveals that what is applied to the body C is the action A of the weight W on the weight lifter's arm C through the tension of his muscles. In turn, the weight has two forces acting on it: its own weight W and the reaction to the weight lifter's hands given by R. The force F is equal and opposite to A, since the two are balanced. The force A is equal and opposite to R, due to the equality of action and reaction. Finally, the force R is equal and opposite to the weight W, since the two are in balance. It is from these three equalities that we deduce the equality of F and the weight W.

Action Equals Reaction

From the argument of the lifter's weights, it is clear that in order to determine the equality of two forces we must also impose the condition of the equality of the forces of action and reaction. Poincaré concludes from a similar argument that Newton's third law is not a matter of experimental observation at all, but rather a matter of the definition of force. We only mention this to demonstrate to the student that these concepts are neither simple nor obvious, and those who believe them to be so are missing the big picture.

Conservation of Momentum Is the Law

A somewhat deeper insight into the third law is gained if we interpret the word *action* to mean the change in *momentum*. In this way, we see that whatever action is lost by one body is gained by another body; that is, whatever the change in momentum induced in one body is compensated for by the change in momentum of the second body. The momentum lost by one body is gained by the other body. This is simply the law of conservation of momentum. This law states that the total momentum in an isolated mechanical system remains unchanged. Of course, no physical system is ever completely isolated and the same is true for the human body. What can be said is that in a conceptually limiting case, one in which the influence of the environment on a body is suppressed as much as possible, momentum will be conserved. We shall have a great deal to say about conservation laws subsequently.

Are Newton's Laws Only Definitions?

If Newton's three laws of motion are, in fact, nothing more than definitions, then why are they so special? The answer to this question lies in the fact that Newton's laws provide a way of logically determining the forces acting on a body. Through the mathematical representation of these forces using Equation (3.1), a way to calculate the future motion of that body is constructed. The three laws model the rate of momentum transfer from one body to another as a force, and enable us to determine the form in which a force may arise in a physical situation that we have not as yet investigated experimentally.

3.1.2 Phenomena Are Not Always Simple

Some Phenomena Have No Mathematics

In an elementary introduction to the physical sciences such as we are trying to lay out here, it is usually assumed that all physical phenomena have relatively simple mathematical descriptions. That is not to say the mathematics is not difficult to understand, but rather that the physical process follows a straightforward mathematical procedure that can be learned by anyone with the desire and willingness to do the work. This is not always the case, however. For example, the swirling water of a mountain stream and the tumbling air currents in the wind are not understood. These phenomena, known as turbulent fluid flow, restrict the speed a plane can travel, determine the maximum speed of the Olympic swimmer, and humble the most valiant of airline passengers. Turbulence is one of those physical phenomena that is not taught in most physics departments, in large part because it is an embarrassment that we do not understand this most fundamental property of fluid flow. Turbulence is taught in engineering departments, however, because in designing structures such as aircraft, bridges, and boats the influence of turbulence must be taken into account even if it is not understood. Turbulence is like the crazy uncle who lives with the family. You do not understand his behavior, but you make allowances for it.

Question 3.1

What did Galileo believe about mathematics and physical reality?

Question 3.2

Name Newton's three laws of motion and give a brief description of what they mean.

Question 3.3

Define inertia in your own words and explain how it is related to mass.

Question 3.4

Define mass, force, and acceleration and explain how they are experimentally determined to be related to one another.

Gravity Is Always There

3.2 SO THIS IS GRAVITY?

The one force with which everyone is familiar is that of gravity. We cannot see gravity, we cannot taste gravity, we cannot smell gravity, but we are all, sometimes painfully, aware of its existence. The understanding of the Universal Law of Gravitation has been called the greatest generalization achieved by the human mind. This is the phenomenon that guides the planets in their orbits, keeps the moon from crashing into the earth, and prevents the bio-shell that encases the earth from drifting into space. It is also the force that breaks the athlete's bones during competition, twists the ligaments at our joints, and causes everything to sag as we get older. It is remarkable that nature follows such an elegantly simple principle as the law of gravitational attraction, but even more remarkable is the fact that we can understand it. So what is the Universal Law of Gravitational Attraction?

3.2.1 Two-Mass Universe

All Objects Attract

Any two masses in the universe attract each other with a certain amount of gravitational force. For example, the bright star Vega and an isolated hydrogen atom between the stars each exert an equal and opposite pull on the other because of the gravitational force. Any two objects attract each other with the same strength; you and your classmates, a football and Betelgeuse, a chair and a table, a runner and the earth, each pulls on the other with equal force. How Newton was able to conceive of such a force is beyond our ken, but he did and we are able to capitalize on his genius.

The Mathematical Model of Gravity's Force

In detail, the Universal Law of Gravitation is that every object in the universe attracts every other object in the universe with a strength that is directly proportional to the product of their respective masses and inversely proportional to the square of the distance between them. If we denote the mass of the first object by m_1, the mass of the second object by m_2, and the distance between them by r, then the Newtonian attractive force F between the two masses might be written as

$$F = \frac{m_1 m_2}{r^2} \qquad (3.3)$$

which is not so bad as far as equations go. How can we determine if Equation (3.3) is correct or not? The answer should be immediate—of course, do an experiment. However, given the limited means available to us in the classroom, how can we test the validity of Equation (3.3)?

A Universal Gravitational Constant

Let us consider a consistency check of Equation (3.3) that involves dimensional analysis. We know that the dimensions of a force are the product of mass and acceleration, so that $[F] = [MLT^{-2}]$, but it is obvious that the right-hand side of Equation (3.3) does not involve the time, but has the form $[M^2 L^{-2}]$. Therefore, something is wrong, because the equation is not dimensionally homogeneous; the dimensions do not agree on the left- and right-hand sides of the equation. This inconsistency suggests that we have overlooked something, and this turns out to be the constant G, not to be confused with the acceleration of gravity g. By including the factor G on the right-hand side of Equation (3.3) we obtain

$$F = G \frac{m_1 m_2}{r^2} \tag{3.4}$$

so that if the dimensions of G are $[L^3 M^{-1} T^{-2}]$, the equation is dimensionally balanced. The constant G is an experimental factor with units and is called the universal gravitational constant. The units of G are chosen such as to make the force come out in Newtons when the masses are in kilograms and the distance is in meters:

$$G = 0.0000000000667 \text{ Nm}^2/\text{kg}^2$$
$$= 6.67 \times 10^{-8} \text{ cm}^3/\text{gm sec}^2 \tag{3.5}$$

The value of G in Equation (3.5) is the best modern value of G and is determined such that the force between two masses of one kilogram each, separated by one meter, is precisely 6.67×10^{-11} Newtons.

Weighing the Earth

The English physicist Henry Cavendish used the value of the gravitational constant G to measure the "weight of the earth," which is to say, to determine the mass of the earth. But he probably believed that the title he chose for his seminal paper, *Weighing the Earth,* was sexier than, say, *Determining the Mass of the Earth,* and he was certainly correct. This experiment was a remarkable feat for the 18th century, or any century for that matter.

The Experiment to Weigh the Earth

Cavendish measured the gravitational constant G in 1798 by means of the delicate torsion balance depicted in Figure 3.3. In the experiment, a light rod with lead balls, two inches in diameter, on each end was suspended in the middle by a slender quartz fiber approximately one yard long. Two other lead balls, eight inches in diameter, were then placed adjacent to the smaller balls on the ends of the rods as indicated in the figure. The gravitational force between the lead balls produced a very tiny twist in the fiber and from the total angle of twist Cavendish could determine the strength of the force between the small and large lead balls. Cavendish was able to measure the force, the two masses, and the distance, and thus determine the magnitude of the gravitational constant G given by Equation (3.5). The force he measured was approximately 5×10^{-11} of a Newton, a value very close to that found using much more refined techniques over the last two centuries.

Converting Force to Mass

The mass of the earth is deduced from the experimental fact that the force exerted on a mass of one kilogram at its surface is 9.8 Newtons. Therefore $F = 9.8$ Newtons in Equation (3.4). The distance between the one kilogram mass ($m_2 = 1$ kg) and the

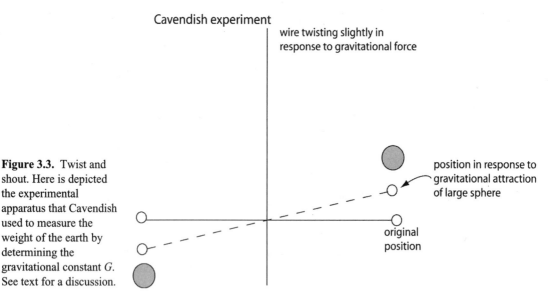

Cavendish experiment

wire twisting slightly in response to gravitational force

position in response to gravitational attraction of large sphere

original position

Figure 3.3. Twist and shout. Here is depicted the experimental apparatus that Cavendish used to measure the weight of the earth by determining the gravitational constant G. See text for a discussion.

center of mass of the earth is 6,400,000 m, the radius of the earth. Therefore, $r = 6.4 \times 10^6$ m in Equation (3.4). If m_1 is the mass of the earth, then by replacing the left-hand side of Equation (3.4) with 9.8 N, the distance r by the radius of the earth, and the mass m_2 by one kilogram, we obtain after some rearrangement

$$m_1 = \frac{9.8 \text{ N}(6.4 \times 10^6 \text{ m})^2}{1 \text{ kg}(6.67 \times 10^{-11} \text{ Nm}^2/\text{kg}^2)} = 6.02 \times 10^{24} \text{ kg} \qquad (3.6)$$

This is the theoretical mass of the earth. Stop for a moment and think about what has been done. Without leaving your chair, you have determined the mass of this rock that is orbiting the sun and upon which all of known civilization has developed. This is truly a remarkable thing!

3.2.2 Why Study Gravity?

Three Reasons For Studying Gravity

But why, in a book on human biodynamics, are we devoting so much effort to the understanding of gravity? The answer has three parts. The first is cultural, and like the advice to a son in *Hamlet*, knowing such things enriches the soul. The second has to do with the fact that most applications of biodynamics involve the force of gravity in one way or another. The impact of the runner's heel, the twisting of the muscle during a fall, and the torque about a joint are all due to gravity. Finally, from a more general perspective, the gravitational law does not possess a characteristic scale, which is to say that it is true on all scales, ignoring quantum mechanics. The form of Equation (3.3) indicates that from the microcosm of atoms to the macrocosm of the stars, the law of gravity remains the same. The form of the equation is self-similar and scale-free, a property that we shall look for in other relations in our study of biodynamics. By scaling we mean that multiplying the distance in the force law [Equation (3.4)] by a constant, say λ, we obtain the same force multiplied by a constant factor, that is, $F(\lambda r) = \lambda^{-2}F(r)$. Note that we shall have some success in finding such scaling properties in biology, particularly in the area of data analysis.

Scale-Free Aspects of the Cosmos

Most of the things we are familiar with have some characteristic time scale, or length scale, associated with them. A runner can only go so far before falling down from exhaustion. That distance for a given runner can be increased with training, but a new limit will be established that the runner will never be able to exceed, regardless of training. In the same way, there is a fastest response time that a given individual will have, say, in applying the brakes of his/her car to avoid an accident. Here again, training might reduce that response time by a factor of two, but it will probably never reduce it by a factor of a hundred or even a factor of ten. There are certain fundamental limitations to what the human body can accomplish. A man can run only so fast, and no faster; a woman can hold her breath under water for only so long, and no longer. Thus, any theoretical description of the processes of running or swimming under water must take these fundamental limitations into account. Any valid theory will have time scales built into it.

Scale-Free Aspects of Locomotion

Gravity does not contain any scale and, therefore, it is valid at both the microscopic and the cosmological scales, whereas we humans are of intermediate size. Therefore, we propose a separation of what we can understand about biodynamics into those parts that have fundamental scales, and thus are of limited application, and those parts, if any in fact exist, that are universal in scope. Are there some aspects of locomotion that do not possess a fundamental scale and are therefore scale-free? We have some indications, based on data processing, that certain characteristics of biodynamics are scale-free. The scale-free nature of such appropriate time series will be discussed subsequently.

Music of the Spheres

The Greeks believed in the beauty of the spheres and in heavenly symmetry, and this tradition was carried into the time of the astronomer Johannes Kepler. The view in Kepler's time (he died 12 years before the birth of Newton, who was born the same year Galileo died) was that nature was simple and orderly and its behavior was regular and necessary. In short, nature acts in accordance with perfect and immutable mathematical laws. In 1609 Kepler published *The New Astronomy* [15] in which he presented his first two laws of planetary motion. The first law displaced the Greek belief that the planets moved on circles, spheres, or epicycles, and asserted that the planets orbit the sun on ellipses, with the sun sitting at one focus of the ellipse. The second law replaced the idea that planets move at constant speed around the sun and stated that they sweep out equal areas in the plane of the ellipse for equal time intervals (see Figure 3.4). This law of equal areas implies that the speed of the planet in traversing its orbit from point P to point Q, a four week period in the spring, is faster than it is in covering the arc length from point P' to point Q', a four week period in the winter.

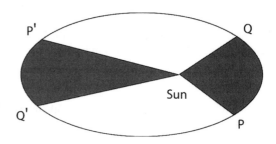

Figure 3.4. Kepler's law of equal areas in equal times. An elliptic planetary orbit is depicted. The shaded regions are equal in area and are swept out in equal time intervals. This implies that the planet has a greater speed in going from P to Q than it does in going from P' to Q'.

Kepler's Third Law of Planetary Motion

Kepler's third and final law was published some 10 years later in his *Harmony of the World* [16]. The law states that if τ is the period of revolution of the planet around the sun and s is the average distance from the planet to the sun during one period of revolution, then $\tau^2 = \chi s^3$, where χ is the same constant for all the planets. Kepler's calculations were designed to reproduce the astronomical observations of Tycho Brahe. This relatively simple algebraic equation describes the harmony of orbital motion, since it not only applies to planets orbiting the sun, but also to moons and any other satellites in orbit around other bodies. It was subsequently shown by Newton that each of Kepler's laws was a consequence of the Universal Law of Gravitation.

3.2.3 The Pendulum

The Physical Properties that Influence the Period of a Pendulum

Now let us come back down to earth. We have used the idea of a pendulum a number of times in our earlier discussions. It is a simple machine that undergoes simple periodic motion and can be used as a timing device. The period of the pendulum is constant, at least up to the point where we can neglect the loss of energy due to friction. If the period of a pendulum were not constant, or nearly so, a pendulum could not function as a clock. In the ideal case of a frictionless pendulum, in which energy is conserved, we want to determine the physical quantities that affect the period, just as we did for planetary orbits. Elements that might influence the period of a pendulum that immediately come to mind are the mass, m, of the bob at the end of the pendulum arm, shown in Figure 3.5; the force w, with which the earth attracts the bob, that is, the weight of the bob; the length of the pendulum arm supporting the bob, l; the length of the arc of the swing of the pendulum, l'; and the radius of the bob, l''. Of course, there is no a priori way to know which of these variables should be included in the dimensional equation. So as a first attempt at scaling the problem, we assume the length of the arc l' is very much less than the length of the pendulum arm and therefore neglect it. The same approximation is made for the size of the bob; as long as it is very much less than the length of the pendulum arm, we neglect it. We emphasize that these are all approximations and their validity can only be determined through experiment.

Pendulum

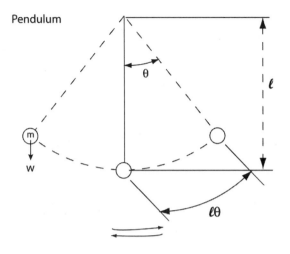

Figure 3.5. Period in terms of physical quantities. A schematic of a frictionless pendulum is depicted. The labeling of the positions of the pendulum is discussed in the text.

Period of the Pendulum from Scaling

Let us use the remaining quantities of mass, m, weight, w, and length, l, to write the period of the pendulum as

$$\tau = Cm^{\alpha}w^{\beta}l^{\delta} \tag{3.7}$$

The corresponding dimensional equation is

$$[T] = [M^{\alpha}(LMT^{-2})^{\beta}L^{\delta}] \tag{3.8}$$

We deduce the equations for the exponents through the principle of homogeneity to be

$$\text{exponent of } L: 0 = \beta + \delta$$
$$\text{exponent of } T: 1 = -2\beta \tag{3.9}$$
$$\text{exponent of } M: 0 = \alpha + \beta$$

These simultaneous equations are solved to yield the parameter values $\beta = -1/2$, $\delta = -\beta = 1/2$ and $\alpha = -\beta = 1/2$. Substituting these values for the exponents into (3.7) yields

$$\tau = Cm^{1/2}w^{-1/2}l^{1/2} \tag{3.10}$$

But we have previously determined that the weight of an object of mass m is given by mg, so that substituting for w in Equation (3.10) yields

$$\tau = C\,\frac{\sqrt{ml}}{\sqrt{mg}} = C\sqrt{\frac{l}{g}} \tag{3.11}$$

and C is a numerical coefficient that is determined by other means to be 2π. Thus, if the frequency of the pendulum, f, is given by the ratio of the number of radians in a circle, 2π, divided by the length of time required to traverse the circle, the period, $2\pi/\tau$, then from Equation (3.11) we obtain

$$f = \sqrt{\frac{g}{l}} \tag{3.12}$$

The frequency of the pendulum is seen to decrease with increasing length and to be independent of the mass of the bob.

Question 3.5

What did Cavendish's experiment demonstrate? What was the end result of his experiment? What was he able to do with his result?

Question 3.6

What are some reasons to study gravity?

Question 3.7

Explain the significance of the gravitational constant.

Question 3.8

Suppose you build a pendulum to measure a dynamic quantity in an experiment, but the timing is too slow. The pendulum does not have the necessary resolution, so you decide to double the pendulum's frequency. What is the simplest way to do this?

3.3 SUMMARY   Remarks Concerning Gravity

We now have one law that reigns supreme throughout the cosmos and on the surface of the earth. The Law of Universal Gravitation does not require any special circumstances, only that two masses be a given distance apart, in which case they will exert a force on one another that is determined by their masses. This is especially true of the earth and everything on it. We pull on the earth and the earth pulls on us. However, the earth is so much bigger than we are that, to our coarse senses, only the influence of the earth on us is perceived by us. We take this into account by saying that the gravitational field of the earth is uniform, giving rise to a constant gravitational acceleration at the earth's surface. We have seen that this factor of g enters into all our considerations regarding weight and mass. It is also important in many other physical phenomena, to which we now turn.

No Initial Choice of Variables Is Wrong

Huntley [11] and others [17] point out that the choice of variables made in constructing the dimensional equation is based on experience. But the novice should take solace in the fact that a choice of variables that is less than optimal will not produce a wrong answer; rather, it will usually produce a result that suggests a better choice of variables. Therefore, one can play with the scaling relations in order to find the proper description of the phenomenon of interest. This strategy for finding equations leads us to the relation for the period of the pendulum that depends on only the acceleration of gravity g and the length of the pendulum, l. We shall subsequently see that this simple equation provides significant insight into how we walk and run.

Apparent Deviations from Newton's Laws

Examples of large-scale motion in the human body are the smooth behavior of the cardiac muscle in the pumping action of the heart, the cyclic behavior of the motion of the leg during walking, the bellows-like behavior of the lungs as we breathe, or any of a number of other physiological systems that operate smoothly as we move around. These are the activities that can reasonably be modeled using Newton's laws for their average motion. The small differences in the time interval from one beat of the heart to the next, from one step to the next, for one breath to the next, are all consequences of the fact that the average behavior does not tell the whole story. But we discuss that later.

PART II

DYNAMICS, THE FUNDAMENTALS OF MOTION

Parts I and II of this book introduce some of the most important concepts in science and in the limited domain of biodynamics. In this chapter, we introduce and discuss the allometric growth law (scaling in biology), conservation of energy, linear response in physical and biological systems, and reductionism. In the following chapters, we extend these discussions to include the traditional notion of scaling as used by D'Arcy Thompson. In this historical approach, there is assumed to be a parameter that can be used to deform one or more of the basic variables describing the organism of interest, but the underlying assumption is that the process is continuous. We contrast that approach with the more modern notion of fractals in which metrics, like the length of a curve, cease to have meaning. We must reexamine certain fundamental concepts, for example, how we measure space and time, in order to understand the phenomenon under study. We use data to show that both the classical and fractal kinds of scaling exist in biology.

The conservation of energy is one of the most mysterious principles in physics. As Feynman argues, what energy is exactly we do not know, but we understand that if we consider a process and keep an eye on the various forms that energy takes as the process changes over time, we will always get the same number. The conservation of energy means that energy cannot be created or destroyed, but can only change form. In physics, we say that a conservation law is implied by a symmetry principle. If a process is symmetric in time, which means that we cannot determine whether the system is moving forward in time or backward in time, then energy is

Biodynamics: Why the Wirewalker Doesn't Fall. By Bruce J. West and Lori A. Griffin
ISBN 0-471-34619-5 © 2004 John Wiley & Sons, Inc.

conserved. If you think about this for a moment, you will see that in order to tell if a system is going in one direction the energy must change. This loss of energy breaks the symmetry in time.

There is even a law about the relationship between the symmetry of a system and the kind of conservation law that is implied by that symmetry. Every symmetry in a system has a corresponding conservation law, involving conjugate or complementary variables. Energy and time are conjugate variables, as are position and momentum. Thus, when space is homogeneous, which is to say there is no way, dynamically, to determine where we are in space along the direction we are going, then linear momentum is conserved. In a similar way, when space is isotropic, spatial rotations are indistinguishable and angular momentum is conserved. Emmy Noether developed a mathematical theorem that proved the relation between conservation laws and symmetries for physical systems.

The reductionistic philosophy has been applied to all manner of phenomena, including something as familiar as white light. The separate frequencies of white light yield the colors of the rainbow, whereas the incoherent superposition of the separate frequencies yield the white light flooding the garden in the morning. Newton was able to separate every component of light from every other component, using a prism to pull all the colors of the spectrum from white light apart. This encouraged scientists to seek the appropriate prism for separating very different kinds of complex phenomena into their constituent parts and, by understanding these parts, influence the process. Scientists and others have worked to make people healthier, faster, smarter, to speak better, and so on, by identifying and sometimes understanding those parts of the process that have apparently kept the individual from being the fastest, the smartest, the best spoken, or the healthiest. In the back of our minds is the assumption that if we only understand the components of a complex process, then we might be able to tinker with these components and thereby change an undesirable outcome into a desirable one.

The linear worldview has been developed so completely because it works for describing so many physical phenomena. However, the stress–strain relation, apparent in linear phenomena like the swinging pendulum, does not properly describe the response of biomaterials. Materials like ligaments respond nonlinearly to applied forces and this would certainly have bothered Hooke, the first scientist to determine the linear stress–strain relation.

In the previous part of this book, we learned that if a physical system has a symmetry, then there is a corresponding conservation law. In the dynamics of the human body, one of the most important of these conservation laws is the conservation of momentum. In developing a devastating punch, the boxer learns the conservation of momentum. Using this simple principle, the momentum of the opponent's attack is transformed into a breathtaking flight through the air by the martial arts expert. In every sport requiring the control of motion, the athlete must master this law and learn how to turn it to his/her advantage.

CHAPTER 4

Measures of Motion

Objectives of Chapter Four

- Understand that the various parts of complex phenomena are interrelated. In the case of biological systems, the allometric relation describes this interdependence.
- The human body acts to maintain its temperature at a constant value and in so doing makes use of a number of physical mechanisms to balance the generation and dissipation of heat.
- Realize that human locomotion cannot be explained by mechanics alone, but requires thermodynamics as well. However, the two are inextricably intertwined.
- Learn how to manipulate data, find an average, and identify error, an error being the deviation of the data from some idealized notion of what the process should be.

Introduction

All Parts of the Body are Interrelated

In this chapter we take the student beyond the rigid confines of mechanics and display the universality and utility of the allometric growth laws of biology. The allometric growth laws may relate the weight of a deer's antlers to its body weight, the length of a fish to its total weight mass and the speed of a runner to the length of his/her leg. These and many other applications show that different organs within a single body do not grow independently of one another. In the same way different characteristics within a body, not necessarily the weight of an organ, do not operate

Biodynamics: Why the Wirewalker Doesn't Fall. By Bruce J. West and Lori A. Griffin
ISBN 0-471-34619-5 © 2004 John Wiley & Sons, Inc.

independently of one another. As we said, the speed of a runner is determined by the length of the runner's leg, but the relation between length and speed is not a strict proportionality.

Why We Need Locometrics

A simple definition of locometrics is the study of the quantitative measures (metrics) that enable us to quantify biological phenomena related to locomotion, and to a certain degree these locometric studies overlap with biophysics. For example, to understand the familiar act of walking requires that we know how energy is generated within the body. Further, we can trace how the chemical energy in food is converted into the mechanical work of the muscles using the air we breathe. Finally, we must understand how we shed the energy generated in this process, in the form of heat, once the work is completed. The human body is a constant temperature reactor. It maintains this temperature by means of a delicate balancing act between the generation and dissipation of heat. The body works as a biochemical engine, one that retains a state of dynamical equilibrium with a corresponding constant temperature. In this chapter, we review some of the mechanical and thermal metrics that have been found useful in understanding the workings of this machine.

Mechanics Explains a Lot, but Not Everything

Because mechanics provides only a partial explanation, we need to understand a great deal more than just how forces act on the joints of our body while we walk or run. We must understand the rest of the biomechanical machine as well. To achieve this understanding, biometrics is used to expand the traditional view of biodynamics. We go beyond the mechanical explanations of the phenomena being investigated, and discuss fields, action-at-a-distance, energy, heat, and other related biophysical concepts. However, whenever possible, we shall reduce our discussions to mechanics, or at least to mechanical analogs, since, in our experience, mechanics is the discipline with which people are the most comfortable. Ropes, levers, pulleys, and pendula are objects familiar to most children; we can base our general understanding of how much of the human body works on them.

Metabolism and Its Rate

Metabolism may be defined as the sum of the physiological processes by which an organism maintains life. The metabolic rate determines how rapidly organic chemicals are processed to yield energy that the body uses to generate new cells and tissues, give off heat, and undertake physical activity [18]. The metabolic rate has traditionally been expressed as a function of an animal's weight, using an allometric relation, as we discuss in this chapter. One of the overarching characteristics of this allometric relation is scaling.

The Importance of Scaling

Scaling is one of those concepts that has been around for a few hundred years, and although used quite often in biology and engineering, its significance has not yet penetrated into the elementary levels of science. One of the arguments we consider is the scaling relation between the volume and surface area of an animal. We do this to understand the regulation of body temperature in warm-blooded animals. Temperature regulation is dependent on the balance between an animal's rate of heat generation (dependent on volume) and the rate of heat loss (dependent on surface area). We find that geometrical scaling is not sufficient to explain the observed relation between volume and surface area. Specifically, we need to introduce noninteger dimensions, so-called fractal dimensions, to understand scaling in animals.

Homeostasis Is Not as Important as We Once Thought

**4.1
BIOTHERMAL
CONCEPTS**

Living systems have traditionally been thought to be homeostatic; that is, being alive is to be in a dynamical steady state that maintains the processes of the system without changing their dynamical characteristics. The degree of an organism's independence from its environment is determined by its ability to maintain a constant internal environment. For example, body temperature in warm-blooded animals, blood pH, blood pressure, pulse rate, and other such physiological measures are all kept more or less constant, independently of what the environment is doing. This is Cannon's principle of homeostasis. Thus, changing environmental temperature does not necessarily change the temperature of the body of a homeothermic animal, but it does trigger a thermoregulatory response whose purpose is to counteract the external change in temperature.

Biostructure Can Be Viewed as Dissipative Structure

This picture of the regulation of bodily functions is consistent with the thermodynamic view of minimum entropy production proposed by the Nobel Laureate Ilya Prigogine [19]. In this latter view, there is a flux through any nonequilibrium system, such as a flow of energy or material, that is necessary to maintain a structure or pattern. The organization required to maintain the pattern is called a dissipative structure, since a certain fraction of the flux is "dissipated" within the system, and it is this dissipated flux that is used to maintain the structure. The pattern or structure in a biological system is the high state of organization required to keep the organism alive. Schrödinger, one of the fathers of quantum mechanics, may have been the first to set down this principle. He called it negentropy [20], referring to the stuff a living body takes from the environment to maintain itself. The local organization, life, with its subsequent decrease in entropy, is paid for by the increase in entropy of the immediate surroundings. In this way, the overall order of the organism plus environment remains the same or increases—a little more order here, a little less order there. This important idea will be elaborated on once we lay the groundwork for such things as entropy and thermodynamics, and we do this emphasizing dynamical and mechanical concepts, without the usually attendant formalism.

What Is Hot, What Is Not?

Even in our discussion of energy transfer processes that involve heat, as in the case of our runner, we shall rely primarily on mechanical considerations. We postpone the general discussion of energy for the time being, in order to carefully introduce this most important concept. We shall be a little looser in our treatment of heat in order to assist the reader in formulating a mechanical interpretation of his/her experience of what is hot and what is not. In parlance, we talk about how hot a day it is, and to quantify how hot it is, we give the temperature, say 100 °F. Most people raised in a temperate climate would agree that 100 °F is indeed hot, but someone raised in the tropics might disagree. The point is that "hot" is a qualitative term subject to interpretation, whereas temperature is quantitative and is presumably objective rather than subjective. So what exactly is being measured with a thermometer? What does temperature mean? To interpret the temperature of the air, we have recourse to large collections of particles and we shall discuss their statistical nature. We do this because the physical meaning of heat is actually contained in the statistics of particle motion. We shall clarify this apparently outrageous statement, subsequently.

Temperature Is What?

In the previous chapter, we introduced the notions of atoms and molecules. The picture that many people share is that these molecules are like tiny billiard balls that do not interact with one another unless they are colliding. These microscopic particles are never stationary. For example, the particles of air in a room move erratically, darting about from place to place, changing their velocities by colliding with one another and with the walls. Each cubic centimeter of air, a volume about the size of a bouillon cube, contains approximately 10^{23} particles. This is an amazingly large number. To get a picture of just how large this number is, assume that a single molecule of air takes up a space as large as our bouillon cube, in which case the original cube of air would be as large as the earth, making the earth a cube rather than a sphere, but this is a minor point. To make matters worse, however, most of this square earth volume would be empty space, that is, there would be no molecules in most of the volume, so there is plenty of room for them to go whizzing around. When particles do strike one another, or strike a surface such as a wall, they exchange momentum, which is to say they exert a force on whatever they hit. The force per unit area of wall is what we call pressure. Similarly, temperature, our colloquial measure of heat, is the average energy of these tireless particles, associated with their motion.

Evaporative Cooling Is Related to Microscopic Particles

Now consider our sweaty runner again. If it is not too hot a day, the runner notices that by ducking out of the sun, into the shade, she cools down. Of course, everyone knows that perspiration evaporates from the body and thereby cools her off. But what exactly does that mean? How does evaporation cool her off? What happens is that the sweat on the skin is, in part, made up of water molecules. These molecules are jumping around like the molecules in the air, and some are sufficiently energetic that they break through the surface barrier of the moisture on the skin and escape into the air. At the same time, however, water molecules (water vapor) in the air are impacting the surface of the skin. Some of these water molecules from the air side are sufficiently energetic that they are absorbed by the sweat on the skin. Thus, there is an ongoing dance of particles leaving and entering the body through the skin and the water on the skin surface. However, since the body of the runner is warmer than the air, on average, the particles leaving the body's surface are on average hotter, that is, they are more energetic, than are those absorbed by the body's surface, thereby lowering the average temperature of the skin surface and eventually cooling the entire body.

Evaporation Is More Particles Out Than In

The "dry heat" of the desert regions of the world can be understood in terms of the absence of water vapor in the air. Since there is less water vapor in the atmosphere in an arid climate than in a temperate one, there are fewer water molecules available to replace those leaving the body. This imbalance in the number of particles leaving and entering the moisture on the skin reduces the amount of water on the skin and reduces the surface temperature that much more rapidly. This is the process of evaporation, which leads to a loss of water from the skin surface and a cooling down of the entire body. This is also how your coffee cools in the morning. The vapor above the hot fluid surface is a consequence of the loss of energy from the escaping particles to the cooler particles of the air. This loss of energy produces centers of condensation where the water molecules collapse into droplets, thereby producing the water vapor we observe.

Question 4.1

Now that we have described evaporation, let us consider condensation, the inverse process of evaporation. Describe, in terms of particle motion, the process of condensation of water on the outside of a cold drink on a warm summer's night. (Answer in 300 words or less).

4.1.1 Metabolic Rate

There Is a Balance in Heat Transfer

The efficiency of the energy exchange process during the evaporation of water is determined by the surface area of the body exposed to the air. The greater the exposed surface area, the more particles that can come in contact with that area, and the more particles that can be involved in the heat exchange process. The greater the surface area of a body, the faster the rate of cooling. The rate of cooling of a body is proportional to the surface area according to Newton's empirical findings and the Stefan–Boltzmann law, both of which we discuss subsequently. In order for the body to maintain a constant temperature, the rate of heat generation through the burning of fuel (heat gain) and the rate of cooling through evaporation (heat loss) must balance one another. Of course, there are other mechanisms for gaining heat other than the burning of fuel, just as there are mechanisms for cooling in addition to evaporation, all of which will be taken up in good time.

Allometric Relations and the Metabolic Rate

The total utilization of chemical energy for the generation of heat by the body is referred to as the metabolic rate, R_{met}. In Figure 4.1, the metabolic rate for mammals

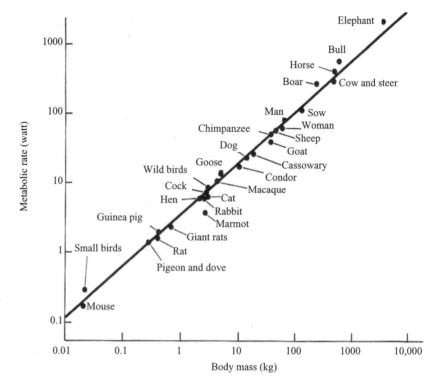

Figure 4.1. Mouse to elephant curve. Metabolic rates for mammals and birds are plotted versus the body weight (mass) on log-log graph paper. The solid line is the best linear regression to the data. Adapted from Schmidt-Neilson [23].

and birds is plotted versus body weight on log-log graph paper. This is often called the "mouse-to-elephant curve" and the first version of it was constructed by Kleiber in 1932 [21]. These data all fall along a single straight line on graph paper that has logarithmic scales for both the ordinate and abscissa. On this graph paper, the logarithm of the animal's metabolic rate R_{met} is graphed versus the logarithm of the animal's body weight W_b. The straight line on such "log-log" graph paper indicates that the metabolic rate is of the power-law form:

$$R_{met} = \alpha W_b^{\beta} \qquad (4.1)$$

where α and β are empirical constants. Equation (4.1) is the first of many "allometric relations" that we discuss in this book. The observed values of the constant α, of course, depend on the units chosen, whereas the value of β is independent of units. The body weight is often replaced with body mass in biological discussions, so that the metabolic rate is related to a mass rather than to a force. Equation (4.1) is, in fact, what is found in Brody's book, *Bioenergetics and Growth* [22], but the difference between weight and mass is often confused in the literature. The value of the observed power-law index obtained from a linear regression of the data in Figure 4.1 is $\beta \approx 0.75$. The term linear regression means that a straight line can be drawn through the data with a minimum of error. The exact value of the power-law index is still open to controversy, however, and the reader is referred to Schmidt-Nielsen [23] for an excellent critique of the experimental literature and the interpretation of this parameter.

Replacing Weight with Length

If we replace weight with volume in Equation (4.1), which we can do since the mass density of most animals is nearly one, we obtain another allometric relation:

$$R_{met} = \alpha'V^{\beta} = \alpha'l^{\beta'} \tag{4.2}$$

where $\beta' = 2.25$, obtained by using $V = l^3$ in Equation (4.1) to get $\beta' = 3\beta$. The metabolic rate is, therefore, a nonlinear function of the characteristic length of the animal, l. It should be clear from Equation (4.2) that the rate at which an animal consumes fuel is neither proportional to its surface area nor to its volume. If the metabolic rate were proportional to the animal's surface area, then β' would be equal to 2, since in that case β would be equal to 2/3. If the metabolic rate were proportional to the animal's volume, then β would be 1 and β' would be 3. Neither of these latter two cases are borne out by experiment, since β lies between these two values, $2/3 < 3/4 < 1$.

An Empirical Law

Here we observe that, although Equation (4.1) is not a law of nature in the sense of Newton's laws in mechanics, it is one of those experimental summaries that captures a constraint on nature's design of biological systems. No matter how efficiently an organism exchanges energy with the environment, the simple fact that the volume increases more rapidly than the surface area containing that volume limits the size of the organism. The notion of growth and balance is a recurrent theme in our discussions.

Comparative Physiology

In our discussion, we often use arguments from comparative physiology that relate biosystem variables such as the metabolic rate, the volume of oxygen exchange in the lung, the average heart rate, and many others to the weight of the body. We review such relations for a number of reasons. First, there is a 100 year history of the successful use of allometric relations that summarize correlations between biovariables and weight. These relations constitute a phenomenological network that guides the theoretical understanding of the biodynamics of the human body. Second, allometric relations emphasize the importance of scaling in the understanding of complex phenomena. Third, biological structures, like the human body, are designed to meet, but not exceed, the maximal requirement to perform its required function. This is the principle of *symmorphosis*, first articulated by Taylor and Weibel [24]. Finally, the discussion of allometric relations, such as given in the excellent monograph by Calder [25], is based on static relations between the mass of the body and a biovariable.

Mass or Weight?

It is probably worth commenting that since these are the first data we have shown, each point on the graph consists of averages over large numbers of measurements. It is, of course, the weight of the animal that is measured, so that the abscissa can be either force units for weight or mass units, the two differing by the gravitational constant g. One gram of mass undergoing an acceleration of one centimeter per second per second experiences a force of one dyne. Thus, the force exerted by gravity on a one gram mass is 981 dynes, and the force exerted by gravity on a one kilogram mass is 9.81×10^5 dynes. Thus, a kg of force is equivalent to 9.81×10^5 dynes, which is equivalent to one Newton of force. We shall have much more to say about the units of force and how they are used as we proceed.

Question 4.2

Describe three different phenomena that are known to scale. Give the experimental basis for this knowledge regarding scaling and explain your understanding of why scaling occurs. (Answer in 300 words or less.)

**4.2
ALLOMETRIC
SCALING**

There Is Scaling in Geometry

We wish to make the arguments regarding allometric relations more transparent, and therefore more useful. To make allometric relations easier to understand, consider the concept of geometrical scaling, also called *isometry,* meaning that two objects or two observables have the same measure. This is another way of saying that the dimensions of two observables are the same, so that whatever ruler we use to measure that dimension, it is applicable to both variables. This sameness is particularly apparent in discussing the geometric similarity of two distinct bodies, say two spheres or two cubes. In this geometrical context, an object may be characterized by means of a linear length scale, say l, so that its surface area, A, increases as

$$A \propto l^2 \qquad (4.3)$$

Here the strange symbol means "proportional to," so that Equation (4.3) shows that the surface area of an object is proportional to the square of a typical length of the object. We do not use an equality because if A is the surface area of a disk of radius l, there would be a numerical factor of π in front of l^2; that is, the area of a disk of radius l is πl^2. For a square of side l, the factor in Equation (4.3) is unity. Therefore, we use the proportionality sign and avoid the use of overall constants. The volume of the above object increases as

$$V \propto l^3 \qquad (4.4)$$

so that if Equation (4.4) is inserted into Equation (4.3), through the length $\propto V^{1/3}$, we obtain the relation between surface area and volume:

$$A \propto V^{2/3} \qquad (4.5)$$

The meaning of the algebraic Equation (4.5) is depicted in Figure 4.2, where the area is graphed as a function of volume, and the nonlinear relation between the area and volume is evident.

Mass is Proportional to Volume

The assumption is often made that mass increases as the volume of a body increases. We assumed this to be true in the discussion of the data in Figure 4.1. The pro-

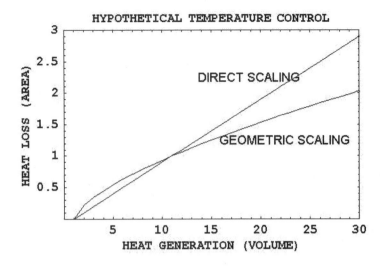

Figure 4.2. Two possible temperature control curves. The surface area of a body is plotted against the volume of the same body.

portionality between weight and volume is probably true in an average sense. However, for the exercise physiologist, the increase depends on whether the mass is muscle or fat, the former being denser than the latter. The mass of both is proportional to the volume, but the proportionality constant would be different between the two. For fat the mass, density is 0.90 gm/cm^3 and for muscle the mass density is 1.10 gm/cm^3, depending on sex, race and so on. On the average, this means that the numerical factor in Equation (4.5) would change, but the general form of the relation would remain the same.

4.2.1 Temperature Control

Control through Geometry

In order to maintain temperature control of the body in homeostasis, the heat the body generates must be balanced by the heat the body dissipates, or else we would either burn up with fever or freeze to death. The curves in Figure 4.2 indicate such possible relations. In Figure 4.2, the curve obtained from Equation (4.5) is a hypothetical temperature control through the balance of heat generation and heat loss. This theoretical relation assumes that the rate of heat loss is proportional to the surface area of the body. Recall the argument for evaporative cooling as a justification for this form of heat loss. Further, the rate of the generation of heat is assumed to be proportional to the volume of the body. Consequently, if we use the above geometric argument to construct a relation between the metabolic rate and the body weight, we would expect that the power-law index β would be 2/3 in Equation (4.1).

Two Kinds of Control

In Figure 4.2, the area increases quadratically with a linear scale and the volume increases cubically with a linear scale. The two curves depict different relations between area and volume. Comparing the two curves we observe a disproportionate increase in one with respect to the other. The straight line indicates a direct proportionality between heat loss (proportional to A) and heat generation (proportional to V).

Surface Law for Temperature Control

The observed value of the power-law index in the metabolic relation [Equation (4.1)] is 3/4 and not the 2/3 given by the geometrical self-similarity argument. The geometrical argument was, in fact, used over a century ago to arrive at the "surface law," which maintained that there was a nearly constant ratio between heat production in a body and its surface area. As Schmidt-Nielsen [23] points out, this assumption led to the extensive but mistaken use of body surface as a base of reference for metabolic rates, most notably used in clinical medicine.

Other Heat Loss Mechanisms

What is being overlooked in the above discussion are two additional mechanisms for heat loss: radiative heat loss and convection. In the latter case, the hot particles leaving the body lose their energy through collisions with neighboring air molecules that are less energetic. The heated air has fewer particles per unit volume than the ambient air because of the increased number of collisions. Therefore, the heated air is lighter than the ambient air, rises away from the body and is replaced by the cooler, denser, ambient air. In this way, cool air comes in contact with the body, is warmed by body heat, and then flows away. This net transfer of heat from the body that is carried away by the moving air is called convection.

Convective Heat Loss

Notice that convective heat transfer requires a difference in temperature between points in space. If there is no temperature difference, there is no convective heat loss from the body because the air does not move. In general, one can define the efficiency with which heat is transferred from the surface of the body to the moving air by the convection coefficient K_c (Kcal per m^2 per hr per degree Centigrade). If the temperature of the surface of the body is T_s in degrees Centigrade and the air temperature is T_a, then for a surface area A exposed to the air, the rate of heat loss due to convection is linearly proportional to the temperature difference:

$$H_c = AK_c(T_s - T_a) \tag{4.6}$$

The thermal convection coefficient is determined by experiment and is found to be a function of air speed, increasing with increasing air speed [26]. The sign convention is such that the exposed area of the body loses heat when the body's surface is warmer than the air, $T_s > T_a$, and the body absorbs heat when the body's surface is cooler than the air, $T_s < T_a$. When the air is moving at a speed below 10 cm/sec, the thermal convection coefficient has a value of approximately 6 Kcal/m^2/hr/°C.

Radiative Heat Loss

Here we must anticipate material that will be introduced later and has to do with electromagnetic radiation. As you probably know, visible light is a form of electromagnetic radiation. However, there is a broad spectrum of wavelengths of electro-magnetic radiation (the colors of the rainbow), most of which we cannot see. This radiation is emitted by the microscopic particles in the light bulb (electrons) being excited by the electricity from the wall socket and de-excited through the emission of radiation at the wavelength of light we see. Thus, radiation is also an energy-gain–energy-loss process and provides another way in which the more energetic particles leaving the body can lose energy. When the escaping particles collide with air particles, they emit radiation, not at a wavelength we can see, but certainly at a wavelength we can feel. The heat we experience standing next to someone who is vigorously exercising is predominantly produced by radiative heat loss. Night vision goggles enable one to see the heat loss due to infrared radiation, even though we cannot see this radiation with the naked eye. Therefore, we can construct a radiative heat loss equation of exactly the same form as Equation (4.6):

$$H_r = A\varepsilon K_r(T_s - T_a) \tag{4.7}$$

where K_r is the radiation coefficient of the air and in typical situations has a value of 7 Kcal/m^2/hr/°C [26]. Notice that T_a does not have to be the ambient air temperature. For example, if the individual is standing on a rock in the desert in the heat of the day, the rock can be much hotter than the air and therefore it is the temperature of the rock and not the air temperature that enters into the temperature difference in Equation (4.7). The additional parameter ε is called the emissivity and is in the range $0 \le \varepsilon \le 1$. This parameter has to do with the electromagnetic properties of an object. An object that is a perfect reflector does not absorb energy and has $\varepsilon = 0$. An object that is a perfect absorber (is also a perfect emitter and is called a black body) has $\varepsilon = 1$. Therefore, ε is a measure of the efficiency with which a body absorbs and reemits radiant energy.

Being Small Is Cool

A consequence of the geometrical scaling relation between surface area and volume of a body is that smaller animals are more efficient in achieving energy balance than are larger animals. This may become clear if we consider the ratio of the unit of surface area to the unit of volume. Using Equation (4.5), we obtain

$$\frac{A}{V} \propto \frac{V^{2/3}}{V} \propto V^{-1/3} \tag{4.8}$$

Therefore, the surface area per unit volume decreases with increasing volume, indicating that a small animal has more surface area relative to its volume than does a larger animal. This geometrical relation supports the statement concerning the increased efficiency of small animals over large animals in dissipating heat by this mechanism, due to their relatively larger surface area.

Scaling in Euclidian Geometry

What is it that the scaling relationship between area and volume tells us? According to this principle, if one species is twice as tall as another, it is likely to be eight (2^3)

times heavier, but to have only four (2^2) times as much surface area. This tells us immediately that the larger plants and animals must compensate for their bulk; respiration depends on surface area for the exchange of gases, as does cooling by evaporation from the skin and nutrition by absorption through membranes. The 2/3 exponent from geometrical scaling is not sufficient to accomplish these tasks physiologically. The data indicate that the surface area of an organ must scale differently with volume. In fact, nature has found more than one way to enhance this scaling law, that is, to increase the exponent above 2/3. One strategy has to do with the geometrical concept of fractals.

Question 4.3

If the ambient air at a temperature of 26 °C, is moving slowly over a nude body of total surface area 1.7 m^2 at a surface temperature of 33 °C, what is the rate of convective heat loss? If a person who is resting quietly has a metabolic rate of 85 Kcal/m^2/hr and is in equilibrium with his/her environment (as much heat is being generated as is being lost), what fraction of the metabolic rate is convective heat loss?

Question 4.4

If the ambient air is at a temperature of 26 °C, what is the radiative heat loss from a nude body of total surface area 1.7 m^2 at a surface temperature of 33 °C? If a person who is resting quietly has a metabolic rate of 85 Kcal/m^2/hr and is in equilibrium with his/her environment, what fraction of the metabolic rate is radiative heat loss?

4.2.2   Scaling Using Fractals

The Term Fractal Was Coined by Mandelbrot

The word fractal was coined by Benoit Mandelbrot in the mid-1970s [27] to describe processes and objects having no characteristic scale length. What this lack of scale implies is that a process or object is self-similar. There are three different ways in which the term fractal is used: geometrically, statistically, and dynamically. A geometrical fractal has a geometrical self-similarity in that any small piece of the object has the same geometrical form as the original object. A statistical fractal has a statistical self-similarity in that any small segment of the process has the same statistical distribution as the original process. A dynamical fractal has a dynamical self-similarity in that any small part of a trajectory has the same dynamical properties as the entire trajectory. There are fractals in space and there are fractals in time, and what distinguishes fractals from other functions is that, in the phenomena they describe, all scales are tied to one another. The outline of the mountains in the distance is like the profile of the foothills in the foreground; the statistics of heartbeats over seconds is the same as those over hours. The point is that fractal processes are peculiar sorts of things that capture the natural variability in space and time of complex phenomena.

Fractals Are Self-Similar

Fractals may be understood through the direct generalization of a few concepts from geometry. In Figure 4.3, a one-dimensional line segment is depicted. A ruler

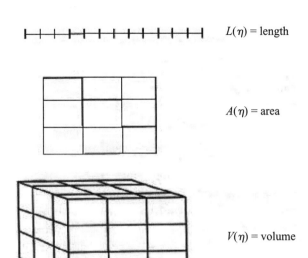

$L(\eta) = $ length

$A(\eta) = $ area

$V(\eta) = $ volume

Figure 4.3. Self-similarity in one, two, and three dimensions. Here we depict the number of hypercubes needed to cover one object in one, two, and three dimensions, with the volume of the hypercube being unity in each case. The size of the hypercube is η^D and its "volume" is $V(\eta) = N\eta^D$ for both integer and noninteger dimension D.

of length η is used to mark off N similar segments, each of length η to obtain N self-similar pieces of the original line. Therefore, we obtain an interval of total length $L = N\eta$ and if we take the total length of the line to be unity, $L = 1$, then the number of times we use the ruler is $N = 1/\eta$. In Figure 4.3 we also show a two-dimensional square. A smaller square of area η^2 is used to trace out a number of smaller areas inside the larger area. In this way, we partition the original square into N self-similar versions of itself. If the smaller square fits into the larger one exactly N times, then the total area is given by $A = N\eta^2$. If the original area is taken to be unity, $A = 1$, the number of smaller squares required to cover the larger square is $N = 1/\eta^2$. Finally, the last figure in Figure 4.3, is a three-dimensional cube. We create N self-similar replicas of the large cube by filling it with smaller cubes of volume η^3. Again, if the number of the smaller cubes fills (covers) the larger one without remainder or excess, then the total volume is given by $V = N\eta^3$. If we take the large cube to be of unit volume, $V = 1$, then we require $N = 1/\eta^3$ smaller cubes to cover the unit volume.

Self-Similar Dimension

Note that in each of the examples shown in Figure 4.3, we constructed smaller objects of the same geometrical shape as the larger objects in order to cover it. This geometrical equivalence is the basis of our notion of self-similarity. In particular, the number of self-similar objects required to cover an object of dimension D is given by

$$N = 1/\eta^D \tag{4.9}$$

where in our discussion D was either 1, 2, or 3. We can reverse this relation by taking the logarithm of each side of the equation and thereby define the dimension of the object by

$$D = \frac{\ln N}{\ln[1/\eta]} \tag{4.10}$$

where η is the linear scale size of the covering objects and N is the number of such objects required to cover the given object. Here we have used the logarithm to the base e, the natural logarithm, rather the logarithm to the base 10. Note that Equation (4.10) is mathematically rigorous only in the limit of the ruler size becoming vanishingly small. Further, although Equation (4.10) defines the dimension of a self-similar object, there is nothing in the definition that guarantees that D has an integer value. In fact, in the general case, D is noninteger. Opening up our minds to noninteger D is the generalization of the above definition of dimension.

Fractal Scaling

It would be possible to devote the remainder of this book to the properties of fractal structures, but that has been done very well elsewhere by a number of people, including Mandelbrot himself [27]. We use only a few examples to emphasize aspects of fractals of interest in biodynamics. Consider the object called a *Menger sponge,* depicted in Figure 4.4. If all the holes were filled, the volume would be given by the geometrical expression l^3, if l is the length of a side. However, the Menger sponge, like a real sponge, has many regions that penetrate from the surface to the interior of the object, thereby increasing the surface area but containing the object in the same overall volume of space. The volume of the space containing the sponge increases as l^3; however as Mandelbrot [28] points out, the volume of the Menger sponge increases as $l^{2.727}$, as determined by direct calculation. The area and volume can be related using two expressions for the length of the side of the sponge:

$$A^{1/D} = V^{1/3} \tag{4.11}$$

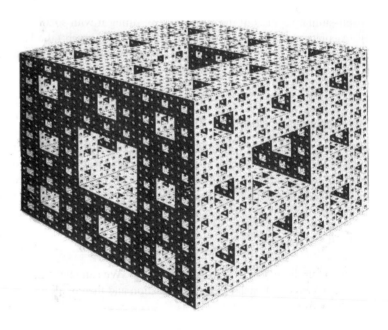

Figure 4.4. The Menger sponge is a fractal. The Menger sponge, with all its penetrations from the surface to the interior, has a fractal dimension of 2.7268. (From Mandelbrot [27], with permission.)

so that using the fractal dimension for the Menger sponge we have (see p. 112 of ref. [27]),

$$A = V^{0.909} \tag{4.12}$$

We see that the exponent 2/3 that arises from a Euclidean geometry argument is replaced by 0.91, in the area–volume relation, using a fractal geometry argument for a particular kind of object. We can again determine the area per unit volume to be

$$\frac{A}{V} \propto V^{-0.09} \tag{4.13}$$

so that the area per unit volume continues to decrease with increasing volume, but much more slowly than in the case of Euclidean geometry. Thus, the geometry of Euclid, for smooth continuous surfaces, is too weak to describe the observed metabolic rate, and the fractal geometry of sponges seems to be too strong. The truth, therefore, lies somewhere between Euclid and Mandelbrot.

Scaling for Different Structures

One way nature adds surface area to a given volume, as we have seen in the case of the sponge, is to make the exterior more irregular. The ratio of surface to volume in an object increases somewhat less strongly in trees, because of their distribution of branches and leaves. The simplest type of branching tree is one for which a single conduit enters a point, called a vertex, and two conduits emerge. This dichotomous process is clearly seen in the design of biological trees such as botanical trees, neuronal trees, bronchial trees, and arteries. An interesting relation, first observed by Leonardo da Vinci for botanical trees [29], involves the relationship between the diameter of an incoming branch with the diameters of the two subsequent branches:

$$d_0^\alpha = d_1^\alpha + d_2^\alpha \tag{4.14}$$

where d_0 is the original diameter and the subsequent diameters are d_1 and d_2. The data of da Vinci yields $\alpha = 2$, whereas the study of Murray [30], some 400 years later, gives $\alpha = 5/2$. The same relation was determined for the lung airway and the arterial system at each generation z [31]:

$$d_0^\alpha(z) = d_1^\alpha(z) + d_2^\alpha(z) \tag{4.15}$$

This is considered to be an efficient mechanism for passing gas or blood, respectively, to where they are needed; see, for example, MacDonald [32]. Suwa et al. [33] found that for blood flow $\alpha = 2.7$, whereas for the bronchial tree Weibel [34] and Wilson [35] determined experimentally that $\alpha = 3$.

The Human Lung

A physiological structure that hollows out the interior of an object, much like the Menger sponge does, is the mammalian lung. The human lung, following the form of a dichotomous tree, starts from the trachea and terminates at about 300 million air sacs. The bronchial tree approaches the more favorable ratio of surface to volume enjoyed by our evolutionary ancestors, the single-celled microbes [30]. For the moment, let us assume that the average diameter of each branch in the bronchial tree is the same after a bifurcation, $d_1(z) = d_2(z) = d(z + 1)$, so that Equation (4.15) reduces to

$$d^\alpha(z) = 2d^\alpha(z + 1) \tag{4.16}$$

This relation allows us to write, using the value $\alpha = 3$,

$$d(z) = 2^{1/3} d(z + 1) \tag{4.17}$$

so that the diameter size is reduced by a factor 0.794 in going from generation to generation. If d_0 is the diameter of the trachea, then the average diameter of the bronchial airway, $d(z)$, at generation z is given by

$$d(z) = (0.74)^z d_0 \tag{4.18}$$

This exhibits a geometric decrease in the diameter of the airway with generation number. At the $z = 23$ generation, the average diameter of the bronchial tube is 0.1% the size of the trachea. We shall discuss this reduction more thoroughly later, including the reasons that one might expect a branching relation of the form of Equation (4.15) in the mammalian lung.

Question 4.5

Explain why there are so many experimental values for the parameter α in above branching laws. (Answer in 300 words or less.)

4.3 CLASSICAL SCALING

Failure of Traditional Scaling

Historically, the classical concept of scaling relies on the key assumption that biological processes, like their physical counterparts, including the physical laws, are continuous, homogeneous, and regular. This concept of scaling, developed by D'Arcy Thompson [36] and others, while of great importance, fail precisely at this point, since most biological systems, and many physical systems, are discontinuous, inhomogeneous, and irregular. Therefore, the traditional scaling is not capable of accounting for the irregular surfaces and structures seen in hearts, lungs, intestines, brains, and even in the dynamical time series generated by these organs. Characterizing these kinds of systems evidently requires new models and new scaling ideas. Here, we discuss how the related concepts of fractals, nonanalytic mathematical functions, and renormalization group transformations provide novel approaches to the study of physiological form and function.

Dimensional Analysis

We discussed dimensional analysis in previous sections. The idea of the analysis was that by balancing the dimensions on both sides of an equation and using an appropriate set of dimensional constants, one can formulate the solutions that describe the evolution over time of the phenomenon being studied. Traditionally, this has been done less in physics than in engineering and biology. In the former case, the equations of motion can often be written down, if not solved exactly, whereas in the latter cases, more often than not, there are no known equations of motion, much less solutions to those unknown equations. In addition, biologists have historically viewed mathematical formalism in their discipline with suspicion. For this reason, those equations that have been developed within biology and which have survived for any extended period of time have an overwhelming base of data support, for example, the allometric relations.

Neither Theory Is Correct

There is a plethora of data in biology, and biologists apparently never tire of testing a mathematical expression that purports to "explain" a biological phenomenon. We have seen how the geometrical scaling of Euclid, along with our intuition and historical precedent, led to the idea of a "surface law." But more importantly, we saw how the data showed that the exponent of 2/3 is not valid. On the other hand, neither is the fractal geometrical argument that leads to an exponent of 0.91. Rather than either of these two exponents, the power-law index is determined by the data to be approximately 0.75. However, we do not abandon our notion of similarity; instead, we extend the similarity argument to the principle of *allometry,* which literally means by a different measure, so that two quantities, with different measures, may be related to one another. We find that allometry combined with the flexibility of fractal geometry can explain many of the observed relations.

4.3.1 Allometric Relations

Allometric Equation

The prototypical allometric equation was constructed by Julian Huxley in 1924, and discussed, along with empirical evidence, in his 1931 book, *Problems of Relative Growth* [37]:

$$Y = aX^b \tag{4.19}$$

where X and Y are biological observables and a and b are constants. In the typical case, X is the magnitude of an animal as determined by some standard linear measurement, such as its weight minus the weight of the organ, and Y is the magnitude of a differentially growing organ. In his book, Huxley records data ranging from the claws of fiddler crabs, to the heterogony of the face relative to the cranium in dogs as well as in Baboons, to the heat of combustion relative to body weight in larval mealworms. The two parameters a and b in Equation (4.19) are important for different reasons, as we shall see. This fact was not always appreciated. In fact, Huxley himself mistakenly dismissed the parameter a as an unimportant constant. Some contemporary scientists continue to hold this view.

All Parameters Are Important

Calder [24] believes that it would be "a bit absurd" to require the coefficient a to incorporate all the dimensions necessary for dimensional consistency of the allometric equation, particularly because b is in most cases not a rational number. He argues that equations based simply on observations are exempted from the dimensional consistency required for homogeneous equations, such as, for example, the kinds of equations we used in the application of dimensional analysis. He asserts:

> The allometric equations ... are empirical descriptions and no more, and therefore qualify for the exemption from dimensional consistency in that they relate some quantitative aspect of an animal's physiology, form, or natural history to its body mass, usually in the absence of theory or prior knowledge of causation.

At first sight, it does appear that to require the overall coefficient in Equation (4.19) to have dimensions with fractional exponents is perhaps "a bit absurd." However, in the years since the first publication of Calder's book the apparent ubiquity of fractal phenomena has become more and more apparent. This fact makes the argument regarding the irrational nature of the dimensions of a, being a justification for ignoring its dimensionality, less and less compelling.

Strange but True

A fractal function, whether representing a process in space and/or in time, has a fractal dimension, a dimension that is not integer valued. Thus, a fractal function might have the dimensions of time to a noninteger power, or a length to a fractional power, or even a mass to such a power. These apparently pathological functions, rather than being the exception, have been found to be the rule in biological phenomena. Examples of geometrical fractals in the human body are the branching of the airways in the lung, the folds on the surface of the brain, the venous system in the kidney, and the His–Purkinje conduction system in the heart to name a few. These and many other such phenomena are reviewed in references [8] and [38]. Examples of statistical fractals in the human body are the statistics of the interbeat interval time series of the heart, the statistics of the interstride interval time series in gait, the statistics of the interbreath intervals, and on and on [8]. We shall examine these dynamical processes somewhat later.

Fractal Dimensions

Equation (4.19) implies that the ratio of the relative growth rate of an organ to the relative growth rate of the body remains constant and is denoted by b. Thus, b is an invariant of the growth process. Even though the two growth rates can change over the age of the body and the organ, the ratio of the growth rates will remain the same over time if this equation is to remain valid at various stages in the animal's development. In this way, biologists discovered the importance of invariant quantities in the nineteenth century.

The Ratio of Growth Rates

The mathematical argument for the form of the allometric equation boils down to this: the percentage change in one biovariable, say $\Delta x/x$ (the ratio of the change in the independent variable, denoted by Δx, to the value of x), is proportional to the percentage change in another biovariable, $\Delta y/y$. The constant of proportionality between the percentage changes is the factor b. Suppose that, rather than a biological system, one was interested in investments in the stock market. Assume you have two stocks with different rates of return, α for stock 1 and β for stock 2. If the two stocks satisfy the allometric relation, the ratio of the two rates of return, α/β, would be constant over time. Thus, the total amount of money in stock 1 at any time would be equal to the total amount of money in stock 2 at the same time raised to the power $b = \alpha/\beta$. In this way, the growth of the two stocks would be related through the complex system known as the stock market.

Question 4.6

Suppose you invested $1000 in a stock that made a 6% return over the year. A friend of yours invested the $500 at the same time, but made the same amount of money as you did in the same period of time. What was your friend's rate of return? Explain how these two investments are allometrically related. (Answer in 300 words or less.)

4.3.2 Data Analysis

Linear Regression of Allometric Equations

Huxley observed that the allometric equation indicates that for any given size of a body, the total amount of mass that can be incorporated into an organ is given by that mass raised to a number. However, it is difficult to manipulate data into the form of the allometric equation and equally difficult to interpret such applications. To overcome this difficulty with the form in which the allometric equation usually appears, consider taking the logarithm of Equation (4.19) and writing

$$\log Y = \log a + b \log X \qquad (4.20)$$

The logarithm of the mass of the organ is determined by the logarithm of the body mass multiplied by the power-law index. On log-log graph paper, this form of the allometric equation is a straight line with a slope given by the power-law index b and the Y-intercept yields the overall coefficient a, as shown in Figure 4.5. Thus, Equation (4.20) is used as a linear regression equation to be fit to the logarithm of the data.

Finally, Some Data

Let us consider some examples of fitting the allometric relation to data from Huxley's book. In Figure 4.6 is depicted data from Huxley's book on the teleost fish, *Orthopristis*. The horizontal axis is the total length of the fish, and the vertical axes are taken to be the lengths of the head, trunk (body), and tail as separate data sets. In addition, the width and depth of the fish are also considered. Each of these data sets satisfy a scaling relation of the growth of all these various parts of the fish are strongly coupled to one another and to the overall length of the fish. Note that the allometric relation is in terms of lengths rather than mass in this example.

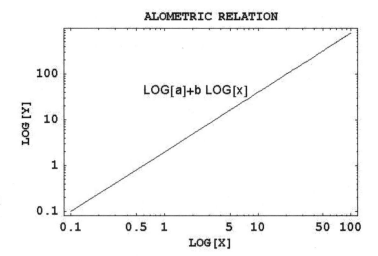

Figure 4.5. Straight line on log-log graph paper. Here is depicted a typical graph for an allometric relation of the form of Equation (4.20) with $a = 2$ and $b = 1.3$.

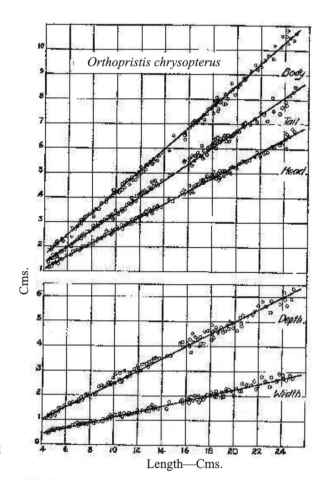

Figure 4.6. Relative growth of various parts of fish. The relative growth of the various parts of the teleost fish, *Orthopristis*; the body, head, tail, width, and depth are graphed as functions of total body length in centimeters. (From Huxley [37].)

Fish Sizes

Equation (4.20) is used as a fitting equation, where the parameters a and b are determined by the data using a fitting procedure that we discuss later. One might observe that the slopes of the various curves in Figure 4.6 for the body, tail, and head, are not parallel to one another; they have different b's. The same is true for the depth and width curves. The range of variation of the slope in these measurements is $0.1 \leq b \leq 0.45$. Note how clustered is the data around the regression line fit by Equation (4.20). Thus, it is possible to have many biovariables dependent on the same independent biovariable. Here, the dependent biovariables are the lengths of the body, tail, and head and the depth and width of the fish, and the independent variable is the overall length of the fish. Of course, we could also have graphed the dependent biovariables versus one another and eliminated the dependent biovariable altogether.

How to Calculate Averages

In looking at Figure 4.6, it should be clear that the empirical relation given by Equation (4.20) does not fit all the data points exactly. Rather, the equation for the straight lines fits the data in some average sense. The horizontal scale is the total length of a fish denoted by X and the measurements constitute a data set $\{X_j\}$ where j denotes the jth fish and $j = 1, 2, \ldots, N$, for a total of N fish used in making the measurements. One cannot rely on the measurements obtained from a single fish, but must concatenate the measurements from many fish in order to determine properties typical of fish in general. In the present case, we define the average length of a fish as the sum over the N measurements divided by the total number of measurements N:

$$\overline{X} = \frac{1}{N} \sum_{j=1}^{M} X_j = \frac{1}{N}(X_1 + X_2 + \cdots + X_N) \tag{4.21}$$

and denote this average with a bar over the variable. This is the definition of an arithmetic average, which we use throughout the book, where the Greek symbol Σ is shorthand for adding up all the terms indicated by the subscript j.

Method of Least Squares

For each of these fish, there is also a data set $\{Y_j\}$, where Y is the length of the fish body, tail, or head in the top graph and either the depth or width of the fish in the lower graph. Therefore, it is safe to say that someone stood around cutting up fish and subsequently measuring the heads, tails, and bodies as well as the depths and widths of the fish parts. That is how data is obtained in the real world. "It is a nasty job, but someone has to do it." The two parameters a and b in Equation (4.20) are intended to characterize the vast amount of data obtained from a collection of such measurements. Their values must be determined from the data set. Let us assume that we take a sequence of measurements $\{X_j, Y_j\}, j = 1, 2, \ldots, N$, in which the x measurement is exact, since this is the independent variable under the control of the experimenter (the total length of the fish), and any error is concentrated in the y measurement, which, presumably, the experimenter observes but cannot control.

With this assumption, we use Equation (4.20) to write the difference in the jth measurement,

$$\xi_j = \log Y_j - [a + b \log X_j] \qquad (4.22)$$

as the deviation in the y measurement from the "anticipated" or "theoretical" expression. If the j data point satisfies the prediction exactly, then the error as defined by Equation (4.22) would be zero. But life is not like that. Things are never quite what you expect and there is always some variation. Graphically, ξ_j is the vertical straight line distance from the point $\{X_j, Y_j\}$ to the optimum straight-line representation of the data. The conditions for optimization are satisfied by selecting a and b in such a way as to minimize the net effect of all the differences between the theoretical straight line and the data. Since the differences are both positive and negative, just adding the errors together would result in some cancellation. To avoid this cancellation, the quantity to be minimized should involve either the square or the absolute value of the deviations from the straight line fit to the data. Thus, the fit is "best" in an overall or average sense. In fact, the fit may not actually pass through any of the data points, but be closest to all the data points on average.

The Lack of Variation

The procedure for minimizing the error, usually called the mean-square error minimization method, was used to obtain the straight lines in Figure 4.6 and the subsequent figures taken from Huxley. What is apparent from the figure is how these data cluster around the regression curves and how the regression curves capture the overall relationship among the data points, with very little variation away from the straight line. The straight line denotes a linear relationship between the independent and the dependent variable in a particular context. These data show how the method of least squares works at its best. We shall also have occasion to see very different quality of fits to data.

Plant Sizes and Range of Data

Of course, it is not only animals that have the growth of their various parts related by the allometric relation; plants do as well. In Figure 4.7, a number of different plants, including cotton, peas, carrots, and turnips have their stem weight plotted against their root weight on log-log graph paper. The scales were not labeled in the original graphs and we leave them that way here. Our intent is merely to show the broad range of application of the allometric relation. There is less data here than for the fish in Figure 4.6, but, here again, the straight lines do intercept most of the data points. Thus, it is clear that the stem and the roots of plants do not grow independently of one another, but the growth of both are tied to one another. The straight line extends for over a factor of ten, in the chosen units, along both axes. This is the range of data we would like to see for any variable of interest; that is to say, we would want the dynamic range of both the dependent and the independent variables to be ten or greater. The axes would then span one to ten, or ten to one hundred, or one hundred to one thousand, so that we would have sufficient data to be confident that the fit is not accidental.

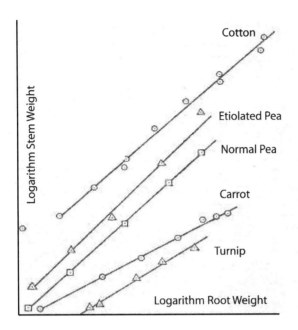

Figure 4.7. Relationship between the weight of the root and the stem. The increase in the weight of the stem is plotted versus the weight of the root for the various plants. (From Huxley [37].)

Heat of Combustion

So far, we have related the parts of an animal to the whole, and the parts of a plant to the whole. We shall also do that for various parts of the human body, subsequently. But now let us examine a different use of the allometric relation. We relate a thermodynamic quantity, the heat of combustion, to the total body weight in larval mealworm, *Tenebrio*. In Figure 4.8, the graph from Huxley's book showing this relation is given. Here we have a little over a factor of ten in both the horizontal and vertical scales, these being calories versus weight. Physically, the heat of combustion is a measure of how much heat the material must absorb in order to induce a

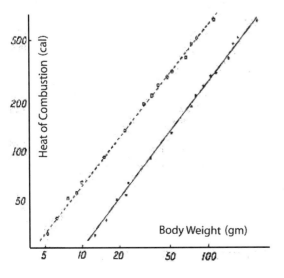

Figure 4.8. Thermodynamic variables also satisfy allometric relations. The heat of combustion is plotted versus the total body weight in the larval mealworm, *Tenebrio,* on log-log graph paper. (From Huxley [37].)

change of phase, that being a change from the solid state to a state of combustion or burning. This is not the metabolic rate that we introduced earlier, but it does relate a thermodynamic variable to a mechanical variable in a biological system.

Relating Different Measures

Thus, the notion of relating different measures through the allometric equation is not an idle boast, but can apparently relate two qualitatively different properties of a complex biological system. The data presented relate length measures, as in the case of fish; weight or mass, as in the case of plants; and thermodynamic measures to weight, as in the heat of combustion and metabolic rates. For the moment, we have no theory to support these "empirical laws," but we shall try to rectify this situation later in this book.

Question 4.6

Explain what is meant by an allometric relation and give (explain) two applications. (Answer in 300 words or less.)

4.4 SUMMARY   Measures of How the Body Is Working

In this chapter, we have looked at some ways of quantifying various functions of the body through the use of allometric relations. The metabolic rate clearly scales with the mass of an animal, but the scaling index in the allometric relation is phenomenological. This is another way of saying that we can describe what occurs, but we lack an in-depth theory of what is going on. The scaling relation is a result of balancing the generation and dissipation of heat by the body. Three mechanisms have been identified for the dissipation of heat by the human body: evaporation, convection, and radiation. The mechanisms for the generation of heat will be taken up in later chapters.

Indicators of Health

The metabolic rate, like the body's temperature, is a gross measure of the proper workings of some underlying complex physiological systems. When the system, that is, the human body, is working properly, all the parameters fall within well-defined ranges of values. What do we do when the parameters fall outside these ranges? How do we treat a fever? How do we handle the inflamation of the leg after an injury? Medicine compiles the empirical information for the treatment of such situations, but we need to develop a theoretical insight into the phenomenological equations in order to better understand the pathological situation.

A Decade of Data Is Needed for a Reliable Measure

The conclusions one should draw from any data set are the following. First, in order to be able to rely on the allometric description of the phenomenon being studied, the data being described must cover one or more decades of scale. Of course, as we have seen, the scale might be length, mass, or even calories. This in itself is not important, but if the scale range of the data begins at one calorie, then it should extend beyond ten calories, or from one centimeter to ten or more centimeters, and so on. Second, the data should be sufficiently dense in each region of the scale, say 100 data points in each decade, to provide a reliable estimate of the biovariable in that region. This would mean that if the measurements were done on a centimeter scale, over a total of one meter in length, then we would want 100 samples randomly distributed over the one meter, with a resolution on the order of a centimeter. If one of the other measurements were done over a kilometer, then we would want 100 samples randomly distributed over the kilometer, with a resolution on the order of tens of meters. This is what would be necessary to provide a "reliable" estimate of the parameters in the fitting equation. The required amount of data would also depend on the degree of variability in the data. We shall introduce such measures somewhat later. Third, the fit to the data by the regression equation from the allometric relation [Equation (4.20)] must be optimal in some well-defined way. This rather cryptic statement means that we define the error for a process and then make that error as small as possible. One measure of the error weighs a positive excursion of the data away from the mean value the same as it does a negative excursion away from the mean. This procedure is often called the method of least squares. But before we get too involved in data processing or data characterization, with which we attempt to fit a phenomenological equation to a data set, we need to develop additional physical concepts in order to "explain" these phenomenological relations. One such fundamental quantity is that of energy, an important but elusive concept.

CHAPTER 5

Energy—The Source of It All

Objectives of Chapter Five

- Understand that energy comes in a variety of forms; it can neither be created nor destroyed, so that the general conservation law applies.
- Understand that mechanical energy is the ability to do work and can be separated into potential and kinetic energy.
- Know how to use dimensional analysis to determine the motion of mechanical systems through an examination of the changes in forms of a system's energy.
- Recognize how scaling can be used to determine the strength of bone and other biological materials.
- Understand how muscles convert chemical energy into mechanical work in simple cases and explain the metabolic scaling relation.

Introduction

The Many Forms of Energy

In this chapter, we set the stage for the analysis of human locomotion using general physical concepts. But in order to do more than just move the furniture around on the stage, we need to understand the underlying assumptions supporting these physical concepts. A useful and subtle physical concept that we will find invaluable is that of energy. Energy is the physical quantity sought by every animal and plant in order to remain alive and healthy. Energy is contained in food, air, and sunshine. Much of the food we eat has no value other than its energy content. Sugar and most fats build no tissue; they are fuel only. However, the amount of energy required to

Biodynamics: Why the Wirewalker Doesn't Fall. By Bruce J. West and Lori A. Griffin
ISBN 0-471-34619-5 © 2004 John Wiley & Sons, Inc.

sustain life is negligible when compared with the energy content of winds, water waves, the radiation from the sun, and most other large-scale physical phenomena. But it is the small amount of energy in foodstuffs that keeps us alive.

Energy Is the Ability to Do Work

Energy is one of the central concepts of the physical and biological sciences, but it is not easily explained. In order to circumvent this difficulty, we use an allegory put forward by the Noble Laureate in Physics Richard Feynman to describe energy and its conservation. Energy comes in many forms: mechanical, electrical, chemical, as well as others. The discussion in this chapter is primarily restricted to the mechanical form, since this is the energy that is the directly discernable by most people and which can be most directly interpreted in terms of the ability to do work. The two types of energy are potential and kinetic, and any given process can change its form of energy from one to the other over time.

There Ought to Be a Law and There Is

What is most remarkable about energy is that physics has a law about it, a law that no phenomenon is known to violate. The law is exact as far as we know. It is the law of the conservation of energy. The total amount of this quantity (a property of food, wind, and sunshine) does not change, regardless of what nature does. It is not like force, which changes for many different reasons, is concrete, and can be directly experienced. Energy is an abstraction, independent of the particular process being examined. Because of its abstract nature, Feynman thought the best method of introducing the concept of energy conservation is by means of analogy or parable. In this we concur and liberally paraphrase his arguments [39].

Muscles Are What Enable Us to Move

The meeting of the biological and the mechanical are seen most clearly in the operation of muscles. Muscles enable the skeleton to push and pull and thereby transmit mechanical forces across space. Muscles also conduct action potentials (electrical signals) from one part of the body to another, thereby being directed to do specific tasks. Biochemical processes provide the energy for muscles to act over long periods of time. In general, all of human locomotion, whether walking or running, depends on muscles and how they are controlled.

The Reductionistic View of Science

The modern view of science is that there are a limited number of laws that describe how phenomena behave. Dynamics, for example, is completely described by Newton's laws. Therefore, if the motions of all the particles in a body are described by Newton's laws, then the motion of the total body made up of those particles must also be subject to those same laws. This is the reductionist view. We will subsequently offer words of caution regarding the applicability of this view to biology specifically, and admonitions as to its general applicability in complex physical phenomena.

The Indestructible Blocks

Suppose a child is given a set of 30 absolutely indestructible blocks for her birthday. The blocks are featureless cubes that are completely equivalent to one another. She takes the set of blocks into her bedroom at the beginning of the day to play with them. In the evening, her mother straightens out the child's room, as she usually does, and in putting the blocks into her child's toy box she notices something remarkable. No matter what she does with the blocks there are always 30 of them. Each day she counts the blocks and each day she finds 30. Then one day the count is 29, but with a little investigation she finds the missing block under the blanket on the bed. She must look everywhere to be sure that the number of blocks has not changed.

You Can't Fool Your Mother

A few days later, the count is again short—there are only 28 blocks. The two blocks are nowhere in the bedroom, but then she sees the open window and, looking outside, sees the two errant blocks lying in the grass. On yet another day, a careful count shows that there are 32 blocks. The mother is very upset until she remembers that a neighborhood friend visited, played with her daughter, and brought an identical set of blocks with her. She must have left two when she went home. After the mother sends the extra blocks back next door, she makes sure the window is closed, and does not let her daughter's friend into the bedroom. Everything is going along fine until, on another day, she counts the blocks and finds only 25. However, she notices the toy box in the room and goes to open it, but the daughter invokes her right to privacy and does not let her mother open the toy box. This is indeed a progressive household, so the mother leaves the toy box closed.

A Mother's Law

The mother, being extremely curious, and somewhat ingenious, invents a scheme for determining if there are any blocks in the toy box without opening it. She knows that each block weighs 3 ounces, so she weighs the toy box at a time when she can see all 30 blocks, and it weighs 1000 ounces. The next time she wishes to check if the toy box contains any blocks, she weighs the toy box again, subtracts 1000 ounces, and divides by three. She discovers the following equation to be true:

$$\text{number of blocks observed} = \frac{\text{weight of box} - 10^3 \text{ oz}}{3 \text{ oz}} = \text{constant} \quad (5.1)$$

In this way, the mother has constructed an "empirical law" regarding her daughter's blocks. The number of blocks she can see, plus those she cannot see but knows from the weight are in the toy box, is always constant, that is, equal to 30.

A Bigger Law

This equation works for a while, but then the mother notices some deviations. In particular, one evening, after her daughter's bath, a careful investigation shows that the dirty water in the bathtub is changing its level. Her daughter has been taking blocks into the water and leaving them there, and the mother cannot see the blocks because the water is so dirty. The mother can find how many blocks are in the water by adding another term to her formula. Since the original height of the bath water was exactly 12 inches and each block raises the water level by one quarter of an inch, the new formula for the number of blocks is

$$\text{number of observed} + \frac{\text{weight of box} - 10^3 \text{ oz}}{3 \text{ oz}} + \frac{\text{height of H}_2\text{O} - 12 \text{ in}}{1/4 \text{ in}} = \text{constant}$$

(5.2)

In the first term, we have the direct experience of the blocks by observation, and the rest of the terms infer location of the blocks by examination of the toy box and the bath water. In the gradual increase in the complexity of the mother's world she finds a series of terms representing ways of calculating how many blocks are in places where she is not allowed to look. As a result, she finds a complex formula, a quantity that has to be computed, and which always yields the same number in her world.

The Conservation of Energy

This picture of the mother counting blocks becomes analogous to the conservation of energy when we make the further stipulation that there are no blocks. This means that we disregard the first terms in the mother of all equations, so there is no direct observation of the blocks; only the inferential terms remain. This reduces the blocks to abstractions that can only be observed indirectly. The salient points of the allegory are:

1. In calculating the energy in a system, it is possible for energy to enter or leave the system. In order to verify the conservation of energy, we must be careful that no energy has entered or left the system, and if it has done so, we must properly account for it.
2. Energy has a large number of different forms and there is a distinct formula for each form. There is gravitational energy, kinetic energy, heat energy, elastic energy, electrical energy, chemical energy, radiant energy, nuclear energy, and mass energy. We shall not consider all of these.
3. Energy is additive, so if we total up the formulas for each of these energy contributions, it will not change. Except for energy going in and out of the various terms, the whole remains constant.

We Do Not Know What Energy Is

Feynman goes on to say that in physics today we do not know what energy is, we just know how to calculate it. Energy is an abstract thing in that it does not tell us the mechanisms or the reasons for the various formulae; we just know that we always get "30" when we do the block calculation. Here again, we see the how of energy, but not the why.

Question 5.1

Give an example of how the conservation of energy might be useful when applied to the human body. Do not be afraid to go into some detail. (Answer in 300 words or less.)

5.1.1   Mechanical Energy

A Formula for Work

Now that we know we need some formulas to calculate the energy contained in a given process, we turn to the definition of mechanical energy. So, to begin, we examine how we put energy into a system or process. Energy injection into a system is accomplished by doing work on the system and storing that work for later use. Work is defined in mechanics as the force multiplied by the distance over which the forces operate:

$$\text{work} = \text{force} \times \text{distance} \tag{5.3}$$

Thus, energy is a property of a system, whereas work is a process, it is something that you do to a system or a system does to you. This determines the sign of the work. If you do work on the system, the sign is positive as shown in Equation (5.3), since the system gains energy. If, on the other hand, the system does work on you, then the sign of the work is negative, since the system loses energy in performing work on you.

Energy From the Earth's Gravitational Field

When you carry a suitcase from the first to the second floor, you are doing work against the pull of the earth's gravity. There are always at least two steps in doing work: 1) exerting a force (here a force is necessary to overcome the weight of the suitcase, an external gravitational force) and 2) moving an object in a direction parallel to the force that is being overcome (the suitcase going up the flight of stairs). In Equation (5.3), we only use the component of the force in the direction the object is being moved. Therefore, if you carry the suitcase the full length of the first floor you do no work in the sense defined here, except in picking the suitcase up and putting it down again. Work is only done on the suitcase when the weight is moved up and down.

What About the Body?

Of course, in the context of this book a number of other things should occur to you. What about the weight of the person going up and down the stairs? That weight should certainly be considered in addition to the suitcase. Of course, in most physics texts, the person carrying the suitcase is assumed to be weightless, and the work done in transporting one's own weight up the stairs is neglected. We try not to neglect such things here, because it is the body in which we are primarily interested. So we can add the weight of the person to that of the suitcase in determining the work done using Equation (5.3), or we can neglect the artificiality of the suitcase altogether and determine the work done as moving your body from the first to the second floor.

Mechanical Work Is Not the Only Kind

Next, consider a fact that anyone who has carried a suitcase across a level floor knows: you are doing work, Equation (5.3) be damned. Here again, we must stress that we are only talking about work associated with mechanical forces. What you experience in your arms, legs, and back in moving the suitcase across the level floor is electrochemical work being done by your muscles. We do not neglect such effects here, but these things must be addressed in some order so that we have time to introduce the reader to the necessary concepts, and for this we ask your indulgence.

Potential Energy

The example given above clearly involves the gravitational field near the surface of the earth. When we do work on an object, against the gravitational field, we are actually increasing the gravitational energy of the object. This is a consequence of the law of conservation of energy. Since we do work by moving the suitcase upstairs, and since work is energy, this energy must be stored in the object in some form. The form of energy we store is called potential energy (PE), as we shall discuss more completely below. Let us agree to separate the work done in carrying the suitcase up the stairs and the work done in carrying our body up the stairs, just to make our discussion less convoluted.

Energy Is Additive

If we carry two identical suitcases up the stairs, we do twice as much work as in carrying up one suitcase, at least in terms of the work done associated with the suitcases. In a similar way, if we carry one suitcase up two flights of stairs, the work is equivalent to carrying two suitcases up one flight of stairs. In the first situation, we have twice the force to overcome in a given distance and in the second situation we have the original force acting for twice the distance, and according to Equation (5.3) these both have the same effect on the work done: they double it. If we included the work done associated with our body, then this separation would not be so simple. The work done on an object (the suitcase or our body) may be stored for future use in the position of the object. Energy storage by position is called potential energy, because once the energy is stored, it then has the potential to do work.

Potential Energy and Work

For example, the balance weights in a grandfather clock store gravitational potential energy when they are raised. As the weights fall under the pull of gravity, they do work against the friction in the workings of the clock and run the time machine. The work done in moving the weight to a height s is given by the product of this force and the height:

$$\text{gravitational potential energy} = \text{weight} \times \text{height} \qquad (5.4)$$

where weight = force. It is interesting to examine the dimensions of the potential energy in Equation (5.4). We can substitute the expressions for the dimensions of the various quantities in this equation—the mass in $[M]$, the acceleration is $[LT^{-2}]$

and the length is [L]—to obtain the dimensional expression

$$[PE] = [M] [LT^{-2}] [L] = [M] [L^2T^{-2}] \tag{5.5}$$

that is also the work done on the object. The units in Equation (5.5) are the mass [M] times the square of a velocity [L/T].

Force, Not Mass

In terms of shorthand symbols, the potential energy in Equation (5.4) can be written, using weight = mg and height = s, as

$$PE = mgs \tag{5.6}$$

and although the two expressions look quite different, their meanings are the same. It is worth emphasizing again that the weight of an object is a force. Therefore, in the previous section, where we presented some of the data biologists have collected on variables as a function of body weight, we were actually comparing those variables to a force. In biology, the force produced by gravity acting on a mass, and the mass itself, are often talked about as if they were the same thing. The situation is made even more confusing by investigators using the same units for both weight and mass, say kilograms. Of course, in static situations the weight and mass are proportional, the proportionality constant being the acceleration of gravity g.

Reference Point for Potential Energy

Note that the potential energy defined by Equation (5.6) is zero when the displacement is zero ($s = 0$). However, the zero point of the displacement s is arbitrary, from which we can conclude that the zero point of the potential energy is also arbitrary. What this actually means is that the absolute level of the potential energy is not important. What is important is how much the potential energy changes between points in space. So we interpret Equation (5.6) to be the increase in potential energy between two points a distance s apart. In lifting a weight, the potential energy at the surface of the earth is taken to be zero and s is the height above the surface to which the weight is lifted. We know that the potential energy at the earth's surface is not really zero. To see this, suppose there were a well nearby, with a rope and pulley. If we tied the mass to the rope and dropped it into the well, it could be used to raise a pail of water and thereby do work. Thus, the potential energy of the mass is not really zero at the earth's surface, or at least not zero with respect to the bottom of the well.

Kinetic Energy

Another simple machine (the rope and pulley constitute a machine) is a pile driver. You have seen these large cantilevers at construction sites; they have massive weights hanging from their ends. The weight is released and, in falling, builds up momentum that drives a pile into the ground upon impact. The amount of energy stored in the moving mass is equal to the work done in lifting the weight. The force used to lift the weight is the mass, m, times the acceleration, where the acceleration is that of gravity, g; therefore, $F = mg$. The units of Equation (5.5) are also the units of a second kind of energy, the kinetic energy (KE), or energy due to motion, as in the case of the pile driver. The equation for the kinetic energy is given, as suggested by Equation (5.5), by

$$\text{kinetic energy} = \tfrac{1}{2} \text{ mass} \times \text{velocity}^2 \tag{5.7}$$

or in symbols, with $u = $ velocity,

$$\text{KE} = \tfrac{1}{2} mu^2 \tag{5.8}$$

Sometimes, when we do work on an object, rather than that work being stored in potential energy (the position of the object), it results in the object moving more rapidly and the energy being stored in the enhanced motion of the object. By this we mean that we can inject into, or extract energy from, a moving object by changing its rate of motion. Further, this change in the rate of motion defines the force acting on the body and/or the force that body imposes on a second body.

The Climber and the Flyer

Recall the circus act of an acrobat being catapulted from a seesaw to the top of a human stack. In this trick, an acrobat stands at one end of the see-saw next to three of four people standing on one another's shoulders, while a second acrobat climbs a tower at the other end of the see-saw. The climber increases his potential energy with the intention of delivering it all to the flyer who remains below. The climber jumps from the top of the tower and lands on his end of the seesaw, slamming it to the ground. All of the climber's potential energy is converted into kinetic energy when he hits the plank of the seesaw. The rapid ascent of the other end of the plank hurls the flyer into the air, parallel to the human stack. Now things get a little tricky. If the climber and the flyer are of equal weight, then the stack of humans on which the flyer lands is approximately equal to the height of the tower. But, for the sake of drama, the flyer is typically a child and the climber is an adult, so their weight differs by a factor of two or more. In this case, the stack of humans can be much higher than the tower and the child does appear to fly before the top person on the stack snatches him from the air.

Under the Influence of Gravity

Let us now use what we know about the force of gravity to construct a useful relation for the kinetic energy of an object falling under the influence of gravity. Using dimensional analysis, we write the speed of an object falling under the influence of gravity for a time t as

$$u = Cm^\alpha g^\beta t^\gamma \tag{5.9}$$

where the mass, gravitational acceleration, and time are put in as test variables. We need the speed in order to calculate the kinetic energy. The dimensional equation corresponding to Equation (5.9) is

$$[LT^{-1}] = [M]^\alpha\,[LT^{-2}]^\beta\,[T]^\gamma \qquad (5.10)$$

so that equating exponents on the right- and left-hand sides of this equation yields

exponent of $[M]$: $0 = \alpha$

exponent of $[L]$: $1 = \beta$

exponent of $[T]$: $-1 = -2\beta + \gamma$

so that we obtain $\alpha = 0$, $\beta = 1$, and $\gamma = 1$. Thus, we may write, with $C = 1$ in Equation (5.9), the velocity in terms of the constant acceleration as

$$u = gt \qquad (5.11)$$

so that the speed increases linearly with time. We can also eliminate time from this equation by using the distance the mass falls, which we calculated previously as $s = \frac{1}{2}gt^2$, from which time can be expressed as $t = \sqrt{2s/g}$, and we obtain after a little manipulation of the variables

$$s = u^2/2g \qquad (5.12)$$

The distance traveled by the falling body s is proportional to the square of the instantaneous velocity u^2.

Work and Kinetic Energy

If we multiply both sides of this last equation by the weight of the falling object, w, we have

$$ws = wu^2/2g \qquad (5.13)$$

Note that on the left-hand side of Equation (5.13) we have weight (force) times a distance, hence work. What we have on the right-hand side can be determined by noting that the weight is given by $w = mg$, which, when substituted into Equation (5.13) yields,

$$\text{work} = \tfrac{1}{2}mu^2 \qquad (5.14)$$

The right-hand side of Equation (5.14) is simply the kinetic energy of the falling body and the equality indicates that this kinetic energy is exactly equal to the amount of work done on the body by gravity due to its falling a distance s, starting from rest. The work that gravity does in accelerating a falling body all goes into, and is equivalent to, the kinetic energy it acquires by falling. Since the speed of the body is dependent only on the vertical height of the fall, independent of the path, this must also be true of the kinetic energy. Thus, the speed attained by a child at the bottom of a helical water slide is the same as that produced by falling the same vertical distance as the height of the water slide.

Question 5.2

Suppose Galileo drops a 10 kgm canonball from the leaning tower of Pisa from a height of 40 m. Sketch how the potential energy changes as a function of time. Also sketch how the kinetic energy changes as a function of time. From the two sketches, what can you conclude? (Answer in 300 words or less.)

5.1.2 Capacity for Work

Stored Energy

When a body possesses energy, it possesses a capacity for doing work, and its energy is measured by the work it can do. Energy is stored work, either in position or in motion. Like money, energy, can be handed around and distributed among other bodies without impairing its value, or it can be banked against a future withdrawal. The body does this with fat. The body saves chemical energy in this form against future needs, when the environment may not be able to supply the body's energy requirements. Thus, using the analogy with which we started this discussion, we can write the conservation of mechanical energy as

$$\text{total mechanical energy} = \text{PE} + \text{KE} = \text{constant} \qquad (5.15)$$

The dynamics of a system are then determined by how energy changes from potential to kinetic and back again, but always subject to the constraint that the total mechanical energy in a conservative system is constant. (Answer in 300 words or less.)

Energy Units

Up to this point, we have not mentioned the units associated with energy. From Equation (5.5), it is clear that energy has the units of $[ML^2T^{-2}]$. In the CGS system of units this would be cm^2/sec^2. Looked at another way, work (energy) is force times distance; therefore, we can also write the units of energy as dynes cm. The work required for one dyne of force to move a mass one centimeter is one erg, so that 1 erg = 1 dyne cm. In the IS system of units, the same reasoning leads to 1 joule = 1 Nm.

Pendulum and Energy Conservation

Let us consider a simple case of the conservation of mechanical energy expressed by Equation (5.15), the frictionless pendulum depicted in Figure 3.5. The mass is initially drawn to one side, the point A in the figure, and in so doing has been raised a distance s from its equilibrium position at M. The point M is called the equilibrium position because if the mass were initially motionless at M, then the pendulum would remain motionless forever. As the pendulum swings from A to M, its potential energy is entirely converted into the kinetic energy form, which is to say that at the point M the potential energy is minimum (zero) and the kinetic energy is maximum. As the mass rises again to B, equal in height to A, the energy has been converted from the kinetic back into potential energy. Thus, the pendulum, without dissipation, continues to swing with undiminished energy, alternately converting its energy back and forth between potential and kinetic. This is done with a period that we calculated in Chapter 3 to be $\sqrt{g/l}$, that is, it depends on the acceleration of gravity and the length of the pendulum arm and nothing else. Further, since by assumption no energy is lost due to friction, the sum of the potential and kinetic energies at any instant is a constant.

Conservative Systems

A system such as this pendulum that is completely isolated from its surroundings (by assumption), such that there is no energy exchange between the pendulum and its environment, is called a conservative system. It is only for such an idealized system that Equation (5.15) holds. Do not confuse Equation (5.15) with the more general principle of conservation of energy discussed earlier. The narrowly defined law for conservation of mechanical energy holds only for ideal or nonexistent systems. The true importance of the law of conservation of mechanical energy lies in its ability to determine the equations of motion for interacting systems. One can determine these equations because of the energy exchange process. A change in momentum is produced by a corresponding change in potential energy, which is to say that a change in potential energy produces a change in momentum, which is a force. This association of a force with a change in potential is a restatement of Newton's laws that was made some time after Newton. We shall have more to say about the relation between forces and potentials later, when we discuss dynamics.

Nonconservative Systems

There are no completely isolated or conservative systems in nature, although celestial mechanics and the orbiting planets come fairly close. We discuss nonconservative effects when we inquire into the various forms of kinetic energy. For example, as we discussed earlier, random molecular motion is experienced as heat, coherent vibration pattern molecules are sensed as sound, and the organized motion of electrons is measured as electrical current. We shall examine each of these various manifestations of energy in an appropriate biodynamical context in due course (see Table 5.1).

Galileo Explains Materials and Mechanics

In 1638, Galileo Galilei published the monograph, *Two New Sciences* [12], in the Platonic literary style of using the Socratic method of dialog between individuals of

Table 5.1. Dimensions of some common physical quantities

Quantity	Description	Dimension
Thermal capacity	heat/mass/degree	$[C] = [L^2 T^{-2} \Theta^{-1}]$
Mechanical energy	force × displacement	$[E] = [F][L]$
Kinetic energy	mass × velocity^2	$[KE] = [M][V]^2$
Potential energy	weight × displacement	$[PE] = [ML^2 T^{-2}]$
Power	energy/time	$[P_w] = [E]/[T]$

Note. the three types of energy listed can all be expressed in the same way in terms of the fundamental units as $[ML^2 T^{-2}]$, but we avoid doing this in order to emphasize the dimensional equivalence of the different ways of expressing the energy.

differing views and, more importantly, with very different levels of understanding of the topic being discussed. We mention this because the two new sciences introduced by Galileo are materials and mechanics, both of which are of direct concern to us. In the case of materials, he related the observed properties of objects to their microscopic structure, often relying on the fact that materials could not be homogeneous, but must change characteristics at different scales. In mechanics, he extended the study from the established realm of statics to dynamics, that is, from stationary, equilibrium processes that were understood at the time, to phenomena that change over time and were not understood by his contemporaries. Galileo's book is the first recorded use of the modern method of reductionism to systematically explain the physical properties of objects in terms of empirical laws, under the assumption that the same laws apply independently of scale size.

Reductionism May Have Outlived Its Usefulness

The physical assumptions of reductionism were succinctly stated by Rosen [40], which we paraphrase as:

1. Reductionism assumes that simplicity is common, whereas complexity, using any reasonable definition, is rare.
2. Reductionism further assumes that simple systems are typical and are entirely independent of any context.
3. The change in going from simplicity to complexity is only a matter of accretion of simple, context-independent parts. In the same way, the analysis of complex systems is merely a matter of inverting the accretion procedure that produced them. In the case of the dynamics of mechanical systems, this idea reduces to the notion that all physical systems obey simple mechanical laws, say Newton's laws of motion. Further, no matter how complicated the motion of the system appears to be, that motion is attained through the application of these simple laws to all parts of the system and superposing their separate effects. In the same way, the inverse process is accomplished by decomposing a complicated dynamical pattern into many smaller parts, each of which is described through the application of Newton's laws.

We argue, subsequently, as does Rosen, that reductionism is a concept that may have outlived its usefulness in science. In large part, its utility has diminished because real complex systems such as biodynamical systems do not satisfy this simple procedure. For example, the cardiovascular control system may not have a reductionist description. Therefore, we use reductionism when it is convenient to do so, but do not require its applicability to all relations we use.

Question 5.3

Suppose your family had a grandfather clock repaired and the watchmaker said the pendulum arm had to be replaced. When the clock was returned home you found that it ran uniformly fast. What might you conclude from this observation? (Answer in 300 words or less.)

Not All Things Scale

One of the many things proposed in Galileo's remarkable book [12] is that physical bodies are not homogeneous:

> . . . [O]ne cannot argue from the small to the large, because many devices which succeed on a small scale do not work on a large scale.

Galileo's argument concerned the fundamental nature of things, not just inanimate objects, but living things as well:

> . . . [F]or every machine and structure, whether artificial or natural, there is set a necessary limit beyond which neither art nor nature can pass; it is here understood, of course, that the material is the same the proportions preserved.

Strength and Weight are Related

Galileo was a practical person, so he did not limit his arguments to abstractions, but gave concrete examples of the general principles he examined. Thus, he not only recognized that the weight of an object increases with the volume, but that the strength of an object such as a column increases in proportion to its cross-sectional area. Therefore, a structure such as a bridge, or a person, can only grow to a certain size, after which it will collapse under its own weight [12]:

> From what has already been demonstrated, you can plainly see the impossibility of increasing the size of structures to vast dimensions either in art or in nature; likewise the impossibility of building ships, palaces, or temples of enormous size in such a way that their oars, yards, beams, iron bolts, and, in short, all their other parts will hold together; nor can nature produce trees of extraordinary size because the branches would break down under their own weight; so also it would be impossible to build up the bony structures of men, horses, or other animals so as to hold together and perform their normal functions if these animals were to be increased enormously in height; for this increase in height can be accomplished only by employing a material which is harder and stronger than usual, or by enlarging the size of the bones, thus changing their shape until the form and appearance of the animals suggest a monstrosity.

Could this be why Godzilla is so ugly?

Galileo Had Great Vision

In the above spirit of Galileo, it was pointed out by Schmidt-Nielsen [23] that as the size of a structure such as a bridge is increased, there are three parameters that can be changed:

1. The size of the materials used in the construction of the bridge
2. The kinds of materials used; for example, changing from stone to steel
3. The design of the bridge, such as changing from the compression of columns to the tension of cables in suspension

These same considerations apply to the human body. In fact, Galileo saw no difference between how animate and inanimate objects respond to the general principles of the physical world in which they exist. From a modern perspective, we see that the scientist or engineer can select the material and the design for the structure being considered, subject to the constraints of the desired outcome. Living entities are different in that evolution has already made all these selections in the past, subject to the constraint of fitness, and we observe the outcome of that selection process. Our purpose in this book is to try to determine the rules that nature has followed in this process; after all, there are many possibilities. For example, which of the three parameters—size, material, or design—is the more important? Or are they of equal importance? Does form follow function? In which case does the design (function) determine the material and size (form).

Failure Modes of the Skeleton

An example of nature's design is the skeletal structure of humans. These are support systems, providing levers on which muscles can act, and, being hard and mechanically rigid, keeping the body from collapsing. We withstand the forces acting on our bodies, such as gravity, by means of our skeleton, or we fail in one way or another. When we can no longer resist the pull of gravity or coordinate our steps, we fall down. This falling is a condition that often afflicts the elderly. Anyone who has lost the cartilage in a joint, so that simple walking entails bone scraping against bone, understands the wonder of this support system. In general, the skeleton must support the weight of an animal, or, as was pointed out by Galileo, it will fail in compression and be crushed. This is partly what occurs during ostheoporosis; the loss of calcium reduces the skeletal strength and leads to an enhanced rate of breaking of bones. It is not just the force of gravity that the skeleton must withstand, however. There are a variety of forces that act when we move in any way. In addition to compression due to our weight, there is bending and torsion during locomotion, which could also produce failure by means of buckling, as happens in a beam when it is twisted.

Elastic Similarity and a Strange Relation

In discussing the collapse of a body under its own weight, we tend to think of the weight as increasing in direct proportion to volume, but this is not always the case. Consider an elastic cylinder of length l and diameter d, say a very crude model of a muscle. The volume of the cylinder is given by the product of the length and the cross-sectional area, $\pi l d^2/4$, so the weight of the cylinder is

$$w \propto l d^2 \qquad (5.16)$$

In the geometrical approximation, this implies that the diameter is proportional to the length and, subsequently

$$l \propto w^{1/3} \qquad (5.17)$$

McMahon [41] argued that because the weight of the column increases in proportion to the volume, we require that the material have sufficient elasticity to keep from buckling. He determines that the relation between the length and diameter of the cylinder, representing the torso or limb of the body, necessary to keep the cylinder from buckling is

$$d \propto l^{3/2} \tag{5.18}$$

See Calder [25] for an excellent synopsis and critique of McMahon's discussion. Note that Equation (5.18) is different from that proposed by Galileo in his famous bone crushing problem, in which he chose $d \propto l^2$. In the elastic similarity model, the increase in the weight of the column is given by inserting Equation (5.18) into Equation (5.16) to obtain

$$w \propto d^{8/3} \tag{5.19}$$

We can see from this result that the weight of the column increases more rapidly than geometry alone would dictate. The strength of the column is determined by its diameter, which increases more rapidly than geometrically in order to keep it from buckling. Inverting the relation of Equation (5.19), we can express the diameter in terms of the mass of the cylinder:

$$d \propto m^{3/8} \tag{5.20}$$

The elastic similarity model of McMahon [41] and Rashevskey [42], therefore, yields a relation between diameter and mass that is quite different from that of geometric scaling, the latter being $d \propto m^{1/3}$. Note that $3/8 = 0.375$ and $1/3 = 0.333$, so that the two results are not so different in terms of the numerical index. However, the implications of the two models are really quite different.

Limits to Growth

Thus, we have known for nearly 400 years that biological organisms can be of a certain size and no larger if they are to function effectively in a gravity-dominated world. However, as Haldane [43] points out in his essay, *On Being the Right Size*, zoologists paid little attention to the sizes of animals until fairly recently. Haldane had a way of making a point using vivid imagery [43]:

> You can drop a mouse down a thousand-yard mine shaft; and, on arriving at the bottom, it gets a slight shock and walks away. A rat is killed, a man is broken, a horse splashes.

When Is Gravity Important and When Is It Not?

The reason for the differing outcomes at the bottom of Haldane's shaft is that air interacts with the falling animal, offering a resistance in direct proportion to the animal's downward facing surface area. On the other hand, the gravitational force causing the animal to fall increases as the volume of the animal, because the animal's mass increases as the volume. Therefore, we can use the ratio of surface area to volume in Equation (5.8) to deduce that the resistance to falling more nearly balances the force of gravity for a small animal than it does for a large animal. In the insect world, for example, gravity is of little or no consequence. We shall address such things more completely when we discuss fluid flow and the movement of bodies through fluids.

Question 5.4

Suppose a body satisfies the elastic similarity model. What would be that body's metabolic rate as a function of a linear dimension of the body? (Answer in 300 words or less.)

5.2.1 The Superposition Principle

A Boy and His Dad

The first principle formulated in the physical sciences was that developed in the letter correspondence of John (father) and Daniel (son) Bernoulli beginning in 1727. In their letters, they established that a system of N point masses has N independent modes of vibration. The number of independent degrees of freedom of the system motion is equal to the number of entities being coupled together. The Bernoullis reasoned that if you have N one-dimensional point masses, then you can have N modes of vibration without making particular assumptions about how to construct the equations of motion as long as the coupling is weak. This was the first statement of the use of characteristic values and characteristic functions in physics, and formed the basis of the principle of superposition.

Phenomena Can be Taken Apart and Put Back Together

Simply stated, the Bernoullis' conclusion means that one can take a complicated system, consisting of N particles or N degrees of freedom, and completely describe it by specifying one function per degree of freedom. If one knows what the characteristic functions are (for waves, they are sinusoidal functions) and one knows what the characteristic values are (for waves, these are the frequencies) then one knows what the motion is. Thus, they had taken a complicated system and broken it up into readily identifiable and tractable pieces, each piece being specified by a characteristic function and value. If one knows the characteristic functions and the characteristic values, then the contention was that one knows everything that can be known about the system. The reduction of the system into these independent pieces, is, of course, in conflict with the view of chaos we mentioned earlier. The conflict notwithstanding, the reductionistic view has been useful for describing many phenomena for over three hundred years.

Normal Modes Are Usual

What about the interaction between the variables? Subsequent investigators were able to show that if the interaction among the degrees of freedom in a system is linear, then there exists a representation, which is to say, there exists a set of variables, in which each degree of freedom is independent of every other degree of freedom. This means that there is a set of variables, often called normal modes, which do not interact with one another, even though, in terms of the original variables of the system, there was an interaction. These normal modes are the characteristic functions discussed by the Bernoullis.

The First Principle in Physics

Later, Daniel Bernoulli (1755) used the normal mode result to formulate the *principle of superposition*: the most general motion of a physical system is given by a linear superposition of the characteristic modes. The importance of this principle cannot be overstated. As pointed out by Brillouin (1946), until this time all general statements in physics were concerned with mechanics, which is applicable to the dynamics of individual particles. The principle of superposition was the first formulation of a general law pertaining to a system of particles. Note that the concepts of energy and momentum were still rather fragile things, and their conservation laws had not yet been enunciated in 1755. Yet, Daniel Bernoulli was able to formulate the notion that the most general motion of complicated systems of particles is nothing more than a linear superposition of the motions of the constituent elements. That is a very powerful and pervasive point of view about how one can understand the dynamics of a complex physical system. In fact, this is how we describe the motion of a rigid body, without taking into account the motion of the millions upon millions of particles that make up the body.

Reductionism Was Assumed Throughout

Over the 150 years after Daniel Bernoulli, the mathematical apparatus to give the principle of superposition a precise interpretation was developed. Superposition was perceived as a systematic way of unraveling complex events in the physical world and giving them relatively simple mathematical descriptions. In this way, science was able to understand the propagation of light, sound, heat, and, eventually, all of quantum mechanics. Thus, at the turn of the last century it was the commonly held belief in the scientific community that a complex process can be decomposed into constituent elements, each element can be studied individually, and the interactive part deduced by reassembling the linear pieces back together.

Superposition Does Not Work in Biology

Physical reality could therefore be segmented, understood piece-wise, and superposed back again to form a new representation of the original system. As attractive as this picture is, and it has helped immensely in the construction of our technological society, it does not work in general. In particular, reductionism does not work in biology in general, nor in biodynamics in particular, as we shall see. But, of course, this general failure of reductionism shall not prevent us from applying the principle in those situations in which it does work.

Different Kinds of Muscle

5.3 MUSCLE

The living machinery of the body is muscle. For example, heart muscle is intrinsically rhythmic, contracting at regular intervals even when denervated. This is quite different from skeletal muscle, with which we are concerned in this section. Skeletal muscle, like heart muscle, is striated, as distinct from the muscle in the walls of hollow organs which is smooth. However, unlike heart muscle, skeletal muscle can be activated voluntarily. We can consciously choose to move our fingers, hands, arms, and so on. By contrast, we cannot consciously change the beating of our heart, or at least most of us cannot. We can, of course, indirectly influence heart rate by thinking of things that arouse or anger us on one hand, or, on the other hand, slow our heart rate by meditating on the waves washing up onto a beach. The dominant function of the skeletal muscles is to contract, and since they are attached to bones by tendons, they act to move these bones with respect to one another. Thus, from one perspective, skeletal muscle is similar to windlasses connecting weight-bearing members (bones) and which transmit forces from one part of the body to another. However, unlike ropes, the forces transmitted by muscle are generated internally, a truly biophysical phenomenon.

Electricity and Muscle Contraction

An impulse generated by the central nervous system (CNS) triggers a contraction of the muscle. The CNS is the elaborate communication network evolved by nature, enabling various organs of the human body to be centrally monitored and controlled by the brain. Some control is voluntary, such as overcoming the pain of running in a long-distance race; some control is involuntary, such as regulating our body temperature; some control is a combination of the two, such as breathing and walking. In each case the control involves the activation and subsequent contraction of muscle. The relationship between electricity and muscle contraction, the basis of neurophysiology, was first observed in 1791 by the Italian physicist Luigi Galvani. His experiments involving the depolarization of frogs legs by touching them with metal rods and causing the detached legs to twitch are well known. We take this subject up again when we discuss electricity.

Two Kinds of Muscle Contraction

In the present context, the word contraction has two meanings. The first meaning has to do with the generation of tension without a shortening of the muscle. This condition is produced by the application of a load sufficiently great that the muscle is unable to move; consequently, the length of the muscle remains fixed and so-called *isometric* contraction takes place. The tension in the muscle under isometric contraction is measurable. The second kind of contraction is *isotonic* contraction, in which the muscle is shortened under a constant load. Isometric contraction does no mechanical work because the load is not moved; however, the muscle does mechanical work under isotonic contraction because, in this case, the load is moved. Let us now consider what a muscle is and a little about how it operates.

Muscles and Metabolism

Previously, we noted that the scaling of the metabolic rate with body mass has an empirical exponent of $\beta = 0.75$ that remains unexplained. The elastic similarity model of McMahon [41] was in fact developed, at least in part, to account for this exponent. The argument begins with the recognition that locomotion requires the contraction of muscles. During contraction, muscles exert a force that increases with the cross-sectional area of the muscle. The power output of the muscle is the energy per unit time and may be equated with the metabolic rate, the product of the force generated that is proportional to the cross-sectional area of the muscle and the velocity of the muscle shortening, u:

$$R \propto \pi d^2 u / 4 \qquad (5.21)$$

Assuming that the velocity of the muscle shortening is a size-independent constant for the species of interest, and using the elastic similarity relation between diameter and mass [Equation (5.20)], Calder [25] obtained for the power output of a particular muscle

$$R \propto m^{3/4} \qquad (5.22)$$

in agreement with the allometric relation for the metabolic rate. As Calder argued, this reasoning would also explain the scaling of the biovariables involved in maintaining the flow of energy to the working muscles. In this way, locomotion is seen to conform to the requirements of symmorphosis [44], by which the formation of structural elements "is regulated to satisfy but not exceed the requirements of the functional system."

Question 5.5

Discuss some of the difficulties with using Equation (5.22) as an "explanation" of how the metabolic rate scales with mass in the mouse-to-elephant curve. (Answer in 300 words or less)

5.3.1 Structure

Function of Muscle

Muscles are very complicated structures having a relatively simple function: to contract and by contracting do work. In the mammalian heart, this work is to pump blood to the rest of the body, and in skeletal muscles this work is generally that of motility and locomotion. But before we turn our attention to what muscles do, let us look, at least superficially, of what muscles consist.

Some Basic Properties of Muscle

Most simply, muscles are made up of protein; in particular, two kinds of protein that make up microtubules: *actin* and *myosin*. Muscles have a fibrous structure, with the diameter of a fiber being from 0.02 mm to 0.08 mm. Each fiber consists of approximately 10^3 thinner fibers called myofibrils, whose diameter is on the order of 10^{-6}m, and is surrounded by a membrane of thickness of approximately 100 Å. These myofibrils consist of protein threads, both thick and thin, with cross-sectional symmetry that is hexagonal, as shown in Figure 5.1. The thick filaments are

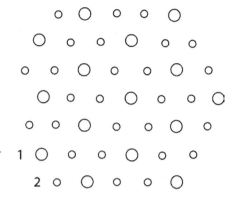

Figure 5.1. Schematic representation of part of the cross section of a myofibril. 1) Thick filaments. 2) Thin filaments. (Suggested by Volkenstein [46]).

formed by the fibrous protein myosin, whereas the thin filaments consist mainly of the protein actin. The thin threads also contain regulatory proteins, such as tropomyosin, troponine, and actinine.

The Motor Unit

Basmajian and De Luca [45] give a clear discussion of the engineering concept of a motor unit, taking cognizance of the fact that muscle fibers do not contract individually but, rather, in small groups that work collectively. Each of these groups of muscle fibers is supplied by the terminal branches of one nerve fiber or axon whose cell body is in the anterior horn of the spinal grey matter. In Figure 5.2 is depicted a

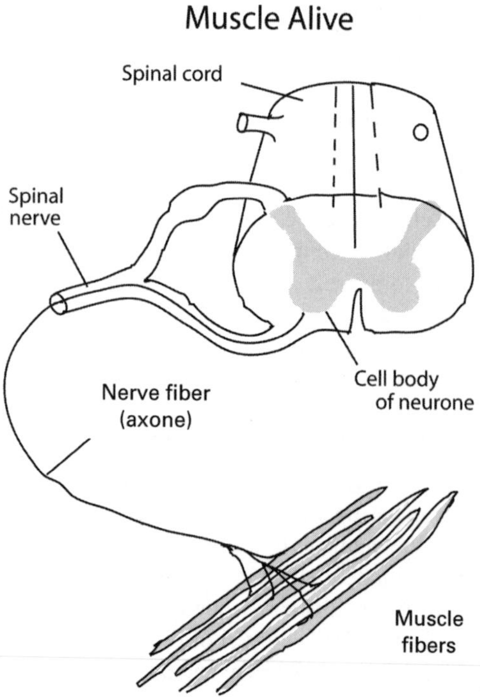

Muscle Alive

Spinal cord

Spinal nerve

Nerve fiber (axone)

Cell body of neurone

Muscle fibers

Figure 5.2. Schematic of a motor unit, indicating the origination point of an impulse that activates the muscle fiber at the end of its journey. (Redrawn from Basmajian and De Luca [44].)

schematic of the motor unit just described. We see that the motor unit consists of the nerve cell body, the long axon running down the motor nerve, the terminal branches, and all the muscle fibers supplied by these branches. This is the functional unit of striated muscle, since an impulse descending the motoneuron causes all the muscle fibers in one motor unit to contract almost simultaneously.

Two Kinds of Excitation Delay

The lack of simultaneity in the contraction of the muscle fibers in a single motor unit has two causes. One cause is due to the finite propagation speed of the activating pulse, which depends on the length and diameter of the individual axon branches innervating the separate muscle fibers. This delay is fixed for each muscle fiber because the axon does not change its structure. The other delay, rather than being static, is dynamic and is produced by the random discharge of acetylcholine packets released at each neuromuscular junction. The random nature of the discharge results in the subsequent excitation of each muscle fiber of a motor unit being a random function of time. Consequently, when the electrical discharge of the muscle fibers is monitored, the resulting electrical signal has a *jitter* or random component. The standard deviation of the jitter in the electrical signal, that is, the strength of the jitter, is approximately 20 μs. We shall see that most biological time series have a random component.

The Shifting Model

It has been observed using electron microscopy that while a muscle is shortening (during contraction of the fiber), the thick filaments (myosin) move into the spaces between the thin filaments (actin), and the overall structure shortens like a telescope. This so-called sliding model of muscular contraction was suggested and confirmed in the works of Huxley, Hanson, and others. Vogel [47] points out that the process of contraction has in fact nothing to do with the familiar concept of contraction, but is actually "a shearing interdigitation of a large number of filaments of fairly fixed length." He goes on to say that because muscle consists mainly of water and is therefore incompressible, the whole notion of contraction may be an illusion. The conservation of volume implies that contraction in one direction implies expansion in an orthogonal direction. "What it does is pull one end toward the other while swelling in the middle."

5.3.2 Muscles Are Machines

Simple Machines

Muscles use the principles of simple machines to gain an advantage over nature. This mechanical advantage is gained through the use of levers. A lever is a simple machine that consists of a rigid arm that can pivot on a fixed point called a fulcrum. Levers are everywhere around us. The gravedigger's shovel, the golfer's putter, and the gardener's hedge clipper are all simple levers. The actin/myosin movement is done by a large number of tiny levers. A lever is used to transmit a force and results in movement of a load. In Figure 5.3, the three classes of levers are shown and are distinguished from one another by the relative positions of the

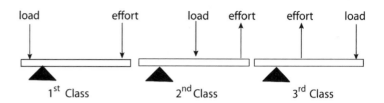

Figure 5.3. The three classes of levers.

load (L), effort or force (E), and fulcrum (Δ). The first class of lever has the form of a child's seesaw with the order $L\Delta E$; the load and effort are on opposite sides of the fulcrum. The second class of lever has the order ΔLE, with the load between the effort and the fulcrum. The third class of lever has the order ΔEL, with the effort between the load and fulcrum. For a lever, it is not the load or the applied force that is important; rather, it is the moments. A moment is the product of the force and the moment arm, that is, the distance from the point of application of the force to the fulcrum.

Mechanical Advantage

A seesaw is in balance when the product of the force of load, F_l, and the load moment arm, d_l, is equal to the product of the force of effort, F_e, and the effort moment arm, d_e, such that $F_l d_l = F_e d_e$. The mechanical advantage, which is defined by the ratio of the load to the effort forces, in a class one lever is given by the ratio of the two moment arms, that is,

$$\frac{F_l}{F_e} = \frac{d_e}{d_l} \tag{5.23}$$

Thus, the minimum effort required to lift a load is given by the ratio of the moment arms. One would think that the mechanical advantage arises when the ratio of the moment arms is greater than one, $[d_e/d_l] > 1$, in which case the effort is amplified in overcoming the load. Assume that the distance from the fulcrum to the effort is one meter and the distance from the load to the fulcrum is ten centimeters. A ten Newton load can then be lifted by a one Newton effort, a mechanical advantage of ten. However, we can also refer to a mechanical advantage that is less than one.

Muscles, Bones, and Joints

Note that the work being done in lifting a load has nothing to do with the moment arms, since work is determined by the displacement parallel to the direction of the force, and the moment arm is perpendicular to the direction of the force. We shall discuss this more fully somewhat later. However, it is useful to point out that bones, serving as moment arms, and the associated joints, serving as fulcrums, act together as levers so that forces applied to the bone to lift weight (at other locations) tend to rotate the bone in the direction needed for movement. Muscles produce the forces that constitute the effort for these biolevers.

Class One Lever

Examples of class one levers (*L∆E*) are claw hammers, pry bars, pliers, and almost anything else found in a carpenter's toolbox. In the human body, the hip is a class one lever since it has the fulcrum between the weight and the applied force. This is the kind of lever of which Archimedes was supposed to have said that, given a lever arm long enough and a fulcrum on which to place it, he could move the earth. Plutarch records that King Hiero, upon hearing of Archimedes boast tested him as follows [48]:

> [Archimedes] fixed accordingly upon a ship of burden . . . which could not be drawn out of the dock without great labor and many men; and loading her with many passengers and a full freight, sitting himself the while far off with no great endeavor, but holding the head of the pulley in his hand and drawing the cords by degrees, he drew the ship in a straight line as smoothly and evenly as if she had been in the sea.

Class Two Lever

A class two lever has the order fulcrum, load, effort (∆*LE*), commonly observed in toilet seats, large paper staplers, wheelbarrows, and some kinds of nutcrackers. The mechanical advantage in this lever is always greater than one because, by design, $d_e > d_l$. The class two levers are not common in the body, but do occur when the muscle supplies the resistance (*L*) to an external force, rather than supplying the force (*E*) to overcome an external load (*L*).

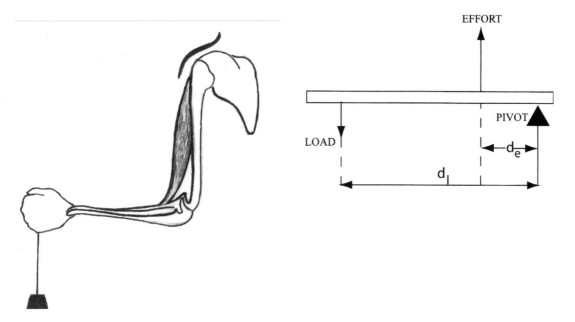

Figure 5.4. A hand-held weight with the attachment of the biceps to the forearm (the attachment of the triceps just above the elbow is not indicated). Adjacent to the schematic of the arm is a class three lever. The comparison between the action of the biceps and the lever is evident.

Class Three Lever

In a class three lever, the fulcrum is at one end and the effort is applied between the fulcrum and the load (ΔEL). An example is a deep-sea fisherman's rig, with the pole tucked into the harness, the fish out at sea, and both hands desperately clutching the pole. The mechanical advantage of this lever is always less than one because, by design, $d_e < d_l$. This is the design of all joints of the upper and lower extremities of the body. There are more third class levers in the human body than any other kind. Consider that our muscles are capable of exerting forces far in excess of any resistance that we have to overcome. The quadriceps, for example, the anterior muscle of the upper leg, is able to apply a force of 200 to 300 kilograms.

Lifting Weights

Vogel [47] gives the illuminating example of a hand-held weight, depicted in Figure 5.4, and compares the muscle–ligament–bone configuration with a class three lever. The lower end of the biceps is attached to the forearm, in which a limited shortening of the biceps causes a wide sweep of the hand much farther from the elbow. He notes that a muscle is an engine for shortening and that is what the biceps does to lift the weight. However, the arm must also be straightened out once the job is done, and to accomplish this straightening out, another muscle is required to oppose the effect of the biceps.

Muscles Acting in Opposition

On the opposite side of the upper arm is the triceps muscle, which attaches to the side of the pivot (elbow) opposite the biceps, just above the elbow. With the elbow as the pivot (fulcrum), the point of attachment of the biceps to the forearm as the load, and the point of attachment of the triceps just above the elbow as the effort, this forms a class one lever. In Figure 5.5, we indicate the attachment of the triceps to the forearm and compare it to a class one lever. Therefore, we have a class three lever (biceps) and a class one lever (triceps) operating antagonistically; that is, each muscle acts as the lengthener of the other. As the biceps shortens, the triceps lengthens and vice versa.

Question 5.6

Suppose you are out in the country and there is an obstacle blocking the road. Identify three different kinds of obstacles and the type of lever that would be best suited to remove the obstacle from your path. (Answer in 300 words or less.)

5.3.3   A Little Excitement

Chemical Energy and Work

In muscle, chemical energy is transformed directly into mechanical work; there is no temperature difference to drive the engine, such as there is in, say, a steam engine. Thus, the only heat generated in muscle is waste; that is, heat is energy that is not available for work. In order to do work, the muscle contracts, as shown

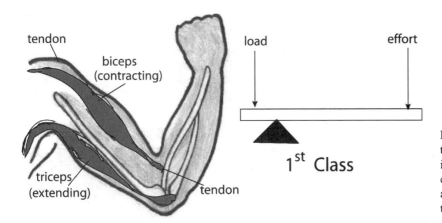

Figure 5.5. Adjacent to the schematic of the arm is a class one lever. The comparison between the action of the triceps and the lever is evident.

schematically in Figure 5.6, where the shortening of a muscle fiber is depicted. The labeling in the figure shows the membrane enclosing the fibril, the thin threads of actin, the thick threads of myosin, and the bridges between the two. We now know that the interdigitation of the protein threads and the pulling or pushing force can only be produced by conformational changes occurring in contractile proteins, that is, in the threads or bridges formed between actin and myosin. The energy source for the conformational changes is the dephosphorylation of ATP. Myosin catalyzes the hydrolysis of ATP and thereby liberates energy. It is this liberated chemical energy that the muscle transforms into the energy of conformational motion. Precisely how the electronic energy of ATP is transformed into the mechanical energy of conformation change goes a little too deeply into chemical mechanisms than we wish to pursue here, but this process constitutes the basis of muscle biochemistry.

How Muscle Contraction Is Done

The electrochemical process in muscle contraction is initiated by a nervous impulse. The contraction can also be initiated by an artificial electrical impulse as well. As Volkenstein [46] explains, a single pulse produces a twitch in a muscle. When sufficiently frequent impulses are transmitted one after another, say on the order of 15 per second, the twitches merge into a tetanic contraction, since each incoming impulse arrives in the refractory (rest) period of the previous one. In other words, the pulses arrive so quickly that the next pulse is present before the influence of the previous pulse has had a chance to decay. The process of contraction begins with an increased

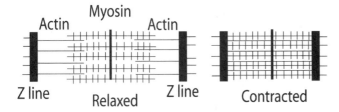

Figure 5.6. Schematic depiction of the shortening of a muscle fiber. The membrane is labeled Z line, actin is depicted by thin threads, myosin by thick threads, and the bridges connecting the actin and myosin are indicated as cross-hatches.

concentration of Ca^{2+} ions appearing in the liquid medium surrounding the protein thread, apparently induced by the impulse. These ions are necessary for actomyosin to function, and precipitates the closing of the bridges between the thick and thin protein threads. The closing of the bridges causes contraction.

Twitches Can Be Quantified

We note that in the present context, the term twitch has a well-defined meaning, since the time course of tension in muscles has a characteristic shape. The functional form for the tension in a muscle twitch is determined experimentally to be

$$F(t) = F_0 \tau e^{-\tau} \tag{5.24}$$

where τ is a dimensionless time defined by the ratio $\tau = t/T$ and T is the excitation time of the muscle. Winter [1] lists a set of five different excitation times for five different muscles. In Figure 5.7 we graph the function of Equation (5.24) for each of these different times. From this figure, we can see two effects of the excitation time. First the maximum tension at a later time as T increases. Second, the width of the pulse increases with increasing excitation time, indicating that certain muscles take longer to reach their maximum tension and remain under tension for a longer time intervals than do others for the same level of excitation.

Excitation Depends on Temperature

It is worth pointing out that the excitation or excitation time of muscles is temperature-dependent. Winter [1] mentions that the biceps brachialis had a contraction time that increases from 54 ms at 37 °C to 124 ms at 23 °C. This decrease in response of the muscle to excitation is reasonable, given the fact that the metabolic rate and indeed all the chemical reaction rates are lower at the lower temperature. In practical terms, this slowing of reaction rate with temperature explains why we are

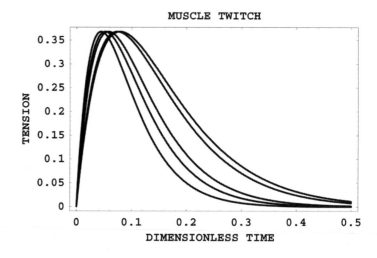

Figure 5.7. Distribution of muscle twitches. The time course of the muscle twitch, modeled by Equation (5.24), is shown for five values of the excitation time. The typical values for the excitation time for triceps bachii is 44.5 ms, biceps brachii is 52.0 ms, tibialis anterior is 58.0 ms, soleus is 74.0 ms, and medial gastrocnemius is 79.0 ms. The amplitude is in units of F_0, which is probably different for each muscle, but is taken to be unity here.

not so agile in winter. Recall fumbling with the car keys on a sub-zero morning or not being able to align the key with the lock to get the car door open.

Speed and Work

It is of interest to determine the dependence of the speed of isotonic contraction on the load, or the tension occurring during isomeric contraction, and the heat exchange of the muscle during these processes. In his classic work, Hill [49] established the main characteristic equation of the mechanics of muscle contraction directly from experimental data. His equation relates the steady-state velocity of isotonic contraction (or shortening) denoted by u with the load F:

$$u = b\left(\frac{F_0 - F}{F + a}\right) \tag{5.25}$$

where F_0 is the maximum force developed by the muscle during isotonic contraction or the maximum load held by the muscle without elongating during isometric contraction, and a and b are constants that differ in the two situations. Note that Equation (5.25) is only valid for $F < F_0$, the situation in which the muscle is shortening.

Work Generated during Contraction

There is a great deal of discussion about the fitting of the parameters in Equation (5.25) to experimental data, and in a book on physiology this would be a proper topic of discussion. However, here we are only interested in the fact that an equation of this form has been developed and refined over the years, since it enables us to calculate the work generated by a muscle during a contraction. We know that work is the product of force and distance, so we can write for the work done by the muscle under a single or tetanic contraction

$$w = Fut \tag{5.26}$$

where ut is the distance traveled by the contraction under the load F. The dependence of the work w on the load F is shown in Figure 5.8 to be bell-shaped. The curve in Figure 5.8 is calculated by inserting Equation (5.25) into Equation (5.26) and using the experimental range of the parameter a, found to be $0.25F_0$ to $0.40F_0$. We see that the work is zero at both extremes of the force, $F = 0$ and $F = F_0$, with a single maximum in between. The highest point on the curve is the value of the load against which the muscle can do the most work. A trainer might look for this load in order to train an athlete at the level of maximum efficiency.

Speed of Twitch Depends on Length

Vogel [47] argues that although the actual speed of shortening depends on the particular muscle, all things being equal, speed depends on the length of the muscle. The argument goes like this. Longer muscles have more contractile units lying end-to-end, so that if the filaments in each unit crawl over each other at the same rate,

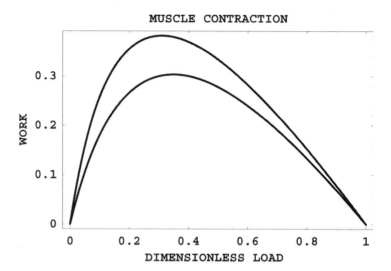

Figure 5.8. Graph of work, W, in arbitrary units, versus load F, normalized to F_0, from Equation (5.26); the overall level has been set to unity, $bt = 1$. The upper curve has $a = 0.25F_0$ and the lower curve is for the parameter value $a = 0.4F_0$.

then the overall shortening will be determined by the total number of units in sequence, that is, the length of the muscle. This is not so different from the dependence of the speed of the crest of a water wave on the length of the wave that we discussed previously.

5.3.4 Running and Walking

The Head Moves Up and Down as We Walk

Alexander, in his *Scientific American* book, *Exploring Biomechanics* [50], gives an excellent review of the distinctive way humans walk. He uses a sequence of still nude photographs the 19th century researcher Muybridge made of himself in his studio in the 1880s [51], to describe how, during walking, we keep each leg almost straight while a foot is on the ground. The head bobs up and down as we walk. The height of the walker changes from the position where the leg supporting the body is vertical (maximum height), to that where the legs are fully extended and the weight is divided equally between both feet (minimum height). This change in height is approximately 40 mm (1.6 in) and oscillates harmonically up and down as we walk.

Force Plates in the Floor

As we walk, our feet exert forces on the ground, which, in turn, as we know from Newton, exerts forces back on our feet. The forces can be measured using a force plate. A force plate is an instrument panel set in the floor that provides a voltage indicating the magnitude and direction of the force exerted by our feet as we walk. The electrical output indicates the components of the force that are downward, backward, forward, and sideways. From such measurements, it has been determined that the force on each foot is always aligned with the leg, or almost always.

The Pendulum Model

One of the more accepted models of walking is based on the simple pendulum, taking into account, as it does, the continuous exchange between potential and kinetic energy as we walk. The sprinter's muscles supply the energy required to rapidly change his kinetic energy from zero to a large value as he crosses the finish line. However, during walking, the energy, like that of the pendulum, is maximally kinetic when the legs are vertical, and maximally potential when the legs have maximum extension. Alexander [50] argues that, like the pendulum, very little work is needed from our leg muscles while we walk, since the legs are straight and the muscles have little tension.

Straight Legs Require Less Energy to Walk

The straight-legged style of walking that we humans have adopted allows us to support our weight without putting a substantial burden on our leg muscles. Alexander [50] explains that when you stand erect, your knees are straight and the line of action of your weight passes quite close to your knee joints. Therefore, relatively little tension in the muscles is required to support your weight. This can be contrasted with a children's game known as the "duck walk," in which you squat down with your butt almost on the floor and you walk around the room in this squashed position. Adults soon find that this causes the quadriceps muscles in their legs to cramp because of the unaccustomed tension required to keep themselves from falling over. From this simple exercise, it is clear that the straighter your legs, the less force your muscles need to exert in walking.

Running Versus Walking

The above argument concerning the efficiency of human walking, must be modified when humans run, since they no longer keep their legs straight when they run. Alexander uses another set of nude photographs taken by Muybridge [51], this time of a young male runner (not Muybridge), to show that when the maximum force is exerted, the knee of the runner's leg in contact with the ground is bent. This bending of the knee in each cycle of the runner's stride requires substantial metabolic energy and, therefore, running has greater costs relative to walking, both in terms of energy expended and in terms of forces exerted. Alexander [50] posed the question that if walking is so efficient relative to running, why do humans ever run? His answer is simplicity itself.

Why Humans Run

The argument, due to Alexander, concerning the transition from walking to running has to do with the stability of locomotion. Assuming the leg remains straight throughout a stride, a person's hip undergoes movement along the arc of a circle with the foot at its center. As we shall review in our consideration of rotational motion, a body that moves on a circle with a speed u has an acceleration point downward along the leg with a magnitude u^2/l, where l is the radius of the circle (length of the leg). The acceleration is necessary to keep the body from moving tangentially from the circle and, in the case of walking, the acceleration is necessary to keep the person from falling over. Because the force acting along the leg is produced by the body's weight, the limiting value of the acceleration is determined by the acceleration of gravity. This implies the relationship

$$u^2/l \leq g \tag{5.27}$$

so that the limiting value of the speed of walking is given by

$$u \leq \sqrt{gl} \tag{5.28}$$

Thus, if the speed of walking is greater than $\sqrt{gl}$, our strides become unstable and we break into a run. Suppose the length of an adult human leg is 0.9 m. Then, using the acceleration of gravity (9.8 m/sec²), we have for the limiting speed of walking $\sqrt{9.8 \text{ m/sec}^2 \times 0.9 \text{ m}} \approx 2.97$ m/sec. This value of walking speed is in agreement with experiment.

Question 5.7

If one divides Equation (5.28) by the length of the person's leg, l, the resulting equation has the dimension of time. Use the equation for the period of a simple pendulum to interpret this equation.

5.4 SUMMARY   Conservation of Energy

The fact that energy is conserved is not obvious. The development of the conservation law took 200 years, both in the formulation of its separate pieces and in the recognition of its universality. The conservation of mechanical energy in an idealized system was recognized fairly early, but it took another century to realize that the heat produced by the friction in a real mechanical system is also a form of energy. It was not until all the various forms of energy were identified that the full articulation of energy conservation was possible. You may have heard that mass is also just another form of energy, a form that is released in nuclear reactors and atom bombs. This form of energy was not fully understood until the second half of the 20th century.

The Two Kinds of Energy

The energy of position is called potential energy and the energy of motion is called kinetic energy. Potential energy is stored in the location of a material body. A watermelon held by a student on the roof of the physics building contains potential energy, which is converted to kinetic energy when the melon is thrown from the building. The amount of potential energy, once contained in the melon, is revealed by the area covered by the splattered remains of the melon when it smashes against the sidewalk below.

Energy Is the Ability to Do Work

Doing problems in the evening is homework, carrying a book bag around campus is hard work, and it is often difficult to get a computer program to work. But it is only the work in physics that we are interested in this chapter, and that is the product of force and the displacement of an object parallel to that force. This is the work that we can identify with the physical energy in mechanical systems.

Energy Was Used Long Before it Was Understood

The waterwheel is a machine used to convert the potential and kinetic energy of flowing water into useful work. In 890 AD, the cam was joined with the waterwheel in the monastery of St. Gall, where the monks used the contraption to make beer. A cam is a piece of wood set on the side of a shaft, so that as the shaft turns, the protruding piece of wood moves anything in its way [52]:

> By the eleventh century there were forge hammers in Bavaria, oil and silk mills in Italy; by the twelfth century there were sugar-cane crushers in Sicily, tanning mills pounding leather in France, water-powered grinding stones for sharpening and polishing arms in Normandy, ore-crushing mills in Austria. From then on the use of water-power spread to almost every conceivable craft: lathes, wire-making, coin-producing, metal-slitting, sawmills and—perhaps most important of all—in Liége, northern France, in 1348 the first water-powered bellows providing the draught for a blast furnace.

The Body Uses Muscle to Do Work

The muscles in the human body connect the macroscopic work the body does with the microscopic electrochemical processes that releases energy within cells. The work is not just lifting your book bag, but is also the pumping of blood, moving the air in and out of the lungs, and locomotion. Each of these functions is interrelated and requires that we know a little about how muscles operate: how muscles receive electrical signals from the central nervous system, how these signals cause the muscles to contract, and how these contractions are controlled and move things around in a useful way. One important function of muscle is to operate as a simple lever.

Stable and Unstable Locomotion

Running and walking are produced by different activities of the muscles in our legs. However, the simplest model of both is that of the simple pendulum. As long as the period of the step is less than that of a corresponding pendulum, walking is stable. If the period increases beyond that of a simple pendulum, walking becomes unstable and a person breaks into a run to regain balance.

CHAPTER 6

Linear Physics

Objectives of Chapter Six

- Take what we understand about simple physical systems and transform that knowledge into some understanding of simple biological systems.
- The tensile strength of a column can be used to estimate the carrying capacity of a weight lifter's skeleton.
- The energy stored in a spring can be used to determine the limit to the elastic energy contained in a pole-vaulter's muscle.
- The torsion in a wire can suggest the maximum torque an ankle can resist before fracturing.
- All these associations and more can be determined using Hooke's law for the proportionality of stress and strain extended to the squishy world of biology.

Introduction

Most Physical Models Are Linear

In previous chapters, where we discussed Newton's laws of motion and mechanics, we did not emphasize the fact that most physical models of the world are linear. This is a profound assumption, since a linear system is one in which the response of the system is proportional to the force applied to that system. Therefore, a system that is unstable cannot be linear because, by definition, instability is an arbitrarily large response of the system to any applied force, even a very small one. Examples of such instabilities might include knocking over a lamp, stumbling on a crack in the sidewalk, a car skidding on ice, losing one's temper, and any of an overwhelming number of familiar phenomena. We shall spend a significant amount of time

Biodynamics: Why the Wirewalker Doesn't Fall. By Bruce J. West and Lori A. Griffin
ISBN 0-471-34619-5 © 2004 John Wiley & Sons, Inc.

discussing linear behavior, because this is what scientists use to understand the evolution of simple systems. Understanding linearity also provides a foundation for stepping off into the world of the nonlinear, which is where we live.

Thinking in More Dimensions

Up until now, we have managed to avoid considering the fact that we do not live in a one-dimensional world. We have restricted our discussions to one vertical coordinate, or one horizontal coordinate, or, indeed, to just a point in time. However, we now need to open up our thinking to three dimensions, because some phenomena require these additional dimensions in order to manifest themselves. For example, the rotation of a diver coming off the high board, a fullback receiving a glancing tackle, or the simple toss of a coin all require at least two spatial dimensions for discussion of their properties. Therefore, we also need to expand our discussion of dimensional analysis to be able to distinguish one coordinate axis from another. This will also allow us to talk separately about the magnitude and direction of a physical quantity like a force or momentum.

Conservation Laws

Another reason to introduce additional spatial dimensions is because we want to discuss conservation laws other than energy, in particular, the conservation of momentum. Like the conservation of energy, the conservation of momentum means there is a physical quantity, called momentum, that is neither created nor destroyed, but is conserved during the interactions in any physical system. Also, like the conservation of energy there are certain conditions under which the conservation law appears to be true and other conditions under which the conservation law appears not to be true. The conservation of momentum is important because it is the change in momentum over time that we experience as a force. In the same way, the conservation of angular momentum is important because it is the change in angular momentum over time that we experience as a torque.

6.1 LINEAR RESPONSE

Linear Restoring Force

A force of overwhelming historical and contemporary importance in modeling both physical and biological systems is due to a contemporary of Newton, Robert Hooke. This is the law of linear response, describing, among other phenomena, the linear spring depicted in Figure 6.1. The force law describing the spring is given by

$$F = -kx \tag{6.1}$$

where x is the displacement of the spring from its equilibrium position, k is called the spring constant, and the minus sign indicates that the force is such as to resist the displacement and pull the spring back to the equilibrium position, $x = 0$. By equilibrium, we mean the condition in which all the forces acting on the spring are in balance so the spring does not move. For example, the spring, depicted as hanging from the armature in Figure 6.1 is at room temperature, so all its atoms are jittering within the material, even though there is no macroscopic motion (displace-

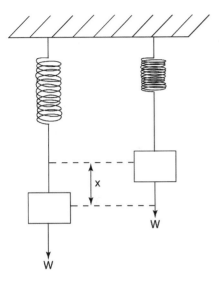

Figure 6.1. The mass, m, or, equivalently, a weight W, hanging from a massless spring is pushed up a distance x from its equilibrium position and then released. Once released, the mass oscillates with a frequency that depends on the physical parameters of the spring.

ment) of the spring. The more the spring is displaced by the application of an external force, the greater is the force that attempts to reclaim equilibrium. Of course, we all know what happens when we pull on a spring and release it. It bobs up and down about the equilibrium position, eventually coming to rest. How long it oscillates depends on the material constant, k (called the stiffness), the mass of the spring, the temperature, and many other variables that are usually neglected in any discussion of the spring's motion.

Control of Sway

Of course, the human body does not consist of springs, but assuming there exists a force proportional to displacement from balance has also been useful in describing phenomena of locomotion. For example, the gymnast who loses his/her balance on the crossbeam moves sharply back and forth, attempting to recapture his/her equilibrium position above the beam. The farther out of balance the gymnasts are when they land, the greater is the force required to pull them back into equilibrium. A less dramatic application of this idea is to explain simple posture control. Stand on one foot and close your eyes. You will soon find yourself swaying back and forth. The human postural control system is highly evolved and complex, but it may be described by a simple mechanism like the linear force law plus noise [53]. Consider the spring hanging in a breeze, where the velocity of the air is turbulent. The spring bounces and dances around in response to the random forces, but it still tries to oscillate as well. This is something like the model of postural control. There is a smooth, slow swaying motion superposed on a much faster jerking behavior attempting to pull yourself upright. This is also the movement of the wirewalker high in the air.

Why Do Springs Oscillate?

After the mass at the end of the massless spring is released, there are two forces acting in opposition to one another. The force of gravity pulls downward on the mass, and the restoring force of the spring pulls the mass up when it is below its equilibrium position, and down when it is above its equilibrium position. Do not confuse these two forces with those in Newton's third law. The forces of action and reaction in the third law act on separate bodies. Here, the two forces are from Newton's second law, with gravity and the restoring influence of the spring being separate forces acting on the same body. But there is another reason for the oscillatory motion of the mass, having to do with the conservation of energy. When we initially pull the spring, we are doing work on the system. This work is stored in a form of potential energy called "elastic energy," which is associated with the relative position of the atoms in the spring. When the spring is released, this elastic energy is converted into kinetic energy, reaching a maximum when the spring passes its equilibrium point at which the elastic energy is zero. As the spring passes its equilibrium point, it compresses and slows down, converting kinetic energy back into elastic energy.

Why Do the Oscillations Stop?

The spring comes to a stop when the compression of the spring is such that all the initial work is back in the form of elastic energy. If no energy were lost during this cycle, things would just repeat themselves and the spring would undergo simple harmonic motion forever. That is not the way things are, however. A little energy is lost during each cycle. In fact, if we had a sufficiently sensitive thermometer, we would find that the temperature of the spring actually increases as it converts mechanical energy into heat during each oscillation. It is the sum of the mechanical energy and heat that is conserved during the motion of the spring. The oscillations stop when all the elastic energy has been converted into heat. The spring therefore heats up, but then it loses heat to its surroundings through radiation and convective heating and eventually comes into equilibrium with the temperature in the room.

Question 6.1

Suppose a spring has an initial displacement of 10 meters and it has a spring constant, k, of 10 Newtons/meter. How much would the temperature of the spring be raised when it stops oscillating, assuming that no energy is lost.

6.1.1 Stress and Strain

Force and Deformation

When a force is applied to a solid body, that body will change in shape as a result of the applied force. Sometimes, the change persists after the force is removed and other times the original shape of the body returns. A diving board will bend under the weight of the athlete, but after the person dives, the board, after a few oscillations, returns to its original horizontal position. A rubber band will stretch when pulled apart and snap back when released, unless, of course, it is pulled too far. Clay can be compressed when our car gets stuck in the mud, and deformed through shearing in the sculpting of a bust. All these various deforming forces can be collectively called stresses. The deformation undergone by the body under *stress* is called *strain*. Note that strain is the ratio of the change in length to the original length of the stressed object and, therefore, is a dimensionless number.

Elasticity Bounces Back

When an object undergoes deformation as the result of stress, it may or may not return to its original position when the stress is removed. As we noted, the diving board returns to its original shape when the diver leaves the platform. The rubber band returns to its flaccid state when the pull on its ends is released. The property of a material to return to its original state with the loss of stress is known as elasticity. It is elasticity that gives a racketball or basketball its bounce, and a golf ball its distance. Elastin is a protein similar to albumin and collagen that is the main constituent of elastic fibers of the body and is responsible for human skin being stretchy and, consequently, smooth.

Plasticity Remains Deformed

A deformed object such as a tennis ball returns to its original shape but clay does not. Clay is said to be plastic rather than elastic; wax is also plastic. However, an elastic material does not return to its original shape when its elastic limit has been exceeded. This failure to return to its initial form means that a certain level of stress produces irreversible deformations of the material that remain after the stress is released, for example the buckling of a beam, the breaking of a bone, or the tearing of a muscle. The time-dependent nonelastic behavior of materials is called *viscoelasticity* and applies to such things as the flow of blood and locomotion using muscles. Even skin manifests this behavior. As one gets older, the skin loses its elastic properties and remains deformed for longer and longer periods of time after a stress is removed. Pinch the skin on the back of your hand and it quickly relaxes back to a smooth surface. The time for the imposed wrinkle to relax becomes longer and longer as one ages. The loss of elastin in the skin accounts, in part, for the wrinkles in the elderly.

Elastic Moduli Are Different in Different Directions

Different material bodies of the same mass and the same geometrical shape respond to the same external forces in different ways. This empirical fact is attributable to the differences in the internal constitutions of the materials. As emphasized by Findley et. al [54], real materials behave in such complex ways that it is presently impossible to characterize this behavior with a single model that is valid over the entire range of possible temperatures and deformations. Thus, separate constitutive equations are used to describe various kinds of idealized material responses. The simplest example, as we mentioned above, is that of a Hookean (linear) elastic solid in which the stress σ (force per unit area) is proportional to the strain ε (deformation per unit length). We ignore the fact that stress and strain are really quite complicated and a given stress can produce different strains in different directions, depending on the material. This level of analysis is beyond our reach; however, we can go quite far in understanding the mechanical operation of the body with relatively simple ideas.

The Response Is Transverse to the Force

The force exerted in a conventional lemon squeezer is in the vertical direction, but the juice squirts out in the horizontal direction, at right angles to the direction of the applied force. The same thing happens when we throw clay against the wall. The force is in the direction thrown, but the response of the clay is transverse to that direction when it encounters the immovable wall. The coupling of the direction of the force in the material to the direction of the response of the material is determined by the *elastic moduli*. The greater the magnitude of the moduli, the stronger is the coupling between the responses to the forces in the various directions.

Steel Versus Wood

Consider two beams, one made of wood and other made of steel. Strike each on the side with a hammer. Which one transmits sound waves better along the beam? If you chose the steel column, you would be right. The reason for the change in sound transmission has to do with the difference in structure of the two materials. The metal has a microscopic lattice that connects the atoms more or less equally in all three directions, like the springs in an idealized bed. The wood column has fibers that interlace one another and define a preferred direction for the movement of these loosely coupled fibers. Thus, the impulse from the hammer sends out a sound wave in the metal, seen as oscillations along the lattice, in all directions. In wood, on the other hand, the oscillations are rapidly dampened by the generation of heat caused by the fibers rubbing against one another. The elastic modulus for the metal would therefore be quite high compared to that of wood. In addition, the elastic moduli of hard woods are greater than that of soft woods, so that hard wood has better tonal quality than does soft wood.

Isotropy Is the Same All Over

An isotropic material is one in which the material properties are the same everywhere. A simple example of an isotropic material is water. Consider a cylinder filled with water and open at the top to receive a piston. A piston with the same diameter as the cylinder pushes downward with a given force and this force is transmitted throughout the water. The force is not just transmitted in the vertical direction but to the sides of the cylinder as well. Thus, the pressure developed in the fluid is equal throughout because the elastic moduli transfer the applied force equally in all directions. Most fluids are isotropic, meaning that they have the same properties in all directions.

Definition of Modulus

We will keep things at their simplest level and treat stress and strain as scalar functions of time. Thus, if the ratio of stress to strain is independent of time,

$$\text{Young's modulus} = \frac{\text{stress}}{\text{strain}} = \text{constant} \tag{6.2}$$

the constant is called the Young's modulus. Within the range of validity of Hooke's law, moduli are constant and characteristic of the type of material. Different types of material have different moduli and therefore different distortions, but independent of the shape and size of the sample and the force applied. The greater the applied force (stress) required for a given material to undergo a specified distortion (strain), the greater is the modulus. Therefore, the size of the modulus indicates the stiffness of the material, just as in the case of the spring constant. Consider the difference between jeans and stretch pants. The stretch pants are elastic; therefore, their modulus is low, since they can be highly deformed and then snap back to their original shape. Jeans, on the other hand, are very stiff and once deformed tend to stay that way, so their modulus is quite high.

Stretch without Breaking

The stress–strain relation for materials is usually given in the form of a graph. In Figure 6.2 is depicted the stress–strain diagram for steel, bone, and saliva. Our experience tells us that saliva is almost infinitely extensible, but that both steel and bone have limits to their responses to stress as reflected in the curves denoting the various stress–strain relations. The tissue in most animals lies between the extreme curves of bone and saliva, and most tissue are inextensible up to strains of 50% or more. Gordon [55] gives the urinary bladder in young people as an example of a membrane tissue that will stretch up to approximately 100% of strain.

Stress–Strain Curves for Tissue Are Not Flat

Arteries and veins generally operate at strains in the vicinity of 50%. If the tissue behaved as an elastic rubber, what would occur is an instability in the swelling of the tube, leading to a spherical protrusion, not unlike an aneurism. Therefore any sort of rubbery elasticity is unsuitable for most internal membranes or for animal tissue in general. Gordon [55] says that the only sort of elasticity that is stable under fluid pressures at high strains, that is, strains greater than 50%, is depicted by the concave curve in Figure 6.2; this will be discussed further in Chapter 12. The shape of this stress–strain curve is very common for animal tissues with only minor modifications. We can see that the response is no longer linear for extreme strains. This would mean that Hooke's force law [Equation (6.1)] would not correctly describe the phenomenon.

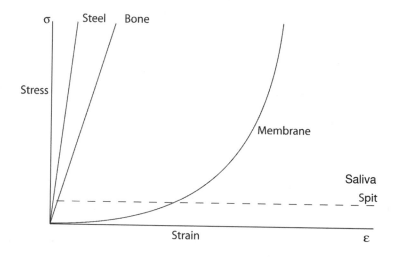

Figure 6.2. Membranes have nonlinear stress–strain relations. Stress–strain diagram for steel, bone, saliva, and typical biological membrane tissue.

Hooke's Anagram

The basic modeling elements of rheology, that is, the study of the reaction of dirt, rocks, and other mundane materials to applied forces, are linear springs that model the reaction force and linear viscous dash pots that model the dissipation, with the additional condition that inertial effects are negligible. If R is the linear spring constant (Young's modulus), then Hooke's law states that the time-dependent stress, $\sigma(t)$, and time-dependent strain, $\varepsilon(t)$, are proportional to one another:

$$\sigma = R\varepsilon \qquad (6.3)$$

Lest we think that human nature has changed very much in the past 300 years, it should be noted that Hooke was very much concerned that he would not receive full credit for his discovery of this relation. Therefore, in 1676 he expressed Equation (6.3) in the form of an anagram (in Latin)—"ceiiinosssttuv"—and challenged the scientific community to decipher it. He gave his rivals, among them Newton, two years to solve the puzzle and perhaps take credit for his discovery regarding springs. At the end of the two years, in 1678, he gave the solution as "ut tensio, sic vis." The translation of the Latin is "as stretch, so force."

Here we read strain as the stretch of the spring and stress as the applied force. It seemed to Hooke and his peers that no matter what was done in science ("natural philosophy"), Newton received the credit, and no doubt a certain amount of that was true. It may be a consequence of the above ploy that Equation (6.3) and its equivalent, Equation (6.1), is still known as Hooke's law today.

Poisson Ratio

It is a familiar observation that applying a tension or compression force to a material body produces a contraction or a bulging of the body in the transverse direction, that is, in the direction at right angles to that at which the force is applied. If σ_1 is the stress applied in direction 1, producing a primary strain ε_1, then a secondary strain ε_2 will be produced in the transverse direction. Poisson found that, for any given material, the ratio of the secondary to the primary strain,

$$\text{Poisson ratio} = \frac{\varepsilon_1}{\varepsilon_2} \qquad (6.4)$$

is a constant for that material. Here again, we are only discussing the linear region of the response curve.

Question 6.2

Suppose a tendon is initially 5 cm long and a stress produces a strain of 0.01. What is the new length of the tendon? Does your answer make sense?

Question 6.3

Assume that an artery in the body is 1 cm in diameter and 1 m long with a Poisson's ratio is 1/2. If the diameter of the artery changes by 1 mm, what would be the total change in length? Does your answer make sense?

6.1.2 Various Forms of Stress

Stressing a Muscle

Consider the stress put on muscles, say by gravity, in lifting a weight. The resulting strain is seen in the bulging of the biceps, which relax back to their undeformed shape when the stress is released. Muscles, therefore, to a certain extent, behave as elastic materials. One of the major differences between muscle and inanimate material is that the strain in an inanimate material is a completely mechanical response to stress. This is not a complete description of the strain in muscle, because of the way stress is transmitted in living tissue. There is no simple analogue to Hooke's law for the dynamics of muscle response. We shall discuss this more fully in later lectures but, for the moment, we remain focused on the linear force law.

Linear Forces Seem to be Everywhere

Hooke's law can take a variety of forms in elastic materials. We considered the case of stretching and/or compressing a spring. The prototype of an elastic body, the bounding of a basketball, could serve just as well. Other examples of the application of Hooke's law include:

1. A wire being pulled from both ends; stretch $\propto$ tension
2. A metal rod being stretched or compressed; change in length $\propto$ force
3. A metal rod being twisted; angle of twist $\propto$ torque
4. A beam being bent under a load; sag of beam $\propto$ load
5. A solid or liquid being compressed; change in volume $\propto$ pressure

From Physics to Biology

Of course, these same ideas, in slightly modified form, apply to the human body:

6. A muscle lifting a weight; stretch $\propto$ tension
7. A bone being stretched or compressed; change in length $\propto$ force
8. An arm or leg joint being twisted; angle of twist $\propto$ torque
9. The air in the lungs, or blood in the heart, being compressed; change in volume $\propto$ pressure

Tension, Torque and Pressure

We have introduced a number of new terms in these examples, such as tension, torque, and pressure. Each of these terms will be taken up at the appropriate time and discussed fully, but we think their intuitive meaning is fairly clear from the context. A tension is the force transmitted by a rod, rope, or other material, such as muscle. Tension transfers force through muscle, from one part of a body to another, say in the lifting of heavy objects using one's legs. Torque is the action of a force over a distance to provide a rotational motion about a point. This is particularly true when we study the action of a force impacting on the knee or elbow and producing a rotational stress. Finally, pressure is a force per unit area. Changes in blood pressure arise from increasing the pumping of the heart, that is, the applied force, or constricting blood vessels and thereby reducing the overall volume of the blood supply system. Both can raise the blood pressure.

These are All Linear Forces

Each of the above linear relations is an example of strain being proportional to stress, that is, the deformation is proportional to the applied force. Of course, this same principle applies to the loading of the skeletal structure of animals and the use of muscles to do work, as we have indicated. The result is that the output is proportional to the input and the relationship is a linear one. The reason why all these materials are Hookean in nature lies in the microscopic domain. In the vicinity of the origin in Figure 6.2, where the microscopic forces of attraction and repulsion of an atomic bond are in balance, the stress–strain curve for a stiff material is linear. The force acting on each bond is the net force applied to the material reduced by the number of bonds sharing the load. In this case, if the area over which the force is applied is doubled, the load carried by each of the bonds is halved. The stress is the natural measure of this microscopic property.

Action and Reaction

We have mentioned repeatedly that for every action there is an equal and opposite reaction. Thus, the force with which a runner's foot strikes the ground is balanced by an equal and opposite force the ground imparts to the foot. When this sharp response of the ground does not occur, the runner is likely to become unbalanced, or be slowed down, perhaps as in running on sand or mud. But what is it about solids that enables them to push back? When we push against a wall, we have every expectation that it will push back against our hands, that is why we are so surprised when a termite-infested structure yields and then crumbles. What does the one wall have that the other has lost?

Hooke's Law and Reactions

The law of action and reaction was clearly known to Hooke and arises from the ability of a material to change shape when a mechanical force is applied to it. Hooke's law states that this change of shape is a linear deformation, that is, a deformation that is directly proportional to the applied force. It is important to understand that this is a property of all materials for some range of forces. It is universal, assuming, of course, that the force is not too large. What is at question is not the response of the material, but whether the material's response is linear or not. But for the moment, we shall not be concerned with such quantitative details.

All Things Deform

Every solid changes its shape to a greater or lesser extent when a force is applied, and the range of possible deformation is enormous. The extension of a rubber band is quite visible, but the change in the length of a tendon or bone as a jogger pounds down the street is much more difficult to see. However, even though we cannot see the deformation of such things, they are nonetheless real, and require special instruments to measure their changes. We shall discuss the measurement of these forces and the anatomical response in due course. Here, we want to get back to how an object responds to an applied force.

Deformation Occurs on All Scales

We must understand that when an object is deflected, that movement is carried down to the molecular level. Thus, when we deform a tendon by pulling on it, the atoms and molecules making up the tendon are displaced, some moving farther apart and others being squashed closer together. The constituent particles in the tendon are held together by stiff chemical bonds. So, as Gordon [55] explains, when the tendon (or any material) as a whole is stretched or compressed, many millions of strong chemical bonds are stretched and compressed in spite of their vigorous opposition. These bonds resist even the smallest deformation, and it is this resistance that constitutes the reactive force to the impulsive loading of the runner's weight.

Elasticity Is Now a Science

The property of most materials to return to their original form after being stressed is called elasticity, as we mentioned earlier. Elasticity as a science did not emerge until 120 years after the death of Hooke. This was in part due to the fact that Hooke and Newton were great enemies and Newton lived an additional 25 years after the death of Hooke. Newton never let an opportunity to denigrate Hooke, Hooke's research, or applied science in general, go by without a sarcastic comment.

Biological Tissue is Non-Hookean

The aorta and principal arteries in our bodies expand and contract elastically with each beat of the heart, but they do not in all respects obey Hooke's law. This is reflected in the fact that the arteries do not move measureably in the longitudinal direction as they expand and contract. Consider a major artery, such as those that supply blood to our legs, and assume them to be a centimeter in diameter and a meter long. Gordon [55] makes that assumption, so for a Poisson's ratio of ½ for the arteries, a change in diameter of 0.5 mm would imply a total change in length of the artery of approximately 25 mm or about ½ inch. This movement of the arteries would have to take place on the order of once every second, which Gordon rightly concludes would be impossible, in addition to the fact that such motion is not observed. Consequently, the materials making up arteries are non-Hookean. They behave elastically in the radial direction, but not along their length. Further, similar arguments can be applied to most biological tissue in animals to reach the same conclusion.

Question 6.4

How did Gordon determine that the change in the length of the artery would be 25 mm in his example?

6.2
DIMENSIONAL
ANALYSIS OF
VECTORS

Scalars Have Only Magnitude

Let us begin our discussion of a general mechanical system by distinguishing between a vector and a scalar. A scalar is a quantity that is specified by a single number, denoting a magnitude. The density of water is a scalar, as is the air pressure in this room and the speed of a car. Indeed, any quantity to which one can associate a single number, without additional qualification, is a scalar. We can multiply scalars by scalars, in which case we get a new scalar. For example, a mass density times a volume yields a mass, all three of which are scalars. Vectors, on the other hand, are a different matter. Their definition is simple, but the way in which they can combine in physical systems can be conceptually quite confusing.

6.2.1   Vectors and Their Products

Vectors Have Magnitude and Direction

A vector is a physical quantity that is determined by at least two numbers. A force is an example of a vector. A force is determined by a number that is its magnitude, or intensity, and at least one other number that is its direction, say the angle from some given reference direction, such as north. This is the case in two dimensions; for example, the velocity of a car moving along a highway. On the other hand, these two numbers would not be sufficient to specify the velocity of an airplane. In addition to the speed of the plane, we need to know the plane's heading and whether it was ascending or descending in order to completely specify its velocity in three dimensions.

Some Properties of Vectors

In Figure 6.3, we show a typical vector in a plane indicated by an arrow. The length of the arrow is the magnitude of the vector and the arrowhead indicates the direction. The x component of the vector is the projection of the vector along the x axis:

$$F_x = F \cos \theta \tag{6.5}$$

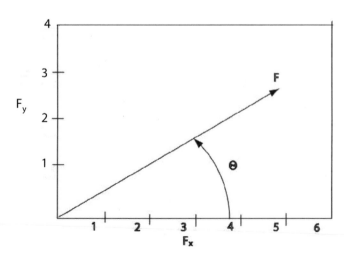

Figure 6.3. A two-dimensional vector. We have sketched a force vector that has components along the x axis and along the y axis. Equivalently, the vector has a magnitude F and a direction θ, as explained in the text.

and the y component of the vector is the projection of the vector along the y axis

$$F_y = F \sin \theta \qquad (6.6)$$

The subscript denotes the direction of the component of the vector. The magnitude, or length, of the vector is therefore given by the Pythagorean theorem, using the trigonometric identity $\cos^2 \theta + \sin^2 \theta = 1$, to be

$$F = \sqrt{F_x^2 + F_y^2} \qquad (6.7)$$

The direction of the vector is determined, by convention, to be the angle θ the vector makes with the x axis:

$$\tan \theta = \frac{F_y}{F_x} \qquad (6.8)$$

We can see from these relations that there are two equivalent representations of the vector. In the plane, one can specify the two numbers (F_x, F_y) as the coordinates of the vector, the size of the force along the x axis and the size of the force along the y axis. Equivalently, one can specify the two numbers (F, θ) as the magnitude of the radius vector and the angle the vector makes with the x axis. The two specifications are referred to as the Cartesian and the polar coordinates, respectively, and are completely equivalent.

Vector Addition

When two forces act on a body, both pushing on it but in opposite directions, the body either moves or is stationary. If the body is stationary, the two forces are equal in magnitude and have nullified one another. But since a force is a vector, it means that we have added two vectors together to get zero; therefore, vectors add. The way to add vectors is by adding together their components. Say the first vector acting on the body is denoted by $F_1 = (F_{1x}, F_{1y})$, indicating a force along the x axis of F_{1x} and a force along the y axis of F_{1y}. The second vector acting on the body is denoted by $F_2 = (F_{2x}, F_{2y})$, showing a force along the x axis of F_{2x} and a force along the y axis of F_{2y}. The fact that the body does not move means that $F_1 + F_2 = 0$. The sum of the two forces vanish, component $F_{1x} + F_{2x} = 0$ by component $F_{1y} + F_{2y} = 0$. Thus, we have equal and opposite components along the x axis, $F_{1x} = -F_{2x}$, and equal and opposite components along the y axis, $F_{1y} = -F_{2y}$. The fact that the vector has a zero in both the x and y directions is the reason that the zero vector is boldfaced, $\boldsymbol{0} = (0, 0)$.

Scalar Product of Two Vectors

We can multiply scalars by vectors, in which case we get a new vector with a different magnitude, but pointing in the same direction as the original vector. We can also multiply vectors by vectors, but here things get a little tricky. Since each vector has a magnitude and direction, there are two ways we can multiply vectors. One way to multiply vectors yields a *scalar product* and is denoted by a dot. Consider vectors A and B; their scalar product is

$$A \cdot B = AB \cos \theta \qquad (6.9)$$

where A and B are the magnitudes of the separate vectors and θ is the angle between the directions in which the two vectors point. Notice that the scalar product of two vectors is a scalar whose magnitude is determined by the product of the magnitudes of the two vectors and the cosine of the angle between them, hence its name.

Work Is a Scalar Product of Two Vectors

One way to use the scalar product is in the definition of mechanical work. You will recall that we defined work as the product of force and the distance moved in a direction parallel to the force. Now we can take into account the vector nature of both force and displacement and define work as the scalar product or dot product:

$$w = F \cdot s = Fs \cos \theta = xF_x + yF_y + zF_z \qquad (6.10)$$

where we have included the components for all three dimensions. Notice that if the two vectors, force and displacement, are perpendicular to one another, $\theta = \pi/2$, and since $\cos(\pi/2) = 0$, the work defined by Equation (6.10) is zero. Since the force and the displacement are at right angles to one another, the force cannot be producing the displacement and, therefore, no work is done by the force. If, on the other hand, the two vectors are parallel to one another, $\theta = 0$, and since $\cos(0) = 1$, the work produced by the force is just the product Fs. The mechanical work done by an applied force is therefore somewhere between the minimum of zero and the maximum of Fs.

Vector Product of Two Vectors

The second kind of vector product yields a vector, not a scalar. This latter product of two vectors is denoted by a multiplication sign, ×. Consider a vector A and a vector B, their vector product is denoted by $A \times B$. We shall discuss the physical applications of this "cross-product" subsequently. For the moment, we merely comment that the cross product means that one vector is crossed into another using the right-hand rule. Consider two vectors, one pointing in the x direction, A_x, say, and the other pointing in the y direction, A_y, say. The right-hand rule requires that if the index finger of your right hand points in the direction of the first vector, and the middle finger of the right hand points in the direction of the second vector, the thumb will point in the direction of their vector product, A_z. This is what we mean by the term "right-hand coordinate system," depicted in Figure 6.4.

Magnitude of the Cross-Product of Two Vectors

The magnitude of the cross-product of two vectors is given by

$$|A \times B| = AB \sin \theta \qquad (6.11)$$

where, again, θ is the angle between the vectors, $0 \leq \theta \leq \pi$. Therefore, when the two vectors are parallel to one another, so that $\theta = 0$, and since $\sin(0) = 0$, the cross-

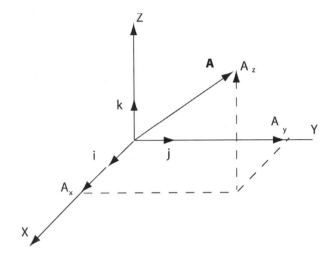

Figure 6.4. Right-hand three-dimensional coordinate system. The horizontal plane is given by x and y axes, and the vertical axis being given by z axis. The so-called unit vectors are indicated by i, j, and k, which are vectors one unit long pointing in the x, y, and z directions, respectively.

product vanishes. If the two vectors are perpendicular to one another, then $\theta = \pi/2$, and since $\sin(\pi/2) = 1$, the magnitude of the cross-product is just the product of the magnitudes of the vectors. Thus, the magnitude of the cross-product falls between the values of zero and the product of the magnitudes of the separate vectors. Before you think that we have lost ourselves in some mathematics, think about the twirling motion of a dancer gliding across the dance floor, the twisting of a high diver in the air above the water's surface, and the gut-wrenching movements of a wrestler throwing his opponent to the mat. All these movements, and infinitely many others, imply not only the application of a force of a given size, but a rotation as well. It is the rotation that is captured by the cross-product. The tighter the diver brings his limbs to his body, the faster he can rotate. But we are getting ahead of ourselves.

Question 6.5

Consider a room 10 ft long by 10 ft wide. We want to specify the location of an object in the center of the room, relative to a coordinate system in one corner. Use a geometrical argument to prove that the Cartesian coordinates of this point are (5 ft, 5 ft) and the polar coordinates are ($\sqrt{50}$ ft, 45°).

Vectors of Interest

In the case of a force, we have the product of the mass (a scalar) by acceleration (a vector) to produce a vector. Further, since acceleration is the time rate of change of velocity, velocity is also a vector. For example, the speed of a runner may be 22 ft/sec, but if we also specify that we are on a street heading north, then we have specified the direction as well as the speed and, therefore, the velocity of the runner. In the same way, we may be only interested in the impact force of a runner's heel with the ground if we want to assess the possible damage done to the heel in a race. However, if we want to determine how to achieve the minimum time in jumping hurdles, then the angle at which the heel comes in contact with the ground might also be as important, or even more important, than the magnitude of the force.

6.2.2   Vector Dimensions

Dimensional Analysis Revisited

Most of our discussions before this chapter concerned vectors, but we were careful in the way we discussed them to avoid problems in which the direction was important. The reason for this care had to do with the choice of the three fundamental units—mass, time, and distance—in our application of dimensional analysis. In discussing only a single dimension, we did not have to worry about such things as the isotropy of space. To properly account for the vector nature of physical problems, we need to generalize the notion of distance denoted by the units $[L]$ to vector units $[L_x]$, $[L_y]$, and $[L_z]$. In this way, we specify with a separate length scale each of the three directions x, y, and z. Thus, we extend the scalar length $[L]$ to the vector lengths $[L_xL_yL_z]$. In this way we generalize all the derived quantities such as the velocity and acceleration to these three components as well. The generalization to multiple indexed dimensions allows us to take into account the fact that biological systems are rarely isotropic in space but, more often than not, have preferred directions. If nothing else, gravity introduces an anisotropy into the dynamics of the human body because we have weight. This is very different from the fact that the body has mass, since space is isotropic relative to the body's mass.

Keeping Track of the Directions

The vector dimensions carry more information than do the scalar dimension. For example, the expressions for the dimension of velocity, $[LT^{-1}]$ and $[L_xT^{-1}]$, really have quite different meanings. The first expression has the dimension of speed, whereas the second has dimension of velocity in the x direction. Most textbook projectile problems have the conditions of an initial uniform velocity horizontally and a uniform acceleration vertically, that being gravity. It is certainly more illuminating to write that the horizontal velocity of a body is $[L_xT^{-1}]$ rather than $[LT^{-1}]$. In the former case, the horizontal direction is quite clearly along the x axis, whereas in the latter case, the body can be moving in any direction. In the same way, the vertical acceleration is better characterized by $[L_zT^{-2}]$ than by $[LT^{-2}]$, since the former directly specifies the vertical direction.

Area Is a Vector

The vector nature of dynamical variables may be familiar, but the vector nature of geometrical quantities is probably less familiar. It may come as a surprise that area is a vector and may be denoted by writing $[L_xL_y]$, $[L_xL_z]$, and/or $[L_zL_y]$, rather than by the single quantity $[L^2]$. Oh, you did not know that area is a vector? Well it is. Consider a quantity like pressure, which is a force per unit area. We take the force to be vertical downward, say the influence of gravity on a column of water in a pipe, so that $[F] = [ML_zT^{-2}]$. The force acting along the z axis is perpendicular to the (x, y) plane, where the cross section of the pipe has an area $[A] = [L_xL_y]$. The pressure is, therefore, $[P] = [ML_zT^{-2}]/[L_x L_y]$. However, if we did not attach the direction to the length dimension, the expression for pressure would reduce to $[P] = [MLT^{-2}]/[L^2] = [ML^{-1}T^{-2}]$, which contains much less information than does the true expression. But enough about equations. How do we use this information?

Scalar Range Problem

Let us consider a projectile fired with an initial velocity $\boldsymbol{u_0}$ in a horizontal direction at a height s above the earth's surface. We want to know how far the projectile will travel before landing on earth. In Figure 6.5, we show a car running off a cliff and we inquire how far from the cliff the car will land. This is, of course, the same problem. The horizontal range of the projectile (car) R is determined by the initial horizontal velocity u_{x0}, the height s, and the acceleration of gravity, g. Therefore, we can write the algebraic equation for the range:

$$R = Cu_{x0}^\alpha s^\beta g^\delta \tag{6.12}$$

where you will recall that C is a dimensionless constant that cannot be determined from dimensional analysis, but must be worked out by other means, such as experiment. The dimensional equation for the range is given in terms of a horizontal distance that is determined by an initial horizontal velocity, a vertical distance, and the vertical acceleration of gravity, as

$$[L_x] = [(L_x T^{-1})^\alpha L_z^\beta (L_z T^{-2})^\delta] \tag{6.13}$$

where x is the horizontal direction and z is the vertical. The simultaneous equations for the exponents in Equation (6.13) are given by

$$\text{exponent of } L_x: 1 = \alpha$$
$$\text{exponent of } L_z: 1 = \beta + \gamma \tag{6.14}$$
$$\text{exponent of } T: 1 = -\alpha - 2\delta$$

so we obtain the parameter values $\alpha = 1$, $\delta = -\frac{1}{2}$, and $\beta = \frac{1}{2}$. Thus, inserting these parameter values into Equation (6.12), the range of the car as it travels from the edge of the cliff to the ground is determined by

$$R = u_{x0}\sqrt{2s/g} \tag{6.15}$$

where $C = \sqrt{2}$ is found with the aid of a different technique, and with this we have the complete solution to the equation of motion. The solution [Equation (6.15)] can also be obtained from our earlier discussion of free-fall, where the distance of the

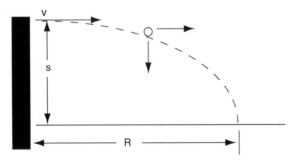

Figure 6.5. A physicist's spherical car going over a cliff. A car drives off a cliff with a given velocity in the horizontal direction. The cliff height is s, so what is the range, R, of the car on its maiden flight?

fall is $s = gt^2/2$. Thus, the time it takes the car to fall through the distance s is given by $\sqrt{2s/g}$ and, therefore, Equation (6.15) can also be written as

$$R = u_{x0}t \tag{6.16}$$

From Equations (6.16) and (6.17), we see that the algebraic expression for the range of the projectile (car) is not unique, but has many forms.

Inserting Some Numbers into the Equations

Let us apply Equation (6.16) to a car traveling 60 miles/hour or 88 feet/sec that plunges at full speed over a sharp cliff 1600 feet above a beach. Fortunately, the driver was able to jump from the car before it made its maiden flight and is uninjured. Using the acceleration of gravity at 32 ft/sec^2 in the vertical direction, we find that the car travels 880 feet in the horizontal direction R before being folded into scrap metal by the impact with the sand.

What the Police Use to Determine the Speed

It is a simple equation like Equation (6.16) that the arresting officer uses to determine how fast your car was really going before you slammed on the brakes. He does this by measuring the length of the skid marks. In the case of the car going off the cliff, the police determine the speed by the range the car traveled before impact. If the driver claimed the car rolled or was pushed off the edge, the officer would be able to determine if the driver had lied (he might use a consultant for this determination). Such was the case of the mother who drowned her two children in a lake, which was recently in the news.

Vector Range Problem

Another typical projectile problem concerns the running broad jump. The runner jumps at an angle with respect to the horizontal and, therefore, has a component of initial velocity in the vertical direction as well as the horizontal; see Figure 6.6. If the initial velocity of the runner is $u_0 = (u_{x0}, u_{z0})$ what is the range of the runner in

Figure 6.6. Standing broad jump. The position of the jumper is depicted in the takeoff, hanging, and landing phases. One can see that the trajectory of the jumper is a parabola, even though we have not drawn one in.

the broad jump? We make use of the vector nature of the velocity to write the range equation as

$$R = Cu_{x0}^{\alpha}u_{z0}^{\beta}g^{\delta} \tag{6.17}$$

The corresponding dimensional equation is given by

$$[L_x] = [(L_xT^{-1})^{\alpha}(L_zT^{-1})^{\beta}(L_zT^{-2})^{\delta}] \tag{6.18}$$

yielding the simultaneous equations for the exponents,

$$
\begin{aligned}
\text{exponent of } L_x&: 1 = \alpha \\
\text{exponent of } L_z&: 1 = \beta + \delta \\
\text{exponent of } T&: 1 = -\alpha - \beta - 2\delta
\end{aligned} \tag{6.19}
$$

with the solution $\alpha = 1$, $\delta = -1$, and $\beta = 1$. Thus, the range is given by

$$R = Cu_{x0}u_{z0}/g \tag{6.20}$$

so that in terms of the angle the initial velocity makes with the horizontal, θ, the horizontal velocity is $u_{x0} = u_0 \cos \theta$ and the vertical velocity is $u_{z0} = u_0 \sin \theta$. Substituting these components of the initial velocity into Equation (6.20), we obtain

$$R = 2\frac{u_0^2}{g} \cos \theta \sin \theta \tag{6.21}$$

where from other considerations we find that $C = 2$. Now, using the trigonometric identity for half angles, $\sin 2\theta = 2 \cos \theta \sin \theta$, the range is determined to be

$$R = \frac{u_0^2}{g} \sin 2\theta \tag{6.22}$$

Note that if we multiply both the numerator and denominator in Equation (6.22) with the mass of the jumper, we have the ratio of the runner's initial kinetic energy, $[\frac{1}{2}]mu_0^2$, to the runner's weight, mg.

How Far Will the Jumper Go?

If the runner pushes off the ground at an angle of 45? with respect to the horizontal, at a speed of 17.89 ft/sec, then using (6.22) we calculate that the range of the runner would be 10 ft. Note that this is assuming that the runner's body is rigid and ignores the additional distance achieved by extending the legs forward beyond the center of gravity while the runner is in mid-flight. We see from Equation (6.22) that the distance covered by the jump depends on the square of the runner's initial speed and not on just the initial horizontal velocity. Stated differently, the distance depends on the ratio of the initial kinetic energy of the runner to the weight of the runner. Thus, all things being equal, two runners achieving the same initial kinetic energy will not have the same range. The lighter of the two will go farther in direct proportion to the ratio of the weights. The range of a runner who is 10% heavier will be 90% of the range of the lighter one.

Question 6.7

Suppose that an arrow is fired directly upward with an initial velocity 60 ft/sec. Use either of the equations you know to determine how long it will take the arrow to reach the ground, or use dimensional analysis.

6.3   SUMMARY   Our Map is Linear

Much of what we know about the world is based on linear ideas. We push a little and the system responds a little. We push harder and the system responds more strongly. It is always unexpected if we push on an object and there is a dramatic response. For example a door falling off its hinges when we push on the doorknob. Such unexpected occurrences are typically the basis of comedy. In this way, much of our operational view of the world is linear. We expect the physical universe to be stable and to respond proportionately to our actions. When the universe is unstable, we can be upset, sometimes confused, but always surprised.

The World Is Nonlinear and Nonisotropic

The stress–strain curves show that most nonorganic materials satisfy Hooke's law over a substantial range of applied stress. Therefore, it is often very useful to treat materials as if they did obey Hooke's law until experiment indicates something to the contrary. We know that biological materials do not satisfy such a simple linear relation through most of their working cycle, so that this approximation has limited value. Another distinction to be made is that biological tissue is not isotropic; there are favored directions for stretching in muscle as well as in arteries and veins. This is manifested in high values of Poisson's ratio for biological tissue.

Always Magnitude and Sometimes Direction

All physical and biological variables can be quantified, which is to say, we can associate a number with the variable. This number is the magnitude of the variable we measure. If that number completely characterizes the variable, such as a density, a heart rate, and a breathing rate, then the quantity is a scalar. Note that this variable may still change over time, but at each instant it has only a single value. This is different from those variables that require more than one number to give a complete specification. The latter variables are vectors, such as forces, momenta, and displacements. The most significant difference between a scalar and a vector is the way they combine. The product of scalars is a scalar. The product of vectors, on the other hand, has two forms: a scalar product and a vector product. As their respective names would imply, the scalar product of two vectors is a scalar; for example, the scalar product of force and displacement is energy. On the other hand, the vector product of two vectors is a vector, a fact we shall use in the discussion of rotational motion.

CHAPTER 7

Biomechanics 1— Rectilinear Motion

Objectives of Chapter Seven

- Make contact between the simple laws of rectilinear motion in mechanics and the traditional arguments of biomechanics.
- Establish that the conservation laws of linear momentum, angular momentum, and energy are the result of symmetry and that these arguments do not require a deep understanding of mathematics.
- Establish, in a preliminary way, how the general physical laws apply to everyday activities such as walking, running, jumping, and swimming.
- Understand that friction is a macroscopic manifestation of microscopic processes and is closely tied to the generation of heat.
- Introduce the notion of center of mass and the use of stick figures to represent what we cannot calculate concerning the locomotion of the human body.

Introduction

Mechanical World View

What we actually see when we examine projectile motion, and the reason that problems examining it are ubiquitous in elementary physics courses, is that it provides the simplest solutions to Newton's equations of motion. Before Galileo and Newton, such motion was usually a mystery, and when it was not a mystery, the explanation was usually wrong. After Newton's enunciation of the three laws of motion,

Biodynamics: Why the Wirewalker Doesn't Fall. By Bruce J. West and Lori A. Griffin
ISBN 0-471-34619-5 © 2004 John Wiley & Sons, Inc.

everything from the planets in their orbits to the motion of springs, pendula, and the locomotion of the human body were all "understood." Understood is put in quotations because it is often a difficult task to determine the motion of a complex body, even though we know what the underlying equations are. This does not detract from the significance of Newton's original discovery of the dynamical laws for massive bodies. At no time, before or since, has any one person had such a profound influence on how we see and understand the physical world.

Knowing as Opposed to Doing

For example, the acrobat knows that at each instant of time there is a point in space at which all her weight is concentrated, and that she rotates faster if she is folded into a tight ball centered on that point. The wirewalker also knows that no matter what is done with arms and legs, this point must remain above the wire if he is to keep from falling. This is knowledge gained from experience, or experiment, and is often more valuable than the knowledge one obtains from books.

7.1 MECHANICAL FORCES

Newton's First Law Defines Inertia

Newton's first law of motion is a restatement of Galileo's law of inertia, which we discussed earlier, but, as with most things of importance, it bears repeating: an isolated object continues to do what it was doing. If it was moving at a constant velocity, it continues to do that; if it was initially stationary, it remains stationary, unless acted upon by some outside force. This last phrase makes the first law useful. In its original form, the law of inertia was applicable only to a completely idealized system, that is to say, a completely isolated system, but in nature nothing is completely isolated. There is always some friction in the system, some loss of energy, some interaction with another body. It is the recognition of this latter fact that motivated Newton to introduce the concept of force into the first law. But once the idea of a force is used to define the change in velocity in the law of inertia, a prescription must be given to calculate how the change in velocity occurs. This is the content of the second law of motion.

Newton's Second Law Defines Force

The second law of motion gives a specific way to determine how the velocity of a body changes due to the application of external influences called forces. These forces are vector quantities and are given by the time rate of change of the momentum of a body. This definition cannot be fully understood without the use of the differential calculus, but it reduces to a somewhat simpler form when the mass is constant, in which case the force is given by the product of the mass and acceleration. In modern parlance, this is called the inertial force, to distinguish it from, say, the gravitational force. However, with so many different kinds of forces, we need to understand how two or more forces influence a body. A stationary object is said to be in equilibrium, in which case the individual vector forces acting on the object exactly balance one another and the object does not move. The study of these forces in balance is called statics and is learned by every design engineer and architect. Manmade structures must be designed to resist the loading by wind, rain, snow, earthquakes, and every other kind of force that nature can apply. Before Newton, a vocation such as architecture was empirical in nature, but now it is a synthesis of art and science, a blend of what can be conceived and what can be realized under the constraints of natural law.

Newton's Third Law Defines the Balance of Nature

If you have ever pushed a car unsuccessfully, you know that your effort is balanced by the stubbornness of the inanimate object. You push against the car and the car pushes with an equal and opposite force against you. This is the content of Newton's third law: every action has an equal and opposite reaction. Note that this is not to be confused with the equilibrium of statics, in which various forces acting on the same body are in balance. The third law refers to forces acting on different bodies. The third law can be seen as the way nature retains its balance. You can sit in your chair without moving because the action of your weight on the chair is balanced by the reaction of the chair back onto your body by an equal and opposite force. The force of the runner's heel at the point of contact with the ground is balanced by the back-reaction from the ground, experienced by the runner throughout her body.

If We Were Alone

Suppose there are two particles alone in the universe. If the first particle exerts a force on the second particle, then the second particle will exert a force on the first particle that is equal in magnitude, opposite in direction, and on a line connecting the two particles. Thus, the push (pull) the first particle exerts on the second is equal and opposite to the push (pull) exerted by the second particle on the first. This equality of action and reaction is true pair-wise for any system of particles. However, we know that forces are vectors and therefore have direction associated with them, so if we introduce a third particle into this two-particle universe, the direction of the net force of reaction will be unclear. The first particle exerts a push (pull) on the second and third particles, and they in turn exert a push (pull) on the first particle, as well as on each other. In this more complex system, the components of the actions and reactions are equal and opposite. For example, the sum of all the actions of the first particle on the second and third particles along the x axis (y axis and z axis) is equal and opposite to the sum of all the reactions of the second and third particles to the first particle along the x axis (y axis and z axis). It should be apparent from this discussion that words are sometimes not adequate to describe the phenomenon in which we are interested.

7.1.1 Two-Player Momenta

An NFL Universe

Let us replace the above two-particle universe with a two-player universe. In many physics texts, the following argument is given in terms of two particles that are considered to be alone in the universe. To make the discussion more acceptable to a biologically oriented audience, we change things and discuss the collision of two football players, rather than two particles. For the purist, we assume each football player to be spherically symmetric and their masses to be uniformly distributed throughout their volumes. We denote the momentum of the first player by the product of his mass m_1 and his velocity $\boldsymbol{u_1}$:

$$\text{momentum of a player} = \text{mass} \times \text{velocity} \tag{7.1}$$

so that, symbolically, the momentum of the first player is

$$\boldsymbol{p_1} = m_1 \boldsymbol{u_1} \tag{7.2}$$

where, as with a force, we again have the product of a scalar and vector quantity. We similarly denote the momentum for the second player with such an equation, but with the indices changed from 1 to 2 in Equation (7.2). Now for this two-player system, a force acting on the second player produces a change in momentum $\boldsymbol{p_2} \rightarrow \boldsymbol{p_2'}$, where we denote the change in momentum by putting a prime on the variable. We use the Greek letter Δ to denote "a change in"; therefore, a change in the momentum of player 2 is indicated as

$$\Delta \boldsymbol{p_2} = \text{change in momentum of player 2} \tag{7.3}$$

so that the relation between the new and old momentum is $\boldsymbol{p_2'} = \boldsymbol{p_2} + \Delta \boldsymbol{p_2}$. If the change in momentum, indicated by Equation (7.3), is produced by player 1, then we know from the third law of motion that the momentum of player 1 is also changed:

$$\Delta \boldsymbol{p_1} = \text{change in momentum of player 1} \tag{7.4}$$

Further, since the action and reaction are equal in magnitude and opposite in direction, the changes in the momenta of players 1 and 2 must be related as

$$\Delta \boldsymbol{p_1} = -\Delta \boldsymbol{p_2} \tag{7.5}$$

Thus, the completely general statement of the third law of motion given by Equation (7.5) can be simplified by considering the following example.

Conservation of Momentum

We can also write the change in momentum as the difference between the momentum before the collision and the momentum after the collision using $\Delta \boldsymbol{p} = \boldsymbol{p} - \boldsymbol{p'}$ so that Equation (7.5) becomes

$$\boldsymbol{p_1} - \boldsymbol{p_1'} = -(\boldsymbol{p_2} - \boldsymbol{p_2'}) \tag{7.6}$$

Transposing terms to the right and left sides of Equation (7.6) we obtain

$$\boldsymbol{p_1} + \boldsymbol{p_2} = \boldsymbol{p_1'} + \boldsymbol{p_2'} \tag{7.7}$$

The left-hand side of this equation denotes the total momentum of the two-player universe before the collision and the right-hand side denotes the total momentum of the two-player universe after the collision. The equality means that the total momentum has not been changed by the collision; the momentum has only been redistributed between the two players. Thus, the total momentum of the two-player system is not changed by the collision and, in general, we have

$$p_1 + p_2 = \textbf{constant} \qquad (7.8)$$

a quantity independent of the kind of force the players exert on one another. This is a statement of the conservation of momentum for the two-player system.

The Bouncing Ball

A child bounces a ball on the floor. In order to reverse the direction of the ball's motion, the floor must absorb twice the momentum with which the ball strikes the floor. The floor first absorbs the momentum of the ball, p, to stop the ball's forward motion, and then another p in order to push the ball away with momentum $-p$. Since the floor as a whole does not move, the momentum that is absorbed by the floor goes into molecular motion. This increased random molecular motion can be measured by an increase in the floor's temperature. This absorption of momentum and increase in temperature is what produces the sting in the catcher's hand as his glove closes around the pitcher's fastball.

7.1.2 Multiple-Player Momenta

Extension of the Momentum Argument

The above argument can be extended to any number of players, subject to conservative forces within a closed universe of players, but only if we continue to assume that the players are spherically symmetric. Therefore, we reluctantly return to the many-particle system, with which we can more directly generalize the above argument. For example, the collision between molecules of gas in a room would satisfy the conservation condition since an increase (decrease) in the momentum of one particle, due to a collision with another particle, is always compensated for by a decrease (increase) in the momentum of the other particle due to the first. The total momentum will always remain fixed. The internal forces in such a system of particles will always balance and, therefore, elastic collisions cannot change the total momentum of the closed system of particles. In its crudest form, this is why a piece of material such as a desk or chair does not fly apart or collapse on itself. All the forces are in balance, so that for a system of N particles,

$$p_1 + p_2 + \ldots + p_{N-1} + p_N = \textbf{constant} \qquad (7.9)$$

where the dots denote a sum over all the momenta not explicitly indicated, so that the left-hand side of the equation indicates the sum over the momenta of all N particles of the new universe.

Forces Violate Momentum Conservation

Thus, according to Equation (7.9) the change in the total momentum of the system vanishes, which, by the second law, means that the net force acting on the system vanishes. Therefore, if the change in the total momentum is not zero, then there must be external forces acting on the system. From this argument for the conservation of momentum, we conclude that the violation of the conservation of momentum is the definition of a force.

Newton's Understanding

Therefore, we cannot push an object in one direction without that object also pushing equally hard on something in the opposite direction, and the same applies to pulling. Every force is therefore two-ended and at each end there is a mass. Forces exist only between bodies, as attractions or repulsions, tensions or compressions, pulling the masses together or pushing them apart. As put so succinctly by Newton himself [56]:

> It is not action by which the sun attracts Jupiter, and another by which Jupiter attracts the sun; but is one action by which the sun and Jupiter mutually endeavor to approach each other.

Our Understanding of the Control of Sway

Another example of mutual action of forces is familiar to everyone who has temporarily lost his/her balance, and shifted body weight to keep from falling. Of course, in stumbling, the mutual forces are the effect of gravity on our distributed body weight and the reaction forces of the ground on our feet. More subtle effects involve the stabilizing forces we internally generate and that are usually detected with the aid of a force plate. By means of such a plate, we find that we move back and forth, ever so slightly, to retain our balance. Even when standing perfectly still, the distribution of your weight between your two feet and across the area of either foot changes in time. In fact, the change in the distribution of force is really quite dramatic. We have only recently begun to understand how we control sway when we are standing still. We shall address this issue when we investigate the phenomenon of walking and when we discuss how to process time series data.

You and the Bar Stool

Suppose you are sitting on a bar stool and you want to turn around, but you do not want to push against anything. Can you do it? The answer is no. You will find that if you rotate your upper body to the right, your lower body will rotate to the left and vice versa. Even if you twist really fast, you will only succeed in hurting your back. This is also a conservation of momentum law, but it is conservation of angular momentum, rather than linear momentum. We shall talk about angular momentum shortly. The point is, if the system is only you and the bar stool, you will be unable to rotate the stool without exerting some external force such as pushing against the bar.

Internal Forces Are in Balance

In this way, we see that the internal forces in a system cannot produce net motion of the system. The conservation of momentum implies that an isolated system must remain stationary. However, external forces can produce unbalanced motion in parts of the system, or in the system as a whole. Therefore, if we view our body as a system, we can ignore the internal forces (for the moment) and consider external forces such as the influence of gravity. In order to investigate how gravity influences our bodies, we introduce the concept of center of mass.

Question 7.1

Suppose you are out hunting with a bow and arrow. It is a lovely day and you decide to cross a frozen lake. As you walk across the polished ice surface, a deer comes out of the brush. You raise your bow, draw back, and release an arrow weighing 2 ounces. If the initial horizontal velocity of the arrow is 100 meters per second, what happens to you, exactly? (Answer in 300 words or less.)

7.1.3 Friction Forces

Friction Comes in Many Forms

The form of friction of particular interest to us in this section is that discovered by Coulomb and which relates the force necessary to slide a body on a flat surface to the weight of the body. The law of friction is empirical and ascribes to a surface, say concrete, a parameter called the coefficient of static friction, λ_f. The force required to slide a box of weight W along a level concrete surface is $F_f = \lambda_f N$, where $N = W$ (the total weight) is the maximum force produced by the weight of the box. The notation, denoting a force by N, is used because, in general, this force is the component of the reactive force *normal* to the surface on which the box rests. The term normal means the force acts at right angles to the surface, as depicted in Figure 7.1. The box shown in Figure 7.1 begins to slide when the applied force F just exceeds the reactive force F_f. Note that the reactive force is always in opposition to the applied force, independent of the horizontal direction in which the external force is applied, assuming the concrete surface is isotropic, that is, uniformly rough.

What Skiers Know

Let us replace the block in Figure 7.1 with a skier. Here again it is well known that snow has a coefficient of static friction and that it takes a certain amount of

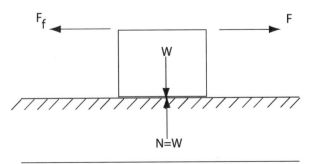

Figure 7.1. Overcoming the frictional force of a surface. A block rests on a surface having a coefficient of static friction λ_f so that the reaction force due to friction is given by $F_f = \lambda_f W$. The applied force must be $F > F_f$ in order for the block to move.

force to initiate the skier's slide. Once the skier begins to move, however, the co-efficient of friction changes from its static value to a dynamic value λ_d and, in general, $\lambda_f > \lambda_d$ so that less force is required to keep the skier moving than is required to get the skier moving. This is a general result regarding the relative sizes of static and dynamic frictional forces. In addition, it is known that it is easier to initiate motion on a slope than on level ground. This is due to two facts. The first fact is that there is a component of the gravity vector pointing downhill that assists the skier. The second fact is that the normal force on which the reaction friction force is based is less than the full weight of the skier, as shown in Figure 7.2. We see that the frictional force is given by $F_f = \lambda W \cos \theta$, where λ can be either the static or dynamic coefficient of friction and θ is the slope of the hill.

Other Kinds of Friction

In general, the coefficient of dynamical friction depends on the relative speed be-tween two surfaces in contact with one another—the skier and snow on the slope, or the box and the concrete surface. We emphasize that the phenomenon of friction is crucial for all aspects of locomotion, as is obvious to anyone who has ever walked, run, and/or driven on ice. The internal friction in a fluid is velocity-dependent, as is the resistance to the motion of a solid in a fluid and the resistance to motion in the joints of our bodies. All these expressions for friction are empirical, because there is no fundamental theory for friction, except to say that it is a macroscopic manifesta-tion of microscopic processes. To determine this for yourself, rub your hands to-gether. The heat you generate results from the direct coupling of the large-scale me-chanical motion of your hands to the microscopic world of cells and molecules in your skin.

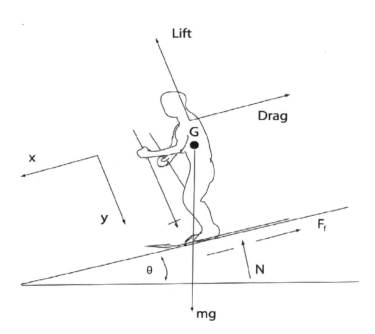

Figure 7.2. The various forces acting on a skier. Here the skier has a weight $W = mg$, but the normal force N is given by $N = mg \cos \theta$ and the frictional force is then $F_f = \lambda mg \cos \theta$.

Question 7.2

You are skating in an ice rink at some speed. Now you want to come to a stop. What must you do? What happens to your momentum? What happens to the kinetic energy you had when you were moving? (Answer in 300 words or less.)

The Real World is Messy

Although true, the arguments we have presented above regarding conservation of momentum and the balance of forces are idealized, and their applications to the world of biology, sports, or medicine are not always immediately evident. This is due, in part, to the fact that we rarely encounter particles or spherically symmetric football players in our daily lives. Things are distributed in space rather than being compact and they are rather floppy as opposed to rigid, so that although we know that the arm of a boxer is made up of countless atoms, this does not help the coach assist the boxer in improving her punch or the sprinter in running faster. Therefore, let us examine how we may extend these fundamental concepts to objects of interest like legs, arms, and torsos and their behavior during movement.

7.2.1 Many Particle Forces

The Human Body is Ungraceful

Physical bodies are typically big clumsy things in terms of how we can describe their motion. The trees you see from your window are attractive, but if you must move one, for some reason, their attractiveness quickly fades. Aesthetics and practicality are not always compatible. The same is true of the most graceful ballet dancer who, having struck his head, lies unconscious on the stage. When we move this person, his body is an awkward connection of sagging limbs and overly heavy torso— a hundred and sixty pound rag doll. It is notoriously difficult to quantitatively describe the way such a body responds to external forces.

Dead Weight Is Easier to Describe

Of course, it is much easier to describe the response of a cadaver to external forces than the response of a live body. This is especially true after rigor mortis has set in. The corpse becomes a truly rigid body, to which the laws of mechanics can be applied in an almost mindless way. A force applied to any point on the dead body is transmitted throughout the rigid structure. This is not true in the case of an unconscious person. A force applied to some part of the unconscious person will most likely cause the region of the body around the point of application of the force to move relative to the rest of the body. But instead of moving as a whole, the body twists, becomes deformed, and otherwise resists any efforts to move it. That is when you call for someone to get the feet.

Some Clever Ideas about the Balance of Forces

Scientists, being the clever people that they are, have devised strategies for describing the motion of even ungainly things like the human body. So now let us start from first principles and see how they do this. The strategy adopted by scientists has to do with applying Newton's laws to complicated physical objects made up of many (10^{23} and more) particles. Remember that this is starting from first principles so we are back to the atoms that make up our bodies. Consider the force acting on a single particle labeled by j so that

$$F_j = m_j a_j \qquad (7.10)$$

In a real physical object, we have so many of these force equations that if we really had to calculate all of them it would be hopeless. One of the results making the concept of force useful, is how many of the forces sum to zero. The net force acting on a complex object is given by the sum of all the forces acting on the individual particles that make up the object:

$$F = F_1 + F_2 + \cdots + F_N = \sum_{j=1}^{N} F_j \qquad (7.11)$$

Notice that the addition of these forces as described by Equation (7.11) actually implies the addition of the components of the vectors in each of the three directions of space. So when we discuss the addition of vectors, we mean the addition of their components. Since F is the net force acting on the body, we know that if F is different from zero, the body will translate in space, that is, it will move in the direction specified by the nonzero force. This is just the content of Newton's second law. However, all the F_j's are forces acting internal to the body, like the forces causing the blood to flow in your veins. The question is whether or not such internal forces can cause the body as a whole to translate in space. To answer that question, let us consider the internal forces in more detail.

Two-Particle Interactions

Let us write the force acting on particle 1 due to particle 2 as F_{12}, and the force acting on particle 2 due to particle 1 as F_{21}, then from Newton's third law, which states that the forces of action and reaction are equal in magnitude and opposite in direction, we have

$$F_{12} = F_{21} \qquad (7.12)$$

Consequently, the net force acting on this two-particle system,

$$F = F_{21} + F_{12} \qquad (7.13)$$

must vanish. Thus, according to the third law, since all the forces in the body arise in pairs, they cancel one another out, and lead us to the inescapable conclusion that the net force on the body due to internal forces is zero. The only forces left after this massive cancellation are those due to interactions of the particles with the environment. But in Equation (7.11) we have only considered internal forces, so $F = 0$ and the body is stationary. We could generalize this to three-body forces and higher, but you get the idea.

Center of Gravity

Suppose the force of gravity is acting on each of the particles in the above body located near the earth's surface. The gravitational pull of the earth will induce the same acceleration in each particle g for every j in Equation (7.11), and using the fact that all the internal forces sum to zero, we obtain the equation for the net force:

$$F = \sum_{j=1}^{N} m_j g = Mg \qquad (7.14)$$

where we have introduced the total mass of the body as the sum over all the individual masses:

$$\sum_{j=1}^{N} m_j = M \qquad (7.15)$$

We can see from Equation (7.14) that gravity influences the body as a whole, not acting on this or that particle separately, but on all the parts of the body together. Also we have included the vector nature of gravity by writing the acceleration of gravity as a boldfaced symbol, g, rather than g.

It Is One Point

Vectors emanate from a single point in space and terminate at a single point in space. Keeping this in mind, if we wanted to specify a single point in an object where the force [Equation (7.14)] effectively operates, it would be the center of gravity. Thus, like the idealization for symmetric objects, even for rather clumsy things like the human body, we can collapse the distribution of weight onto a single point, the *center of gravity* (COG), at which point the force of gravity for the entire body effectively operates. We literally feel this point, when after standing for too long a time, our back muscles have cramped from the continual fight to keep our center of gravity from collapsing under the pull of gravity. You may have experienced a similar reaction after winning a hard-fought tennis match, or in the shower after a particularly strenuous run, or in some other way. So although we know that the force of gravity does not act at a point in our body, it is often a useful fiction for us to employ.

Inertial Forces Hold Us Together

Note that the gravitational force acting on the body is a linear additive process. The net gravitational force is the sum of the gravitational forces acting on all the little parts of the body. But we would not call this a body if it were not for the internal forces that enable these parts to maintain their relation to one another. A mist hanging in a valley is held there by gravity but, eventually, the apparent solidity is lost as the mist dissipates. In other words, the forces that hold the particles of bone together in the skeleton, the cells that make up the blood and the vessels in which they flow, and so on, are the internal forces that enable the live human body to maintain its integrity. Therefore, even though these internal forces do not produce translational motion of the body, they are quite necessary to properly characterize the overall response of the body to applied external forces. If we introduce another external force, not gravity, but one that is essentially uniform over the body's mass, we can still write

$$F = Ma \tag{7.16}$$

where M is the total mass of the body and a is the acceleration of the body treated as a rigid object. In Equation (7.16), we have introduced a fictitious point at which all the mass of the body appears to be located and it is this point that has the acceleration a. This fictitious point is often called the center of mass (COM), which, for the gravitational force, coincides with the center of gravity. This is only true for rigid bodies; floppy bodies, like those of most humans, have a COM in one place and a COG in quite another place.

Distribution of Mass

In Figure 7.3, we present a schematic of the distribution of mass in a human arm. The arm has been segmented (conceptually) into N pieces. The jth piece has a mass m_j and when the arm is fully extended, as shown in the figure, this mass is a distance x_j from the shoulder. What we want to determine is the COM of the arm. Note that this is different from the center of gravity of the arm, since the latter requires a gravitational field, whereas the COM would exist in space where the influence of gravity could be negligible. The COM can be expressed as a fraction of the total length of the arm from either the distal end, farthest from the shoulder, or proximal end, closest to the shoulder.

The COM of the Arm

The total mass of the arm is given by the sum over the masses in all the segments, as in Equation (7.15). The location of the COM, denoted in Figure 7.3 by X, is defined by the relation

$$\sum_{j=1}^{N} m_j x_j = MX \tag{7.17}$$

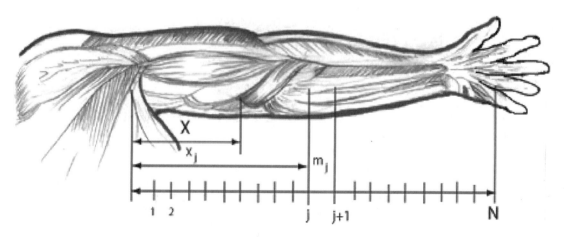

Figure 7.3. Sketch of the center of mass of the arm. Here we have divided an arm into N segments. The mass of the jth segment is m_j, and it is a distance x_j from the proximal end of the shoulder. The location of the center of mass is denoted by the total mass M, when the arm is fully extended, and the COM location is denoted by X.

where M is the total mass of the arm. The distance from the shoulder (proximal point) to the COM is

$$X = \frac{1}{M} \sum_{j=1}^{N} m_j x_j = \sum_{j=1}^{N} f_j x_j \tag{7.18}$$

where f_j is the fraction of the total mass of the jth segment, m_j/M. In this way, in certain calculations, the distribution of mass over the arm may be replaced by two quantities: the total mass of the arm M, and the distance from the origin of the coordinate system (here, the shoulder) to the COM, X. The COM of the upper arm and forearm is simply the point at which the weight balances between the two sides. Since the upper arm is larger than the lower arm, we expect the COM to be closer to the proximal point than to the distal point. This has been determined to be the case experimentally, using cadavers. In particular, COM estimates for the various limbs have traditionally been made from male cadavers. These estimates are known to be incorrect in particular cases.

The COM Can Move

We note from Equation (7.18) that the COM does not remain constant in time, but moves as the structure changes its distribution, that is, its orientation in space. After all, the arm is not a rigid structure. For example, if the arm in Figure 7.3 were to bend at the elbow, this would change all the distances in Equation (7.17) and, therefore, the COM would shift. It is this fact that suggested the use of the COM to represent the dynamics of complex objects like the human body. Thus, rather than following the often tortuous motion of an athlete in competition, we introduce a simplified picture involving COM coordinates. Consider, for example, the motion of a pitcher's arm. The coach may use slow motion pictures to track the motion of the arm's COM to determine what, if anything, can be done about the pitcher's lack of a curve ball. This assumes, of course, that a curve ball is generated by a particular trajectory of the arm's COM.

COM Coordinates

The human body is expressed in COM coordinates by replacing the body's limbs by straight lines, with the appropriate aggregate masses for arms, legs, and torso, resulting in a stick figure. The motion of the limbs is characterized using the appropriate motion of the corresponding COMs. In Figure 7.4 is depicted a three-segment system, with each segment joined to another at one of its endpoints. The connections of the segments denote joints, so the three segments can move with respect to one another and Figure 7.4 is a snapshot of the system in time. The mass of each segment is shown, as are the coordinates of the COM of each of the segments. The total mass, M, is, of course, just the sum of the three masses, and the COM of the entire system is given as

$$X_{\text{COM}} = \frac{m_1 x_1 + m_2 x_2 + m_3 x_3}{m_1 + m_2 + m_3} \tag{7.19}$$

$$Z_{\text{COM}} = \frac{m_1 z_1 + m_2 z_2 + m_3 z_3}{m_1 + m_2 + m_3} \tag{7.20}$$

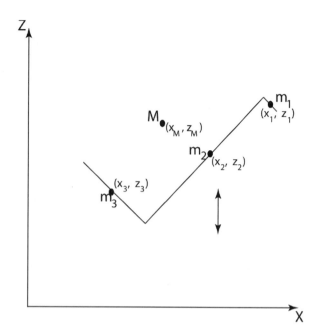

Figure 7.4. A three-segment system is depicted, with the mass and location of the COM of each segment shown. The total mass, M, and location of the center of mass of the entire system (X_M, Z_M) is also shown. These three segments could represent the arm in Figure 7.3, with the shoulder, elbow, and wrist being the joints.

Note that we have restricted our three-segment system to lie in a plane so that only the x and z coordinates are required to specify its location. In the case of a real system, we would have to specify all three spatial coordinates, except under very special conditions, say, in modeling a soldier crawling under barbwire.

Question 7.3

Assume an outstretched arm at an angle of 45° from the horizontal, whose segment m_1 is 1 kgm and is located at (5 cm, 5 cm), $m_2 = 0.5$ kgm located at (15 cm, 15 cm) and $m_3 = 0.01$ kgm located at (17 cm, 17 cm). Find the coordinates of the COM. If the person lifts the arm to a horizontal position, what are the new COM corrdinates?

Stick Figures Replace People

The three-segment system in Figure 7.4 could be used to represent the arm in Figure 7.3. The connecting points would then represent the elbow and the wrist, and the joint at the top would be the shoulder. This coarse model of a real arm is only as good as the application one makes of it. The model might be adequate to study the mechanical work done by the arm. However, it probably would not be adequate to study the heat loss generated by the arm while this work was being done.

7.2.2 What Good Is the COM?

COM Is Useful for Sports

Winter [1] points out that the COM of the total body is a quantity that is frequently calculated, but which is of limited value in assessing human movement. Its major importance seems to be in the analysis of sporting events, particularly in jumping events in which the initial takeoff often determines the path of the jumper's COM and, therefore, the outcome of the event. Examples of where manipulation of COM knowledge has been used to improve performance are in the high jump and in the prevention of injury due to positioning of the body, such as in weight lifting. In the *Fosbury flop* technique, the athlete's COM passes under or through the bar, while the trunk and limbs go over the bar [2]; see the stick figure diagram in Figure 7.5. Because of the approach to the jump and the use of the gluteal muscles, less effort is required to jump higher than was required using previous techniques. Furthermore, because of the utilization of larger muscle groups and the position of the bottom of the body at the end of the jump, there is less injury using this relatively new technique.

Internal Forces Do Not Move the COM

We should emphasize further that the action of internal forces cannot change the position of the COM of a body. In Figure 7.6, we show the analogy between a coiled spring and a diver in the middle of a dive (suspended in mid-air). When the spring is released, it uncoils, but its ends snap out symmetrically and its COM, denoted by the circled x, does not move. The same is true of the diver. When she opens her dive, the position of her COM does not change. It should also be noted that the diver and her COM are moving under the influence of gravity, so the remarks about the internal forces indicate that such forces cannot modify the COM trajectory of the diver determined by her initial state and the force of gravity.

Monitor Motion in Time

As the human body moves, so too does the COM. The time history of the COM has been used by a number of investigators to track and analyze the translational motion of a person through space. If the motion is rectilinear, that is, essentially straight-line motion, then Newton's second law can be used without adornment. The mass of the person will provide the proper resistance to changes in motion. On the other hand, if there are rotations involved, such as turning a corner, or going over a bar and twisting one's body, then a new type of inertia must be introduced and that is called rotational inertia. We now turn our attention to this new concept.

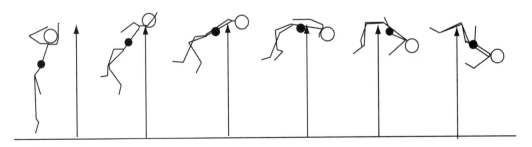

Figure 7.5. A stick figure depicts the various stages of the Fosbury flop.

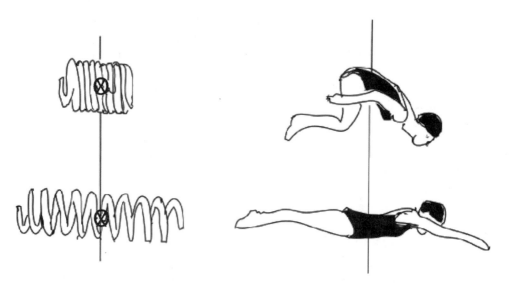

Figure 7.6. On the left, a spring is depicted in both its coiled and uncoiled positions. It is clear that the location of the COM remains unchanged as the spring opens. On the right, a diver is depicted in mid-air. Here again, it is clear that the position of the COM of the diver does not change as she begins her dive.

Question 7.4

Assume you run into a horizontal pole that impacts your body above or below the body's COM. What is the difference in your body's response if the pole is above or below the COM? Explain why these responses are different. (Answer in 300 words or less.)

7.3 SUMMARY   Newton's Three Laws of Motion

You are probably bored with the repetition of Newton's three laws. So, for brevity, let us call them inertia, force, and action–reaction. The measure of inertia is mass and it corresponds to an object's reluctance to change what it is doing. To affect change in an object's inertial state requires a force, one that changes the momentum of the object in time. A stationary object is one in which all the applied forces cancel one another. Standing still, our muscles resist the force of gravity, so we remain upright. When our muscles relax, gravity wins and pulls us into a heap. Finally, there is the most subtle of the mechanical laws, in which a force acting on a body elicits an equal and opposite reaction from the body. The action of the bat on the ball causes the ball to exit the stadium, and also induces a reaction in the bat that splits it into two unequal pieces.

Friction Ties The Large and Small Scales Together

One force with which we are all familiar is that of friction. We established that friction depends on two things: the normal force of an object on a surface, and the coefficient of friction. The coefficient of friction is an empirical parameter characteristic of the interface between the object to be moved and the surface on which it rests. The coefficient of static friction is greater than the coefficient of dynamic friction, since it requires greater force to initiate movement than it does to maintain movement. Anyone who has pushed a car from a standing start knows the truth of this statement.

Center of Mass

We presented a rather detailed argument to convince the reader that Newton's laws of motion for point particles can be used to describe the motion of particle aggregates like our bodies. The argument we used is an example of reductionism, whereby the laws that apply at the smallest scales are seen to apply at the largest scales as well. In the case of mechanical forces, reductionism is empirically true, that is, the superposition of forces is determined to be true experimentally.

The Principle of Least Action

Newton postulated the existence of his three laws and from calculation deduced their correctness through comparison of the results of these calculations with observation and experiment. However, the question often arises as to whether Newton's laws are themselves the fundamental rules for the motion of matter. It turns out that they are not. Hamilton, a century or so after Newton, determined that Newton's laws are a consequence of a more fundamental principle, the principle of least action. The principle states that the path of a particle between a given point in space at a given time, and another point in space at a later time is such that a quantity determined by the particle's energy and called the action, is a minimum. This does not sound like Newton's force law, but it is, in fact, equivalent to it in every way. In formulating this principle, Hamilton drew on a well-known principle from optics: light follows the shortest distance between two points. One consequence of this rule is that the angle of incidence of light on a surface, say a reflecting pool, is equal to the angle of reflection of the light from the pool. This rule is exhibited by a ball elastically bouncing from a rigid wall. The angle with which it strikes the wall is equal to the angle with which it bounces from the wall.

Light Consists of Particles

Thus, although light does not obey Newton's law, both the propagation of light and the motion of material particles are determined by the same deep truth, a truth that has to do with the variation of energy in time and space. The variation in the energy determines motion in physical systems. Whether the energy is that of a mechanical system or that of light, the way the energy changes at different places in space and time determines the motion of material particles and/or light particles (photons). It is interesting that Newton thought of light as being corpuscular in character and understood light in the same way he did mechanical systems.

CHAPTER 8

Biomechanics 2— Rotational Motion

Objectives of Chapter Eight

- Understand how to use the center of mass to describe the motion of an extended object, including both rotational and rectilinear motion.
- Replace the inertial mass with the moment of inertia for rotational motion and understand the laws of motion for rotations.
- Provide guidance on how to jump higher, run faster, or generally improve how we and athletes move about.
- Learn to use stick figure representations of graceful maneuvers from biomechanics in order to offer guidance to the coach and trainer in their attempts to improve the performance of athletes.
- Introduce the notion that the variability in movement is a better indicator of normality and health than is regularity.
- Understand that the unpredictability of variability in locomotion may be one of the best indicators of health; this can be important in rehabilitation after severe trauma.

Introduction

Locomotion of Complex Bodies

Up to now, we have restricted our discussion of motion to situations measured in feet/second or miles/hour, indicating that the moving object covers some net distance in a given interval of time. This motion of a falling object or ball rolling on the floor is called translational or rectilinear motion. Newton's laws have been used to determine the positions and velocities of simple point particles and their rectilinear motions. We now turn our attention to more complex systems, made of aggregates of such particles, and discover that, although Newton's laws still reign supreme, they lead to some kinds of motion that can be quite striking. For the moment, we restrict our focus to phenomena involving rigid bodies, that is, bodies that move as a whole and turn as they move. Later, we consider other more complex phenomena, such as the flow of fluids such as blood and the locomotion of real animals.

Head over Heels

The description of the rigid motion of the human body not only requires the rectilinear translation (uniform motion in straight lines) discussed so far, but also the effect of rotation. When bodies rotate, forces are replaced by torques and linear momentum is replaced by angular momentum. Perhaps the most subtle replacement is that of mass (the measure of inertia) with the moment of inertia. The moment of inertia is introduced because mass is distributed in space, and during rotations this distribution is important. It is no longer adequate to treat a body as a point mass, so we replace it with a stick figure. This is still an approximation, but it is one that is closer to reality.

Rotational Motion

A high diver springs from the platform into the air, arching forward toward the water. Suddenly, she tucks her knees into her chest, bows her head, and rapidly rotates forward once, twice, three times; finally, she extends her arms and legs to enter the water fingers first, without so much as a ripple on the water's surface. This ability of the diver to rotate in the air is as much determined by Newton's laws as it is by the projectile's motion we discussed earlier. Once in the air, no motion of the diver can impede or accelerate her fall to the water, but she can make the descent look spectacular. It is the description of her control of the fall, through rotations, spins, and twists, that concerns us here.

**8.1
ROTATIONAL
ACCELERATION**

Little Annoyances Can Be Revealing

I find that I am always straightening the throw rug on my kitchen floor. It is not that I am fanatical about symmetry, but in a small apartment, rugs lying at random angles are disorienting and annoy me. The point is, when the rug was first put down it was parallel to the sink and looked as it should: four-square parallel to the sink and stove and perpendicular to the windows. But as I walk in and out of the kitchen, each step on the rug imparts a small horizontal force, thereby causing it to move. If the forces were directly in line with the COM of the rug, then it would be translated parallel to its original position. However, in the more usual case, the applied force does not pass through the COM, and the horizontal force imparts a rotation to the rug. These rotations are random both in magnitude and direction (clockwise and counterclockwise), and they add up as I walk in and out of the kitchen. Finally, the rotation is sufficient to irritate me to the point of straightening the rug again. So how can we understand these little annoyances in life using Newton's laws?

Torque Replaces Force

The validity of Newton's three laws of motion do not depend on whether motion is translational or rotational; the laws operate just as well in either case. Rotational inertia is an extension of the first law, the law of inertia, from rectilinear motion to rotational motion: an object rotating about an axis tends to remain rotating about that axis unless acted upon by some external influence. In the case of my kitchen rug, the force that stops the rotation is the friction between the rug and the floor. I once had a throw rug in a dining room with a highly polished hard wood floor. There the rug did not stop rotating, due to the reduced friction, with the result that I landed flat on by back more than once. Since not having the floor so highly polished was not an option, I eliminated the throw rugs.

Torque Is the Cross Product of Force and Displacement

We, therefore, extend our discussion of the first law to encompass rotational motion. In the same way, we extend the concept of force to rotational systems and introduce the notion of *torque*. The latter is the external influence that disrupts the first law extended to rotations. Torque is the rotational analog of force and just as it takes a force to set a body at rest into translational motion, it requires a torque to put the body into rotational motion. In the case of my throw rug, my foot supplies the applied force, and the distance from the point of contact of my foot to the COM of the rug is called the lever arm, so the resulting torque is

$$\text{torque} = \text{lever arm} \times \text{applied force} \tag{8.1}$$

In symbols, using T for the torque, F for the applied force, and s for the lever arm, we have

$$T = s \times F \tag{8.2}$$

We point out that the multiplication sign in Equation (8.2) is the cross-product notation for vectors we introduced earlier. The cross product is a notation for determining rotating quantities, so that since both the force and the lever arms are vectors in the plane of my kitchen floor, the torque is a vector pointing upward from the floor. Recall that we introduced the ideas underlying these quantities when we discussed the structure and function of muscles, including the notion of simple machines.

Real Things Always Rotate

Thus, when a force is applied to a "point mass" (a point mass is what a physicist calls a very small, rigid object that has a well-defined COM), only translational motion is produced, since no rotation can result for a point. For the purist, we mention that we do not discuss the quantum notion of spin. In the real world, however, massive bodies are not concentrated at a point, but are extended in space, and forces do more than produce translations. If the force F is applied to a body along a line a distance s from a parallel line passing through its COM, then the magnitude of the torque is $Fs \sin \theta$, producing a rotation of the body in addition to the translation produced by the force. Here, θ is the angle between the direction of the force and the direction of the displacement, and there is a torque about the COM. This is what happened to my rug, as we saw in the definition of the cross product of two vectors. Just as there is translational inertia leading to the conservation of linear momentum, there is also rotational inertia leading to the conservation of another kind of momentum, called angular momentum.

Question 8.1

Suppose you change a flat tire on your car. In loosening the nuts on the wheel, you apply a force to the wrench, but one of the nuts does not rotate. What are some of the ways you might cause the stubborn nut to turn? (Answer in 300 words or less.)

The Rotating Ball

Let us look a little closer at the phenomenon of rotation before finding the equation for the torque. We know from Galileo's experiments that if we push the leading face of a rotating ball, it will slow down. Further, if we push the trailing face of a rotating ball, it will speed up. However, if we push on the side of the rotating ball, it will be deflected in the direction that we push, but its speed will remain unchanged. How does this change in direction occur?

Going in Circles

Assume that our ball is replaced with a roller skate, which we connect with a string to a nail in the floor. If the string can rotate freely around the nail, the skate will move in a circle with the nail at its center of motion. The string is taut, with an inward-pointing force along the string to keep the skate from flying off as it circles the nail. If the distance from the nail to the skate is r and the skate has a mass m and moves with a speed u, then the force acting on the skate, through the tension in the string, is given the candidate dimensional equation

$$F = Cm^{\beta}r^{\delta}u^{\gamma} \tag{8.3}$$

The corresponding dimensional equation is, therefore,

$$[MLT^{-2}] = [M^{\beta}][L^{\delta}][LT^{-1}]^{\gamma} \tag{8.4}$$

so that, equating exponents of the dimensions,

$$\text{exponent of } M: 1 = \beta$$

$$\text{exponent of } L: 1 = \delta + \gamma \tag{8.5}$$

$$\text{exponent of } T: -2 = -\delta$$

we obtain $\beta = 1$, $\gamma = 2$, and $\delta = -1$ so that the acceleration takes the form

$$a = \frac{u^2}{r} \tag{8.6}$$

where, again, r is the distance from the COR (nail) to the point at which acceleration is being measured. Thus, from Newton's second law of motion we have that the force being applied at the contact point is

$$F = \frac{mu^2}{r} \tag{8.7}$$

Question 8.2

Assume we have a table with a hole in it and a string extends though the hole. On one end of the string, a distance r from the hole, is our skate moving in a circle at a velocity u. On the other end of the string, freely hanging under the table, is a weight that slowly pulls the skate towards the hole. What happens to the skate? What can one conclude if the radius of the circle does not change? (Answer in 300 words or less.)

Equation for Torque

In the above argument, we have used the concept of force in rotational motion. Now let us replace the string with a rigid rod, and replace the skate with an applied force at right angles to the end of the rod in the plane of the table. Then there is a torque experienced by the pivot point (nail). The application of a torque defined by Equation (8.2) introduces an angular acceleration, which changes the angular velocity of the body. We denote the magnitude of the angular, or rotational, acceleration by α. The units of angular acceleration are radians per second per second (rads/sec^2), since the units of angle are either degrees or radians. We again use dimensional analysis to express linear acceleration in terms of angular acceleration and displacement and thereby obtain the candidate equation for acceleration

$$a = Cr^\beta \alpha^\delta \tag{8.8}$$

The corresponding dimensional equation is, therefore,

$$[LT^{-2}] = [L]^\beta \, [T^{-2}]^\delta \tag{8.9}$$

so that equating exponents of the dimensions,

$$\begin{aligned} \text{exponent of } L: 1 &= \beta \\ \text{exponent of } T: -2 &= -2\delta \end{aligned} \tag{8.10}$$

we obtain $\beta = 1$ and $\delta = 1$. Thus, the relation between the two kinds of acceleration takes the form

$$a = r\alpha \tag{8.11}$$

where we have again set the unknown constant to unity. The force associated with a mass m moving in a circle of radius r is, therefore, given by

$$F = ma = mr\alpha \qquad (8.12)$$

and the torque being applied at the COR is

$$T = mr^2\alpha \qquad (8.13)$$

This leads to a new concept. Recall that the ratio of the magnitude of the force to the magnitude of the linear acceleration, F/a, could be used to define the concept of mass; that is, $m = F/a$. In Equation (8.13) we can take the ratio of the magnitude of the torque to the magnitude of the angular acceleration to obtain

$$\frac{T}{\alpha} = mr^2 \qquad (8.14)$$

In rotational motion, the product of mass and the square of the distance to the COR replaces the mass from translational motion and is called the moment of inertia of the body. The moment of inertia is usually denoted by the letter I so that the torque is expressed as

$$T = I\alpha \qquad (8.15)$$

So the product of mass and acceleration is replaced with the product of the moment of inertia and the angular acceleration.

Rotational Analogue of Mass Time Acceleration

Consider a rotating sphere made up of a large number of particles of equal mass. Some of these particles are close to the axis of rotation and some are far away from the axis. If r is the distance from the axis of the sphere to the mass, then those particles close to the axis have a small r and, therefore, a small mr^2, whereas those far from the axis have a large r and, therefore, a large mr^2. The sphere as a whole has a total mass M that is an effective distance R from the axis, and an average MR^2. It is this average quantity that is called the moment of inertia. In vector form, the torque can be written

$$\mathbf{T} = I\mathbf{\alpha} \qquad (8.16)$$

The torque or twist required to stop a rotating sphere in a given interval of time depends not on the mass of the sphere, but on its moment of inertia. The value of the moment of inertia can be changed by redistributing the mass within a body, but only if the redistribution is done without changing the total mass. For example, a sphere with a uniform distribution of mass in its interior has a different moment of inertia than does a hollow sphere with all its mass concentrated in a thin shell forming the sphere. To construct the hollow sphere, mass must be moved from near the COM (also the center of rotation) to near the surface of the sphere, thereby increasing the average value of R. Thus, the torque necessary to stop a rotating hollow sphere in a given time interval is significantly greater than that required to stop a uniformly dense sphere of the same total mass and the same radius, spinning with the same angular velocity.

8.1.1 Rotational Speed

Different Parts Move at Different Speeds

Consider a line of skaters, skating in a circle, with hands joined, around a pivot skater who essentially turns in a circle. The speed at which a skater travels depends on how far he/she is from the pivot person; those farther away move much faster than those close in, if they are to maintain a straight line. Using dimensional analysis, we determine the speed of the various skaters u in terms of the distance from the pivot person s and the angular speed ω:

$$u = Cs^\alpha \omega^\beta \qquad (8.17)$$

where C is a constant. The dimensional equation corresponding to Equation (8.17) is

$$[LT^{-1}] = [L^\alpha T^{-\beta}] \qquad (8.18)$$

where we have taken cognizance of the fact that the angular speed has the dimensions of inverse time. Equating the indices on both sides of the equation we obtain

$$
\begin{aligned}
\text{exponent of } L: 1 &= \alpha \\
\text{exponent of } T: 1 &= \beta
\end{aligned}
\qquad (8.19)
$$

Therefore, dimensional homogeneity determines that the speed of a skater in the line depends linearly on the distance from the pivot person, since the exponent of $[L]$ is one, and linearly on the angular speed

$$u = \omega s \qquad (8.20)$$

where the constant $C = 1$ is found by other means.

The Period of Revolution

Equation (8.20) tells us that different people along the line of skaters move at different speeds u. Those farther from the axis move more rapidly than those closer to the axis, since they must cover a greater distance (arc length) in a given interval of time. However, all points along a radial line complete one revolution at the same time, if the line is to remain straight, and that time is the period of revolution. Since there are 2π radians in a circle, and the frequency with which the line turns through the circle is ω, the length of time it takes to complete a revolution is given by $2\pi/\omega$, that is, the period of revolution. Of course, this does not always happen with our line of skaters. We can see a bend develop in the line if the pivot person tries to get the line to rotate too quickly and the outermost skaters cannot keep up.

Coupling Together Different Types of Motion

Let us replace the line of skaters with the spokes of a bicycle wheel. The bicycle wheel, while it is spinning on the rack in the bicycle shop, undergoes rotational motion rather than translational motion. However, the two types of motion become coupled when the wheel is put back on the bicycle and the rider takes it down the street. While on the rack, the precise center of the turning wheel remains motionless. Therefore, any attempt to characterize the motion of the wheel by saying it is spinning at so many centimeters/second has no meaning, unless we specify the exact radius, the distance along a spoke, to which we are referring. Therefore, we want a quantity to refer to the motion of the wheel as a whole. We require a measure of the rotational speed that applies to the entire rotating body at once. The measure we choose is the number of revolutions of the wheel per unit time. Though various points on the wheel move at different speeds, the wheel rotates as a whole, so that every point on the wheel completes one revolution in one period. We may therefore speak of the wheel or any rotating rigid object as having so many revolutions per minute (rpm). This is what the odometer in the car does. It measures the revolutions per minute of the car's tires.

Various Angular Units

Notice that ω does not involve a length scale, only a time scale, that being the period of revolution of the solid body. If we divide a period into 360 equal parts, called degrees, denoted by the superscript symbol °, then one revolution per minute would correspond to 360° per minute, or 6° per second. As a dancer turns through these degrees, a point on the dancer's body marks the total distance covered in angles, or the fraction of a revolution completed. A speed in rpm's or degrees per second is therefore referred to as angular speed. The constant in Equation (8.18) is equal to unity when the angular speed is measured in a new unit called radians. A radian is an angle that marks out on the rim of the wheel an arc that is just equal in length to the radius of the wheel. The circumference of the wheel is 2π times the length of the arc marked out by one radian. The entire circumference is marked out in one revolution, so one revolution equals 2π radians or 360°. It follows that one radian equals $360°/2\pi$, and since π equals 3.14158 one radian is approximately equal to 57.3°. In these units, the angular frequency ω is in radians/sec.

Round and Round You Go

Thus, whether we consider a dancer pirouetting on the stage, a diver rotating in midair, or a wheel on the triathalon leader's bicycle, the speed of rotation is dependent on how far the moving mass is from the axis of rotation. In the same way, the time rate of change of this angular velocity determines the angular acceleration, which in turn determines the torque acting on this moving mass. Further, according to Equation (8.16), what we need to determine is the moment of inertia associated with this motion, as any field goal kicker in the NFL knows all too well.

Question 8.3

A gymnast runs across a mat, tumbles forward, leaps forward, does a somersault, and lands standing on her feet. Give a discussion of each of the force laws as they apply to her separate movements. (Answer in 300 words or less)

8.1.2 The Moment of Inertia

Changing the Dynamics

We discussed the difference in the moments of inertia of hollow spheres and filled spheres, and pointed out that because of this change in mass distribution inside the spheres, the torque required to change the rotation of the two objects is quite different. There is also a child's toy that looks like an ordinary ball, but when you roll it, instead of moving smoothly, it starts and stops, making lurching movements as it proceeds. This is due to the movement of mass inside the ball, changing its COM and its moment of inertia as the ball is rolling. Gymnasts also do this by pulling in their arms and/or legs during a tumble, thereby changing their moment of inertia and modifying the body's resistance to changes in rotational motion.

How to Calculate the Moment of Inertia

The moment of inertia of a body that does not have the symmetry of a sphere may be calculated by segmenting the body as we did with the arm in Figure 7.3. The moment of inertia, I, of the arm relative to the shoulder is

$$I = \sum_{j=1}^{N} m_j x_j^2 \tag{8.21}$$

where we use the square of the distance from the shoulder to the mass point, as distinct from the calculation of the center of mass coordinates, and where the linear distance to the mass point is used and not the square. Here again, we can see that the influence of a given element of mass increases with increasing distance, a property we use to distinguish between the COM and the center of rotation (COR) in an arbitrarily shaped body.

Twisting a Joint

In judo, there is a technique called *kansetsu-waza*, which consists of locks directed against the opponent's joints, which are twisted, stretched, or bent with the hands, arms, or legs. In our vocabulary, we would say that a torque is being applied to the joints. These techniques are only taught to students over the age of sixteen because they are quite painful and have the potential for great damage. These torques end football and basketball careers, incapacitate runners and gymnasts, and break the bones of ordinary people every day. Therefore, it is useful for the professional and lay person alike to understand the concept of moment of inertia and how this information may be used to strengthen and protect the body.

Center of Rotation (COR)

Like the mass in Newton's second law, it is the moment of inertia that determines how responsive a body will be to an applied force or, in this case, an applied torque. However, unlike mass, which is an experimentally determined quantity, the moment of inertia is calculated from the spatial distribution of mass. In our example of the rotating sphere, the COM and the COR coincide, because the distribution of mass is spherically symmetric, but this is usually not the case. For our limbs, the COR is a joint, and our joints are typically not located at the COM of a limb, assuming that everything is working properly. Therefore, the more we know about how to determine the moment of inertia of a complicated object like the human body, the more use we can make of it.

Parallel-Axis Theorem

The moment of inertia of a body about any given axis consists of two parts: the moment of inertia about an axis parallel to the axis through the COM, plus the moment of inertia about the axis with all the mass of the body concentrated at the COM. Since this is a very important theorem in biomechanics, we present a demonstration of it here. Do not confuse this demonstration with a proof. Consider a two-segment system, in three-dimensional space, fastened at one end, the origin, which we call the COR. We intend for this system to simulate a limb. Further assume that the mass of each of the segments is $m/2$, located a distance R from the COR. The COM of the separate segments are symmetrically placed on either side of the total COM, as shown in Figure 8.1, and independently located by the spatial vectors r_1 and r_2. The moment of inertia of the two segments is, therefore, given by the product of the mass and the square of the distances to that mass from the COR,

$$I_{COR} = m_1 r_1^2 + m_2 r_2^2 \qquad (8.22)$$

Here, we see that the total moment of inertia with respect to the COR consists of two pieces. Now we introduce the COM distance locating the two masses, $m_1 = m_2 = m/2$, using

$$\begin{aligned} r_1 &= R - r \\ r_2 &= R + r \end{aligned} \qquad (8.23)$$

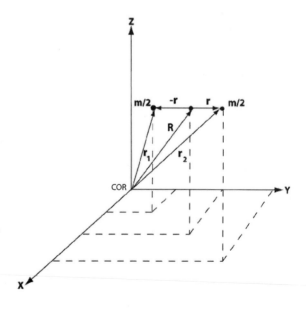

Figure 8.1. Two masses in three-dimensional space, a two-segment system with a COM of the first segment of mass $m/2$ at $R - r$ and the COM of the second segment of mass $m/2$ at $R + r$. The COM of the total system is a distance R from the COR.

so that substituting Equation (8.23) into Equation (8.22) yields

$$I_{COR} = \frac{m}{2}(\mathbf{R} - \mathbf{r})^2 + \frac{m}{2}(\mathbf{R} + \mathbf{r})^2 = mr^2 + mR^2 \qquad (8.24)$$

Again, the moment of inertia has separated into two pieces. The first piece, call it $I_{COM} = mr^2$, is the moment of inertia about the COM, that is, the moment of inertia that would be obtained if the coordinate system had its origin at the total COM. The second piece of the total moment of inertia is mR^2, the moment of inertia of the COM relative to the COR. This second piece is the moment of inertia that would be obtained if the total mass were, in fact, just one mass located at $\mathbf{R}$, rather than being the two masses separated symmetrically about this position. Note that the COR, that is to say, $\mathbf{R}$, can be any distance along a parallel line joining the two segments. This is the so-called parallel-axis theorem, often seen in the form

$$I_{COR} = I_{COM} + mR^2 \qquad (8.25)$$

where R can be in either direction along the line on which I_{COM} was calculated.

A large number of examples of the implementation of Equation (8.25) are given in Winter [1], Chapter 3. We shall consider some of these applications as well as others subsequently.

Energy Transfer Determines Torque

In our study of rectilinear motion, we found that by following how energy was transformed from kinetic to potential and back again, we could deduce the force that would produce such an energy transfer. A similar kind of analysis can be done for rotational motion, where another expression for the torque can be obtained from the definition of the change in energy. It is worth pointing out that this procedure of determining the equations of motion for a mechanical system by investigating how energy changes in form is quite general. The change in energy produced by a force moving a body through a small distance $(\Delta x, \Delta y)$ is given by the equation for the mechanical energy,

$$\Delta E = F_x \Delta x + F_y \Delta y \qquad (8.26)$$

where the components of the force are along the directions of the displacements. If this motion also produces a rotation through a small angle $\Delta\theta$, such that $\sin \Delta\theta \approx \Delta\theta$, then we have for the displacements (see Figure 8.2),

$$\Delta x = -y\Delta\theta \text{ and } \Delta y = x\Delta\theta, \qquad (8.27)$$

where the displacement along the x axis is in the negative direction. Inserting the small displacements determined by Equation (8.27) into Equation (8.26) yields the remarkable-looking expression

$$\Delta E = (xF_y - yF_x)\Delta\theta \qquad (8.28)$$

Recall that we said that the torque is the rotational analogue of the force, so that just as the work done is the distance over which the force acts, so too the work done is

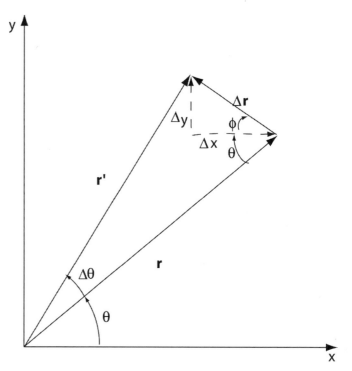

Figure 8.2. Small rotation of a vector in the two-dimensional plane. The original vector is rotated through a small angle $\Delta\theta$ so that the sine of the angle and the angle are equal. Using the fact that $\theta + \phi = \pi/2$ and some trigonometric identities, we arrive at Equation (8.27).

the angle through which the torque acts. Here again, just as the displacement is along the direction of the force, so too the angle is along the rotation direction of the torque. The change in energy is, therefore, given by

$$\Delta E = T_z \Delta\theta$$

The torque about the z axis, T_z, involves the forces in the (x, y) plane of rotation and, therefore, is given by the expression

$$T_z = xF_y - yF_x \qquad (8.29)$$

We emphasize here that the torque is expressed relative to an axis of rotation, so if for some reason we change the axis of rotation, then the torque will change as well.

Power Is the Time Rate of Change of Energy

If the small change in angle, $\Delta\theta$, obtained by a small rotation of the force vector in space, takes place in a small time interval, Δt, then the change of energy during this time interval is given by the ratio of small incremental changes:

$$\frac{\Delta E}{\Delta t} = T_z \frac{\Delta\theta}{\Delta t} \qquad (8.30)$$

The time rate of change in the energy is the power, P_ω; the time rate of change of the angle is the frequency, ω; so that from Equation (8.30) we can write for the power expended in rotating the body

$$P_\omega = T_z \omega \tag{8.31}$$

This expression can be generalized to an arbitrary torque T using a general rate of rotation ω such that the power is given by

$$P_\omega = T_x \omega_x + T_y \omega_y + T_z \omega_z = \omega \cdot T \tag{8.32}$$

where the subscripts denote the axes of rotation. For example, T_x is the torque on the body about the x axis and ω_x is the rate of angular rotation of the body about the x axis. Thus, the total power expended is the sum of the powers expended relative to each of the three coordinate axes.

Torque Causes Rotation

Let us close by noting that the concept of equilibrium we discussed earlier, in which all the forces in a body are in balance, also applies to torques, that is, to rotational motions, as well. A body does not rotate without the application of a torque and the attendant expenditure of energy to do work. Thus, in equilibrium, the change in energy denoted in Equation (8.28) is zero. In equilibrium, the sum of the forces on the particles is zero and the sum of the torques on the particles is also zero.

Rotations Give Rise to Angular Momentum

Recall that the force law could be expressed in terms of the change of the linear momentum of a particle over time, so that when no force is applied, the linear momentum is conserved, that is, it does not change in time. In the same way, the torque can be expressed in terms of the change in angular momentum of a collection of particles over time. If we denote the angular momentum about the z axis as L_z, then we can define

$$L_z = x p_y - y p_x \tag{8.33}$$

where p_x is the linear momentum along the x axis and p_y is the linear momentum along the y axis. The change in the angular momentum over time yields the torque, just as the changes in the components of the linear momentum over time yields the component of the force. In this way, Equation (8.29) can be derived from Equation (8.33). Further, although Equation (8.33) is only true for the angular motion in the rotational plane, the relation between the angular momentum and the torque is completely general. In fact, the time rate of change of the angular momentum about any axis is equal to the torque about that axis. Consider a car going around a corner at high speed. Since the car is not traveling in a straight line, the linear momenta p_x and p_y change, and, therefore, the angular momentum L_z changes, as the car negotiates the curve. It is this change in angular momentum that we experience as torque in sliding to the side of the car as we negotiate the turn.

Different Equations of Motion for Angular Momentum

A relation for the angular momentum that is probably more useful in the study of the human body is obtained by using the small changes in linear displacements due to small changes in angle [Equation (8.27)]. The definition of the x component of linear momentum that arises due to the small displacement along the x axis is

$$p_x = m\frac{\Delta x}{\Delta t} = -my\frac{\Delta\theta}{\Delta t} \tag{8.34}$$

and the definition of the y component of linear momentum that arises due to the small displacement along the y axis is

$$p_y = m\frac{\Delta y}{\Delta t} = mx\frac{\Delta\theta}{\Delta t} \tag{8.35}$$

Substituting these last two equations into the definition of the angular momentum [Equation (8.33)] yields

$$L_x = mx^2\frac{\Delta\theta}{\Delta t} + my^2\frac{\Delta\theta}{\Delta t} = I\omega \tag{8.36}$$

where it is clear that the moment of inertia in Equation (8.36) is given by

$$I = mx^2 + my^2 = mr^2 \tag{8.37}$$

Therefore, the angular momentum of a rotating body is the product of its moment of inertia and its angular speed. Note the similarity with the linear momentum, in which the mass is replaced by the moment of inertia and velocity is replaced by the angular velocity.

Rotating Divers

Consider a diver leaving the platform with a given angular momentum. When her arms are fully extended, the moment of inertia is maximum and the angular velocity is minimum. Then when the diver tucks her knees and chin into her chest, the moment of inertia is greatly reduced, but because the angular momentum in Equation (8.36) is conserved, the rate of rotation of the diver must increase. In the same way, the timing of a tumble by a gymnast is determined by the magnitude of the moment of inertia she can generate at the beginning of the maneuver.

Table 8.1. Dimension of some of the physical quantities associated with rotational motion

Quantity	Description	Dimension	Units
Moment of inertia	mass × displacement²	$[I] = [ML^2]$	kg·m²
Torque	displacement × force	$[T] = [ML^2T^{-2}]$	m · N
Angular velocity	angle/time	$[\omega] = [T^{-1}]$	rad/sec
Angular acceleration	angle/time²	$[\alpha] = [T^{-2}]$	rad/sec²
Power	torque/time	$[P_w] = [ML^2T^{-3}]$	watts

Spinning Skaters

Consider an ice skater, say the gold medal winner, Ms. Yamaguchi, spinning in place with her head and shoulders laid back and her arms outstretched. This configuration of her limbs gives rise to a particular I_{COR} emphasizing the magnitude of the mR^2 in Equation (8.25). When she straightens her back, squares her shoulders, and draws in her arms, $R = 0$ in Equation (8.25), so that I_{COR} reduces to I_{COM}, thereby substantially reducing the moment of inertia. Thus, in order to maintain a constant angular momentum, the rate of Ms. Yamaguchi's rotation dramatically increases and she becomes a flickering diamond on the ice.

Modeling Anatomical Structures

8.2 LINK-SEGMENT MODEL

Let us apply some of the ideas we have introduced in developing a model that is popular in biomechanics, the link-segment model. This model is based on replacing anatomical structures with stick figures having the appropriately defined COM and torques. Thus, just as in physics when we collapse a spherically symmetric ball into a single point at the COM, concentrating the total mass of the original ball at this point and calling this a particle, so too we replace the distributed mass of a limb by a line segment having the proper length, mass, COM, and moment of inertia. Winter [1] lists the assumptions underlying this modeling that we paraphrase as:

1. Each segment has a fixed mass modeled as a point mass at its COM of the segment.
2. The location of each segment's COM remains fixed during the movement.
3. The joints are considered to be hinge (or ball-and-socket) joints.
4. The mass moment of inertia of each segment about its COM, or about either proximal or distal joints, is constant during the movement.
5. The length of each segment remains constant during the movement.

Three-Component Model of the Leg

To show how to use the link-segment model, as well as to elucidate the meaning of the parallel-axis theorem, consider the link-segment model of a leg depicted in Figure 8.3. This three-segment model consists of three elements having masses: the thigh, with mass $m_1 = 7.2$ kg; the calf, with mass $m_2 = 3$ kg; and the foot, with mass $m_3 = 1$ kg. Further, suppose this is the leg of a sprinter, as considered by Hay [2]. The moments of inertia of each of the three segments relative to their own COMs have been determined previously from measurements done on cadavers and are given by $I_1 = 0.1$ kg·m^2 for the thigh, $I_2 = 0.05$ kg·m^2 for the calf, and $I_3 = 0.004$ kg·m^2 for the foot. Therefore, using the distances from the COM of the hip to the COMs of the separate parts of the leg yields $R_1 = 0.3$ m to the COM of the thigh, $R_2 = 0.45$ m to the COM of the calf, and $R_3 = 0.5$ m to the COM of the foot.

Moment of Inertia of the Leg Relative to the Hip

The thigh has a mass m_1 located at the COM a distance R_1 from the axis of rotation about the hip. The calf has a mass m_2 located at the COM a distance R_2 from the

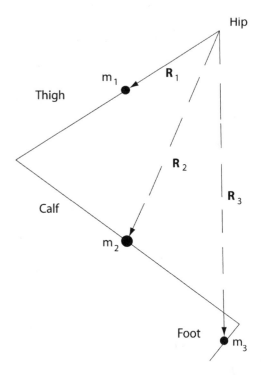

Figure 8.3. A three-segment representation of a runner's leg. The thigh with mass m_1 at the COM distance R_1, the calf with mass m_2 at the COM R_2 and the foot with m_3 at the COM R_3.

axis of rotation about the hip. The foot has a mass m_3 located at the COM a distance R_3 from the axis of rotation about the hip. We can determine the moment of inertia relative to the hip COM using these values of the coordinates in Figure 8.3. Using the mass of the thigh and the distance from the hip, we calculate the moment of inertia of the thigh relative to the hip:

$$\text{Thigh: } I_{\text{hip}}^{(1)} = I_1 + m_1 R_1^2$$

$$= 0.1 \text{ kg m}^2 + (7.2 \text{ kg})(0.3 \text{ m})^2 \qquad (8.38)$$

$$= 0.748 \text{ kg m}^2$$

In the same way, the moment of inertia of the calf relative to the hip is determined by

$$\text{Calf: } I_{\text{hip}}^{(2)} = I_2 + m_2 R_2^2$$

$$= 0.05 \text{ kg m}^2 + (3 \text{ kg})(0.45 \text{ m})^2 \qquad (8.39)$$

$$= 0.655 \text{ kg m}^2$$

Finally, the moment of inertia of the foot relative to the hip is determined using the mass of the foot and the distance to the COM of the foot to obtain,

$$\text{Thigh: } I_{\text{hip}}^{(3)} = I_3 + m_3 R_3^2$$

$$= 0.004 \text{ kg m}^2 + (1 \text{ kg})(0.5 \text{ m})^2 \qquad (8.40)$$

$$= 0.254 \text{ kg m}^2$$

The moment of inertia of the entire lower limb with respect to the hip axis is then determined, using the parallel axis theorem again, and adding the three numbers together:

$$I_{\text{hip}} = I_{\text{hip}}^{(1)} + I_{\text{hip}}^{(2)} + I_{\text{hip}}^{(3)}$$
$$= 1.657 \text{ kg m}^2$$

(8.41)

Exactly how to interpret this number is a matter of experience, which is sort of like saying you will understand it when you get older.

The Procedure Is Important

The number for the moment of inertia calculated above is a little different from that determined by Hay [2] because we used slightly different input numbers. For our purposes here, the exact number for the total moment of inertia is not important. It is the procedure, which can be extended to include all segments of the human body, that is important. Of course, if one is calculating the moment of inertia for a particular diver or gymnast, the number obtained can be quite important. This calculation could be very useful for correcting flaws in technique that would be difficult to pinpoint otherwise.

Taller Means More Power

One thing we can deduce from the form of the moment of inertia is that distance is more heavily weighted than is mass. For example, consider two individuals with the same amount of mass in their legs, but one taller than the other. The taller person has a greater moment of inertia. Consequently, for a given angular velocity, the power generated by the taller person is greater than that generated by the shorter one. What this means is that the taller person is expending more energy to maintain the same pace as the shorter person, even though they have the same mass. Is this what you would have guessed?

Gravity Is Ubiquitous

The most common force encountered in biodynamics is the gravitational force. Gravity is a force we have spent a great deal of effort to understand and which forms the backdrop for many of the other forces observed. Two important categories of forces are external and internal. Examples of internal forces are those produced in muscle and ligaments, because they are internal to the body, and often act in opposition to the external forces. In such activities as maintaining balance while standing, the external forces include the ground reaction forces, and are usually idealized for purposes of both measurement and computation. The idealization of a force acting at a point in space is associated with the fact that in biomechanics, forces are invariably distributed over an area of the body and, as such, are actually pressures rather than forces, that is to say, forces per unit area. To convert the applied pressure back to an external force, we must find a center of pressure and assume that the applied force acts at this point with the appropriate magnitude and direction. Force plates are designed to measure the center of pressure so that we may apply this transformation back to forces.

8.2.1 Separate Forces into Components

Examples of Vector Forces

Let us now consider some examples designed to illustrate the vector nature of forces. Consider a single segment, modeling part of an arm or leg, say, under the influence of gravity, along with external forces acting at both ends of the segment. The forces acting at the ends might be due to joints connecting this segment to other segments, or the distal point might be a ground reaction force, and so on. Figure 8.4 depicts the various variables necessary to completely describe the dynamics of this segment.

Horizontal Balance of Forces

The equations of motion for this line segment are obtained by separating the force equation into their various spatial components. In the horizontal direction, the sum of the force components operating at the proximal and distal points, denoted by an appropriate subscript, equals the x component of the net force:

$$F_{xd} - F_{xp} = ma_x \tag{8.42}$$

The fact that the force does not have to operate directly at the COM is indicative of the fact that we are dealing with a rigid body. By definition, a rigid body transfers the force from the point of contact to the COM. Ask a friend to stand in front of you and hold his body rigid while you push horizontally on his forehead with your finger. If you push sufficiently hard your friend will be forced off balance. If the body were not held rigid, then we would have to account for a certain amount of bending in the body. In the less idealized models we shall discuss later, the deformation of the body will be taken into account. Note that in order to specify how the body responds to applied forces, using Equation (8.42), we must have explicit forms for F_{xd} and F_{xp}.

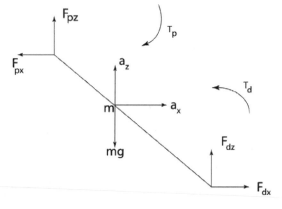

Figure 8.4. The forces acting on a single segment of mass m are depicted. The forces acting on the distal end are $\boldsymbol{F}_d = (F_{xd}, F_{zd})$, the forces acting on the proximal end are $\boldsymbol{F}_p = (F_{xp}, F_{zp})$, and gravity acts vertically downward. The torque acting in a clockwise direction at the proximal point is denoted $\boldsymbol{T}_p$, and the torque acting in a counterclockwise direction at the distal point is $\boldsymbol{T}_d$.

Balance the Forces Vertically

In the vertical direction, the sum of the force components operating at the proximal and distal points plus the weight of the segment at the COM equals the z component of the net force:

$$F_{zd} + F_{zp} - mg = ma_z \qquad (8.43)$$

To proceed further with the rectilinear equations, we need to know what the external forces are in a particular situation.

Balance the Torques

At a formal level, we are not done, however. We must also take into account the rotational factors, as determined by the torques operating at the proximal and distal points. The sum of the torques (clockwise rotation is taken as positive and counterclockwise rotation is taken as negative) must equal the static moment of inertia times the angular acceleration:

$$\boldsymbol{T_p} + \boldsymbol{T_d} = I_0 \boldsymbol{\alpha} \qquad (8.44)$$

The vector nature of torque takes the sense of rotation into account, so that the relative sign between the torques is taken as a convention.

The Rotation of the Diver

Let us again consider a diver, such as the one depicted in Figure 8.5. The torque acts about an axis passing through her feet and has a magnitude given by the product of

Figure 8.5. A diver is shown with her COM (center of gravity) on a line a distance x from the axis of rotation (the diver's feet). Her weight is W.

her weight and the distance x, the distance from her feet to the line of action of her center of gravity. The farther she leans out just prior to her dive, the greater the torque and the faster the rate of rotation with which she will leave the platform. In this way, she can adjust how much torque she has available to carry out difficult maneuvers like the rotation in the forward two-and-one-half.

8.3 SUMMARY   All Motion Is the Same

Netwon's equations of motion describe dynamics of all kinds, both the rectilinear motion discussed in the last chapter and the curvilinear motion examined in this chapter. The concepts of torque, angular momentum, angular acceleration, and angular velocity were introduced to facilitate the general discussion of the motion of nonrigid bodies. For example, it is easier to talk about angular dynamics in terms of angular quantities, which are dimensionally neutral, than it is to discuss them using dimensional quantities. For a spinning skater, or the spinning earth for that matter, only the rate of rotation is important, not the size of the object rotating. The rate of rotation is controlled by the moment of inertia, through which distance plays a role due to mass being distributed at different distances from the COR.

Rotation Is a Vector

The vector or cross product of two vectors is a difficult concept to understand because it involves rotations. More accurately, we should say that rotation is a difficult concept to understand because it is a vector; the direction of the vector is the axis about which the rotation takes place. The cross product of a displacement and a force yields a torque. When the two vectors, the displacement and the force, lie in a plane, the torque acts in a direction perpendicular to that plane. The torque defines the axis about which the rotation takes place. This is the principle by which a wrench operates. The force is delivered across the wrench by your hand, the displacement is the lever arm of the wrench, and the torque is the amount of twist that is delivered to the nut at the end.

The Mechanical View of the World

The mechanical view of the world is that things must come into direct physical contact in order to be moved around; that is, they must either be pushed or pulled. In this process of moving things, a force must be applied to the object being moved, and in so doing, the momentum of that object is changed in time. This change in momentum of the moving object is exactly compensated for by the change in momentum of the object applying the force. This is recognition of the fact that linear momentum is conserved and that every action has an equal and opposite reaction. The resistance of an object to being deflected from whatever it was doing before an external force was applied to it we call inertia. The measure of inertia is the mass of an object. The greater an object's mass, the greater is its resistance to imposed changes in motion. The concepts of mass and momentum lead us to our definition of force, that being either the mass of an object times its acceleration, or the time rate of change of the momentum of an object. These two definitions are equivalent when the mass of an object is constant in time.

Energy Goes To and Fro

Another mechanical concept that we used to form our view of the world is that of energy. Mechanical energy comes in two forms: potential and kinetic. Potential energy is the energy a body possesses by virtue of its position in space. Kinetic energy is the energy a body possesses by virtue of its motion. In a conservative system, the mechanical energy is constant, so that the dynamics of the bodies in such a system consist of the interchange of potential and kinetic energy. The swinging arm of a friction-free pendulum has a certain total energy, some of which is potential and some of which is kinetic at any given time. At the extremes of the swing (maximum displacement of the arm), the arm stops moving so that the kinetic energy is zero and the total energy is potential. At the mid-point of the swing, when the arm is vertically aligned, the potential energy is zero and the kinetic energy is maximum, that is, the speed of the arm is greatest. To a good approximation, this conservative picture of mechanical motion also describes the planets in orbit around the sun, and the moons in their various orbits about their respective planets.

Action at a Distance

The motion of celestial bodies brings into our picture another concept: action at a distance. The invisible force of gravity acts between any two massive bodies, and attracts one to the other with a strength given by the inverse square of the distance between the objects. Newton's Universal Law of Gravitation maintains that the influence of gravity is instantaneous, that is, as one body moves, there is no time delay between the body's motion and the response of the second body to the change in position of the first body. Many scientists did not believe in this action at a distance and thought it was unphysical, but it was not until they understood the electromagnetic field that a possible resolution of this problem became apparent.

PART III

Fields—How Complex Systems Move

Charged particles are found to have an interaction force that is a billion billion times stronger than the gravitational force. This force, like that of gravity, is also invisible (most of the time) and does not require direct physical contact of the charged particles in order to be experienced. The force acts over a distance and has a strength that decreases as the inverse square of the distance between the charges. Faraday determined that it is not this force that acts over a distance with nothing between; instead, this action is due to an entity called a field. This field is produced by the charged particles and fills all of the space between and beyond the charged particles. It is called the electric field.

The electric field is not just an invention of scientists driven by the desire to have nice neat theories that fit into their mathematical structures, although they do want that. An electric field has measurable properties, such as the ability to retain energy and to do work. We will see this in our discussion of electric currents and the generation of heat in material bodies due to the flow of electrons; for example, the conversion of chemical energy in a battery into electrical energy, producing a flow of electrons and the subsequent conversion of some of that electrical energy into heat. The ability of an electric field to do work is measured by Ohm's law and the relation between power and the strength of the current, the potential difference and the resistance.

Thus, the view of the universe that had started with Galileo was made complete with the efforts of Maxwell. By the end of the nineteenth century, it was apparent that there is nothing in the universe but matter in the form of atoms and a mysteri-

Biodynamics: Why the Wirewalker Doesn't Fall. By Bruce J. West and Lori A. Griffin
ISBN 0-471-34619-5 © 2004 John Wiley & Sons, Inc.

ous substance called electrical charge. Both of these quantities generate fields, electrical and gravitational, which have a reality that is of a different kind than that of matter alone. These fields produce the forces to which matter responds in accordance with Newton's laws of motion. Thus, it is no wonder that it was believed that once the properties of matter were deduced from those of atoms, of which matter is composed, all natural phenomena would be completely understood.

The prevailing attitude in the last part of the nineteenth century was reflected in an 1873 note in the prestigious English journal *Nature* that maintained that the age of science was over and only its application in the form of technology remained. Many of the best scientists of the day were advising young men not to go into the physical sciences, as nature held no further secrets of importance to humanity. It was thought that human beings could at last dream of a final understanding of nature at its most fundamental level, a belief not unlike that in a number of books that have appeared in the 1990s. Perhaps this has something to do with the coming end of a century, both then and now.

We have restricted the above discussion to the deterministic view of the universe in which the world consists of a sequence of events that are a consequence of the deterministic mechanical laws of motion. As we mentioned earlier, this view dates back some 2500 years to the work of Democritus, who introduced the concept of the atom into the scientific lexicon. He also believed that the events of atomic collisions were the result of deterministic causal relations. In this view of the universe, there is no room for free will, spontaneity is predetermined, and the rise and fall of empires is recorded at the origin of time. The deterministic view was challenged by Epicurus, who taught that some actions are not a consequence of the laws of nature, but occur spontaneously. Thus, whether or not an event occurs is of no consequence, since the symmetry of time is not disturbed by events. Actions are different, however. After an action, the universe is changed forever. An action is not predictable beforehand and after its occurrence, the symmetry of time is lost forever. Epicurus' view of the universe included determinism and randomness, predictability along with uncertainty. It allowed for the existence of free will.

We touch on the fact that science introduces the elements of randomness into the deterministic view of the universe through both the description of the motion of microscopic bodies using quantum mechanics and the description of the motion of macroscopic bodies using nonlinear dynamics and chaos. For the purposes of our discussion here, we ignore quantum mechanics. However, the uncertainty in the evolution of physical phenomena comes about, not through the subjective notion of the knowledge of the observer, as in the resolution of some of the problems in thermodynamics, but as an implicit property of the interaction of the particles in matter and the fields that matter and charge create. The modern theories disrupt the relentless character of the clockwork universe, making the distinction between what can be known in principle and what is actually knowable by observation a crucial one.

At the end of the nineteenth century, a tremendous range of natural phenomena was understood in terms of a remarkably small number of principles and laws. Classical physics became the paradigm for the acquisition and modeling of human knowledge. It was this flawed quantitative view of the universe that many scientists outside the physical sciences attempted to apply within their separate fields of study. This paradigm was applied to biology, physiology, the interpretation of history and dreams, various economic theories, social organizations, and so on. We are only now coming to the realization of just how wrong these theories were and are. It is only through the analysis of data that the fundamental limitations of the traditional worldview reveals itself.

In this section, we argue that in order to understand how the body uses energy we must first have a perspective on how molecules behave as, say, in the burning of fuel. We begin with the application of Newton's laws to billiard balls (molecules) and show how simple mechanical models can explain the thermodynamic concepts of pressure and temperature. The physical basis of temperature is not simple, since it involves the average kinetic energy of the molecules, but we are able to show that this effect is manifest in the perfect gas law, $PV = k_B NT$. Pressure is used to explain the large-scale transport of fluids in the body, such as blood flow through the pumping of the heart and air flow through the bellows action of the lungs. The small-scale transport of molecules, such as oxygen exchange with carbon dioxide at the interface of the respiratory and circulatory systems, are explained using the mechanism of diffusion. Diffusion is a ubiquitous phenomenon that dominates the transport of microscopic particles in both physical and biological phenomena. This is the level at which the heat of the body is used to actually do work, even though we cannot define the average speed of the molecules being transported.

So how do we summarize in a few pithy remarks what is perhaps one of the most misunderstood areas of physics that has direct application to the human body: thermodynamics and its four laws? We might first remark that thermodynamics only concerns itself with the behavior of average quantities such as pressure and temperature, and how such measurable quantities are related to one another, say through the perfect gas law. These relationships are empirical in nature in that they represent the results of three centuries of experiments. However, we are able to go beyond the empirical relations and explain their theoretical form using the molecular theory of gases, which is to say, the relations between the average quantities are the result of the atomic theory of matter.

Of course, thermodynamics means the motion of heat, and one of the things we investigate is when heat can be used to do work and when it cannot. We know that a fire generates heat. Heat can change water to steam, steam can be used to operate a steam engine, and a steam engine can be used to do work. Therefore, heat can do work. But this is only true when there is a difference in temperature. It is the temperature difference that determines the amount of energy that is available to do work. The heat generated by the human body cannot do work under most circumstances. It is the waste product of chemical reactions and serves no purpose other than to keep us alive by maintaining the body temperature at a constant value.

CHAPTER 9

Bioelectricity— Signals in the Body

Objectives of Chapter Nine

- Understand the physical/experimental basis of electric forces, through the introduction of two new physical concepts: the charge and the electric field.
- The electrical basis of muscle contraction through electrochemical interactions should become familiar if not clear.
- The central nervous system is seen to be an electrochemical network for the transfer of information within the human body.
- The basic laws of electrical activity are seen to be consistent with those of mechanics; in fact, electrical forces are shown to be the basis for mechanical forces.
- Erratic fluctuations in physiological time series are seen to mask underlying patterns.
- Simple standing, without locomotion, is seen to be not so simple.

Introduction

Fields Can Store Energy and Do Work

The human body is one large electrical circuit, an interconnected network of neurons. In this chapter, we examine the behavioral consequences of the movement of charged particles in the human body along these neurons. We do this through a study of the fundamental laws of electricity and the dynamical properties of the electric field. Electrical fields can store energy. They can also be used to do work through the con-

Biodynamics: Why the Wirewalker Doesn't Fall. By Bruce J. West and Lori A. Griffin
ISBN 0-471-34619-5 © 2004 John Wiley & Sons, Inc.

trolled release of that stored energy. In fact, the mechanical forces with which we are the most familiar are manifestations of these electric fields. Thus, the most macroscopic of human behaviors, standing still without falling over, is a consequence of the electric fields generated by the microscopic particles within our body.

Electricity Is in the Body

In our discussion of the contraction of muscle, we introduced the idea that the central nervous system supplies impulses that stimulate muscle contraction. The relationship between muscle and electricity was first established by Luigi Galvani, whose conclusions were, in part, due to observations first made in 1791. Others had observed that an electrical spark would cause the thigh muscles of a dissected frog to contract, even though there was no life in them. Galvani observed that when a muscle was touched with a scalpel and a spark was drawn nearby, but not in direct physical contact with the scalpel, the contraction was still observed. After a sequence of experiments, he finally determined that merely by simultaneously touching the muscle with two different metals, a contraction would result. Since this process could be repeated over and over, Galvani mistakenly concluded that the source of the electricity producing the contraction in muscle was "animal electricity."

No Life Force

Other scientists viewed animal electricity as a kind of vitalism and looked for more physical sources to explain the effect. In particular, the junction between two metals was seen as a strong candidate. In 1800, Alessandro Volta studied the interfaces of dissimilar metals, connected not by muscle tissue but by simple chemical solutions. His results clearly showed the superfluous nature of the "life force" in explaining the contraction of muscle and indicated the existence of a nonmechanical source of forces.

The Body's Communication System

Galvani's and Volta's early experiments on the movement of frog legs made it clear that the impulses that stimulate the activity of muscle are electrical in nature. Therefore, it is worthwhile to understand a little of the nature of electricity and magnetism, in order to better understand the way in which the human body communicates with and coordinates the operations of its separate parts.

Another Unseen Force Like Gravity, Only Much Stronger

The laws of the mechanical world, although detailed and elegant, do not exhaust the physical phenomena that we observe. Gravity describes the balance of the cosmos, with the planets in their orbits, due to the inverse square law of gravitational attraction. This, however, is not the only unseen force in the universe. There is another that dwarfs gravity into insignificance, one that is a billion billion billion (yes three factors of a billion) times stronger, but which also has the inverse square form, that is, whose strength decreases inversely as the square of the distance between the interacting entities. This other force does not depend on the mass of the objects that are interacting, but on another property of matter called charge. Unlike mass, which is always positive, charge is of two kinds: positive and negative. Of course, how we choose which charge is positive and which is negative is arbitrary. The convention is that the charge of the electron is negative and the charge of the proton is positive. Therefore, the electrical force can be either attractive or repulsive, depending on whether the two charges are of the same kind or are of different kinds. If the charges are of the same kind, then the electrical force is repulsive, and if the two charges are of different kinds, then the electrical force is attractive. The first systematic observations of the properties of electricity were made by Benjamin Franklin in 1751. Franklin's book, *Experiments and Observations in Electricity,* was such a success that he became a colonial resource, was, consequently, forced into the service of the American colonial cause, and was never again able to find time to pursue his scientific interests.

Electrical Fields and Signals

The existence of charged particles, with the corresponding electric fields and forces, implies that there is an energy we can associate with the electric field. As in the case of mechanical phenomena, electrical phenomena also separate energy into kinetic and potential parts. These quantities have mechanical analogs, but they must be recalculated in this new context. We examine their importance for the propagation of information from the brain, through the central nervous system, to the point of response in a muscle.

Some Things Are the Same, but Other Things Are Different

9.1 ELECTRICAL SIGNALS

In our striving to be individuals we forget that certain aspects of human behavior are universal, being the same for everyone. Such phenomena as swallowing, blinking, vomiting, coughing, breathing, orgasm, coordination of eye movement, and being startled when surprised remain invariant from person to person. As important as these things are, our primary focus remains on those actions over which we have control. It is not that we cannot consciously control the activities mentioned, for we know that we can. We can, for example, hold our breath until we pass out. Once we pass out, however, we relinquish conscious control and start breathing again. In a more subtle way, we can learn to control our breathing in such a way that it remains different even while we are asleep. Athletes, professional singers, and even yogis do this. Therefore, there is a normal, genetically controlled set of activities that we humans share. There is another set of activities that we may or may not learn, but which certainly differs from person to person. These might include the way we talk, our manner of walking, smiling, or singing, and infinitely many other things. For either class of activity, those that are learned and those that are inherited, in order to understand how behavior unfolds, we must have at least a cursory knowledge of the body's nervous system.

Electrical Impulses Excite Membranes

We pull our hand back from a hot stove and it is seconds later before we actually feel the pain of the burn to our fingers. We can drive in the car for hours before we suddenly slam on the brakes to avoid a thoughtless driver who has cut us off and, in so doing, returned our consciousness to the here and now. These everyday examples indicate how the separate parts of the body must communicate with one another to ensure our survival. Communication within the body is accomplished by means of impulses that travel along neurons. These nervous impulses are electrical or, more precisely, electrochemical phenomena that propagate along axons (parts of neurons) with a velocity on the order of 20 m/sec in humans. The generation and propagation of nervous impulses was first understood theoretically at the turn of the 20th century. According to theory, these excitations are determined by electrochemical processes in membranes and produced by the movement of small ions, these being molecules that have charge.

A Typical Message in the Central Nervous System

The nervous system of humans is historically divided into the central and peripheral systems. The peripheral system contains axons, used to transmit efferent messages from the sense organs to the central nervous system (CNS) and from the CNS to the muscles. Pulling one's hand back from the stove involves the contraction of arm muscle, just as the slamming on of the brake involves the contraction of leg muscle, and both are initiated before the CNS gets involved in the process. In the car near-accident example, the immediate reaction, after many years of driving, is to send a signal to the right leg to contract certain muscles in order to step on the brake pedal. This is a *reflex arc* that may be established through learning. When we see the car turning into our lane without warning, a signal is transmitted from the periphery, which in this case is our eyes, to the brain. The brain activates both the reaction of the leg and the adrenaline rush that results in shaking hands on the wheel of the car after the incident is over.

The Spinal Cord

The two kinds of reactions discussed above are mediated by the spinal cord. The spinal cord has two major functions. One is the conduction of nervous impulses to and from the brain and the other is the integration of reflex behavior occurring in the trunk and limbs. As stated by Dethier and Stellar [57], a reflex is a simple response to a simple stimulus, such as the knee jerk or withdrawing one's hand from the hot stove, above. The basic spinal mechanism for integrating reflexes is called a reflex arc, which sounds rather simple, but is not. The many experiences we have, say, in driving, guarding the goal in soccer, diving from a platform, wrestling, and so on, develop many connections of the associational neurons within the spinal cord. These connections provide the pathways to organize and integrate the information coming over various sensory neurons so that a pattern of activity is set up in the motor neurons. All this leads to an organized pattern of response that we refer to as a reflex arc. The difference between a great athlete and a good athlete may, in fact, lie in a person's ability to fine-tune such arcs.

Question 9.1

Why do you think a pitcher goes through all those little movements before pitching the ball? Are these movements psychological in origin, or do you think there might be a biological basis for them? (Answer in 300 words or less.)

9.1.1 Electrostatics

The Underlying Mechanism

In our discussion of the activity of neurons within the body, we repeatedly mention the propagation, synthesis, and storage of information. However, we never really did mention what was being propagated or how it was done. We now know that it is all a consequence of the electrical force studied two and a half centuries ago by Benjamin Franklin. However, the communication system within the human body is so complicated that the body does not necessarily use these forces in the form found in the laboratory, but rather in some modified form resulting from the body's complexity. This is not unlike the applications of Newton's laws to the human body. Some investigators have used rigid body motion equations to describe the body's dynamics, but we know that for such large floppy objects such as our bodies, a more realistic construction is required. But since we have the laws of mechanics, we can construct as complicated a description of the body's dynamics as we are capable of understanding. So now let us examine the laws of electricity and try to do the same with electrical phenomena within the body.

Coulomb Force Acts Between Electric Charges

On the macroscopic level (the level of our five senses), the form of the electrical force was first experimentally established in 1789 by the French engineer Charles Coulomb. He used the torsion balance for his experiment, just as Cavendish had done to weigh the earth. Coulomb, using this device, was able to prove that, like gravity, the electrical force acting between two charges q_1 and q_2 a distance r apart is

$$F = k\frac{q_1 q_2}{r^2} \tag{9.1}$$

that is, the electrical force varies as the inverse square of the distance of separation between the two charges. There is no theoretical justification for Equation (9.1), so do not look for one; it is the result of observation. For example, the overall constant k is determined from experiment. This equation for the electrical force is an experimental fact that must be memorized, not understood. The understanding comes in what the electrical force law can explain.

New Physical Unit

Notice that we introduced a new physical concept in formulating Equation (9.1) and that is the unit of charge, denoted by q with a subscript. The subscripts are used to distinguish between the separate bodies. Given our orientation, toward units and dimensions in this book, we should spend a little time with this new idea. In the cgs system of units, distances are measured in centimeters and forces are measured in dynes. By convention, which is to say, by an agreement reached by a group of senior scientists with long grey beards at some time in the distant past, if two equal charges separated by a distance of one centimeter exert one dyne of force upon each other, then they are said to have a charge of one electrostatic unit (esu) in magnitude. The esu is, therefore, the unit of electric charge in the cgs system of units.

The Electron Is Very Small

We now know that the smallest possible charge is that of a single electron, which is equal and opposite to that of a single proton. The measured charge of an electron has been determined to be, in terms of our new unit of charge, -4.8×10^{-10} esu, where the minus sign, by convention (again, the agreement made by the same old men), indicates a negative charge. A body containing one esu of negative charge actually contains an excess of approximately two billion billion electrons, whereas a body containing one positive esu of charge contains a deficiency of approximately two billion billion electrons relative to an uncharged body. Here, we see that in a physical system, a lack of a negative charge (electron) is effectively the same as the presence of a positive charge.

The Coulomb Is One Unit of Charge

The *coulomb* (C) is another commonly used unit of charge. In the SI system of units, the coulomb is equal to three billion billion esu. Therefore, in our above example, we can see that a body containing one coulomb of negative charge has an excess of six billion billion electrons (actually 6.25 billion billion). This is an astronomical number, but it represents the number of electrons passing through the filament of a 100 watt light bulb in a little over one second.

The Strength of the Electrical Force

Let us compare the magnitudes of the electrical and gravitational forces between two groups of electrons one meter apart. Suppose each group has one coulomb of charge and, since both groups are the same, the electrical force is repulsive. The total repulsive force between them, using Equation (9.1), is determined using the experimental value for the overall constant $k = 9 \times 10^9$ Nm^2/C^2, to be

$$F_{elec} = \text{electric constant} \frac{(\text{charge 1})(\text{charge 2})}{(\text{separation})^2} = 9 \times 10^9 \frac{\text{Nm}^2}{\text{C}^2} \frac{(1 \text{ C})^2}{(1 \text{ m})^2} = 9 \times 10^9 \text{ N}$$

$$(9.2)$$

The gravitational force between the same two clusters of electrons, noting that each electron has a mass of 9.1×10^{-28} gm, is given by

$$F_{grav} = \text{gravity constant} \frac{(\text{mass 1})(\text{mass 2})}{(\text{distance})^2} = 6.67 \times 10^{-11} \frac{\text{Nm}^2}{\text{kg}^2} \frac{1}{(1 \text{ m})^2}$$
$$= 2.16 \times 10^{-33} \text{ N} \qquad (9.3)$$

where we have used the universal gravitational constant from our earlier discussion. The ratio of the gravitational and electrical forces, F_{elec}/F_{grav}, is, therefore, given by approximately 4×10^{42}. The difference in the strength of the gravitational and electrical forces is like comparing a spring breeze to a nuclear explosion, or a pinprick to the destruction of the earth.

The Symmetry of the Electrical Force

From our discussion of mechanics, we know that forces are vectors and have direction in addition to magnitude. The direction of the electrical force, again, like gravity, is along a straight line joining the two interacting charges. If we define F_1 as the force acting on charge one due to charge two and F_2 as the force acting on charge two due to charge one, we find

$$F_1 = -F_2 \qquad (9.4)$$

Thus, we find the astounding fact that Newton's law of equal and opposite forces is not only true of mechanical and gravitational forces, but it is true of electrical forces as well. Here we have uncovered a deep truth. It is not only significant that electrical forces obey Newton's law, which they do, but, more importantly, that the law of action and reaction for mechanical objects is a consequence of the electrical force. It is the electrical nature of matter that ultimately gives rise to mechanical forces. This may not be clear now, but perhaps later it will be.

Action at a Distance

We can see that the coulomb force acting between two charged particles does not require an intermediary. If one of the particles moves, then, through the change in the distance between the particles, the force acting on each of the particles is modified. In particular, the response is instantaneous, which means that it does not take any time for the movement of one particle to influence the force on the other particle. This action at a distance bothered most physicists in the last century, since the dynamical response of one particle to the other is instantaneous, regardless of how far apart the two particles are. This did not make sense to them. The problem was at least partially resolved by the experimental physicist Michael Faraday, who introduced the concept of an electric field into science. The notion of a field is an extension of the idea that matter and force need not be separate concepts, but that matter might well be just the center of a force, and the space between objects that exert forces on one another is filled with something Faraday called a field. The function of a field is to transmit a force. As an aid to visualizing the field, Faraday developed a pictorial scheme called lines of force, as shown in Figure 9.1.

The Electric Field

The direction of the electric field is indicated by an arrow. In this scheme, the lines of the electric field emanate from a single positive charge, as depicted in Figure

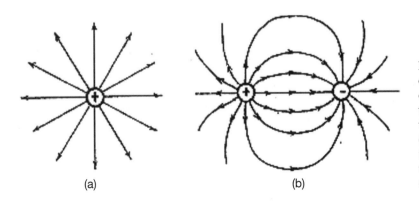

(a) (b)

Figure 9.1. The lines of the electric field around two charge configurations. In (a), the field lines emanate from the single positive charge. In (b), the field lines emanate from the positive charge and terminate on the negative charge.

9.1a. In the same way, electric field lines terminate on a single negative charge. For a single charge, the lines of the field radiate uniformly from the charge, either inward or outward. The strength of the electric field is determined by the number of filed lines in a region of space; the more lines per unit volume, the greater the field strength. However, the absolute number of lines is arbitrary. It is only the relative number of lines that has meaning. Lines that converge, and, therefore, increase in number in a region of space, correspond to an increasing electric field in that region. The field lines that diverge, becoming sparser in a region of space, correspond to decreasing electric field strength in that region of space.

Nonuniform Electric Field

Things get interesting when a second charge is present. In Figure 9.1b, we see that the lines of the field leave the positive charge and enter the negative charge, no longer just emanating or terminating radially from each of the charges, but bending from one to the other. The lines converge in the region between the two charges and, therefore, the force along the line joining the two charges has the greatest field strength. Therefore, another way to represent the force on the charge q_1 is through the equation

$$\boldsymbol{F_1} = q_1 \boldsymbol{E_1} \tag{9.5}$$

where the force acting on particle one is the product of the charge of particle one and the electric field at the same point in space as particle one. The introduction of the electric field dispelled the notion of action at a distance; the force was now understood to be a consequence of the physical field that is generated by the charges and which permeate space. These fields are generated by the charged particles contained within the neurons in the body and are, in fact, passed along by charged particles called ions. The fields constitute the signal that is used to carry information in the body.

Experimental Electrical Field Strength

The field lines drawn in Figure 9.1 are conceptual rather than actual. So to convince you that these electrical fields do, in fact, exist, we show in Figure 9.2 the fields generated by a positive and negative charge (a), as well as by two like charges (b). To visualize the electric field, slender threads were suspended in an oil bath surrounding charged conductors. These threads line up, end to end, along the field lines. It is clear from the figure that the real world behaves in just the way we described above. More elaborate experiments have been carried out to verify the field concept; see, for example, Hewitt [58].

Diminishing Field Strength

We can see from Equations (9.4) and (9.5) that the electric field at charge 1 produced by charge 2 has the inverse power-law form:

$$\boldsymbol{E_1} = \frac{1}{4\pi\varepsilon_0} \frac{q_2}{r^2} \boldsymbol{e_{12}} \tag{9.6}$$

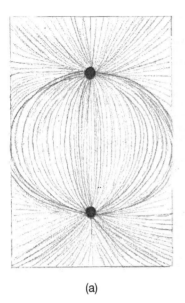

(a)

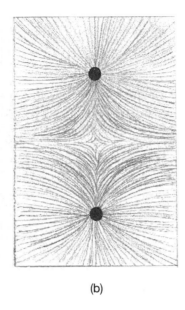

(b)

Figure 9.2. (a) Field lines produced by two oppositely charged particles. (b) Field lines produced by two like charged particles. The fields are visualized by suspending fibers in an oil bath surrounding appropriately charged conductors. Adapted from [57].

where we have replaced the overall constant in Equation (9.1) with a new parameter, the dielectric constant ε_0. We have also introduced a unit vector, e_{12}, to remind us that the electric field, like the electric force, is a vector pointing from one particle to the other. The dielectric constant is a parameter that characterizes the medium in which the electric field exits and is in the denominator of Equation (9.6). Using this form for the electric field, we can draw some conclusions about the size of forces within material objects.

How Things Dissolve in Water

The dielectric constant ε_0 is a property of matter and not of the electric field. This constant provides a measure of a material's ability to resist the formation of an electric field within it. This constant is larger in insulators than it is in a vacuum, and since we are dividing by this constant in Equation (9.6), the electric field and, therefore, the electric force for a given charge is smaller in an insulator than it would be in a vacuum. Consider common table salt, sodium (Na) chloride (Cl). The constituents of table salt are held together by electrical attraction. In the salt-shaker, there is air between the Na and Cl atoms, and the dielectric constant for this air is 1.00054. On the other hand, when the salt is put into water, the water enters the space between the constituent particles. The dielectric constant for water is 78. Therefore the forces between the Na and Cl atoms of table salt are correspondingly decreased, by almost a factor of 100, due to the higher dielectric constant of water with respect to that of air. The decrease in the forces holding the atoms together, associated with the increase in the dielectric constant, is one reason why salt dissolves readily in water; the salt literally falls apart. This is also why water is generally such a good solvent. The water gets between the atoms in the material and reduces the electric forces holding the particles of material together.

Electromechanical Forces

It is now possible to synthesize the laws of mechanics with this new entity, the electric field. From Equation (9.5), we see the relation between the two concepts. If we also introduce mass times acceleration into the left-hand side of Equation (9.5), we could determine the motion of a massive charged particle acting under the influence of an electrical field, just as we could determine the motion of a mass acting under the influence of gravity. For example, if the electric field is constant, then the equation of motion for a charged particle is of exactly the same form as that obtained earlier for a massive object falling in a gravitational field. To obtain the equation for the momentum of a charged particle under the influence of a constant electric field, we use the dimensional equation

$$p = C(qE)^{\alpha}t^{\beta} \tag{9.7}$$

where the product of the particle's charge and the electric field strength, qE, is a force according to Coulomb's law, t is the time, and C is that bothersome overall constant that we cannot determine by dimensional analysis. The dimensional equation corresponding to Equation (9.7) is

$$[MLT^{-1}] = [MLT^{-2}]^{\alpha}[T]^{\beta} \tag{9.8}$$

where we obtain by equating exponents,

$$\text{exponent of } M\text{: } 1 = \alpha$$
$$\text{exponent of } L\text{: } 1 = \alpha \tag{9.9}$$
$$\text{exponent of } T\text{: } -1 = -2\alpha + \beta$$

yielding $\alpha = 1$ and $\beta = 1$. Thus, we can write the momentum increase induced by the constant electric field for an initially stationary charged particle,

$$p = qEt \tag{9.10}$$

as a linear function of time. In the same way, the distance traveled by the particle in a time t, assuming that the particle starts from zero velocity, is given by

$$2s = qEt^2 \tag{9.11}$$

a quadratic function of time. If the applied electric field were not constant, but changed over time, the response of the particle would be more complicated.

Electrical Potential Energy

Today we talk about a gravitational field in the same way that we talk about an electrical field. Faraday's field concept has become one of the more useful ideas in science. Another correspondence of electricity with gravity is that of electrical potential energy, in analogy with gravitational potential energy. Just as one adds energy to move one mass away from another mass, working against gravitational attraction, such as carrying a suitcase up the stairs, one must add energy to move a negatively charged body away from a positively charged body to overcome the mutual attraction of the two charged bodies. The energy that is added to the charge system is then stored in the relative position of the two charges. The energy is stored in the potential energy of the electric field. However, the existence of positive and negative charges means that work must be done to bring like charges closer together, in order to overcome their mutual repulsion, for which there is no gravitational analog, since there is no such thing as gravitational repulsion.

Electrical Potential Differences

Electrical potential differences are quite common and today can be measured in terms of the energy that must be added to a unit of charge to move it a given distance. In the SI system of units, the unit of charge is the coulomb, so that the unit of electrical potential difference is the joule per coulomb. This unit is used so often that it is given a special name in honor of Alessandro Volta—the volt—a joule/coulomb. Consequently, the electrical potential difference is often referred to as "voltage." For example, a nine volt battery has a potential difference of nine volts between its terminals.

The Electromotive Force

To understand the force produced by the electric field, we again use what we have already learned about gravitational fields: a mass will spontaneously move from a region of higher potential energy to a region of lower potential energy, unless constrained not to do so. This is a pedantic way of saying that if you hold a rock over a cliff and let go, it will fall. Similarly, a charged particle will move from a region of higher electrical potential to a region of lower electrical potential, unless constrained not to do so. Since it is the electrical potential difference that induces motion of an electrical charge, that potential difference is often called an electromotive force and is abbreviated emf. You may have observed that fluids flow from regions of higher pressure to regions of lower pressure, and heat flows from regions of higher temperature to regions of lower temperature. In each and every case, it is the inhomogeneity that causes the entity of interest to move. In this regard, whenever you see something moving, look at the background to see what is changing, and perhaps inducing the motion that you observe.

9.1.2 The Action of Neurons

A Typical Neuron

The human nervous system consists of approximately 10^{11} neurons, the basic building blocks, which come in many different shapes and sizes. The spinal motor neuron, depicted in Figure 9.3, exemplifies the various parts that one finds in the typical neuron. It consists of a soma or body that contains the nucleus. From the soma is extended a long fibrous axon that divides into terminal branches, each ending in a number of synaptic knobs. These knobs contain vesicles in which the synaptic transmitters secreted by the nerves are stored. Outside the CNS, the axon acquires a sheath of myelin, a protein–lipid complex made up of many layers of cell mem-

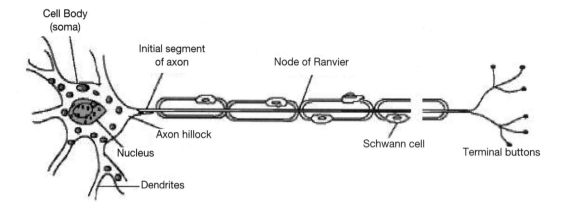

Figure 9.3. Spinal motor neuron with myelinated axon.

branes. The myelin sheath envelops the axon except at its ending and at the nodes of Ranvier, periodic 1 μm constrictions that are approximately 1–2 mm apart. This sheath is an electrical insulator, which means that it does not conduct electricity. But even though the sheath is an insulator, it does leak quite badly, which is to say that charged ions leak out of the neuron.

The Connectedness of the Body

Each neuron receives messages from many other neurons through the thin branches that form contacts (synapses) with the body of the cell and with its short appendages (dendrites). But these contacts are not only made with other cells through the synapses; they are also made in the region of the nodes, where the axon's membrane is no longer insulated and makes contact with the surrounding medium directly.

The Neuron Is Like Other Cells

Nerve excitation is transmitted throughout the nerve fiber, which itself is part of the neuron. The neuron is in most respects quite similar to other cells in that it contains a nucleus and cytoplasm. It differs from other cells in that long, threadlike tendrils emerge from the cell body, and these numerous projections branch out into still finer extensions. These are the dendrites that form a branching tree of ever more slender threads, not unlike the fractal trees discussed with regard to other physiological networks [38]. One such thread does not branch and often extends for several meters, even though it is still part of a single cell. This is the axon, which is the nerve fiber in the typical nerve. Excitations in the dendrites essentially always travel toward the cell body in a living system, whereas in the axon, excitations always travel away from the cell body.

All Neurons Are Similar

Because of the difficulty in examining patterns of interconnections in the human brain, there has been a major effort on the part of neurologists to develop animal models for studying how interacting systems of neurons give rise to behavior [59]. There appears, for example, to be no fundamental difference in structure, chemistry, or function between the neurons and their interconnectedness in man and those of a squid, a snail, or a leech. However, neurons do vary in size, position, shape, pigmentation, firing pattern, and in the chemical substance by which they transmit information to other cells. Here, we comment on the differences in the firing patterns taken, for example, from the abdominal ganglion of *Aphysia*. As Kandel [59] points out, certain cells are normally "silent," whereas others are spontaneously active. As shown in Figure 9.4, some of the active neurons fire regular action potentials, or nerve impulses, and others fire in recurrent brief bursts or pulse trains. These different patterns result from differences in the types of ionic currents generated by the membrane of the cell body of the neuron.

The Reflex Arc as an Example

We mentioned above that the reflex arc is a basic mechanism of the spinal cord for integrating reflexes. Let us see how it works. A typical reflex arc has the following components [57]:

1. *Receptors* in the skin, muscles, and joints that are selectively sensitive to various stimuli.
2. These receptors form one end of *afferent neurons*, which enter the dorsal part of the spinal cord.

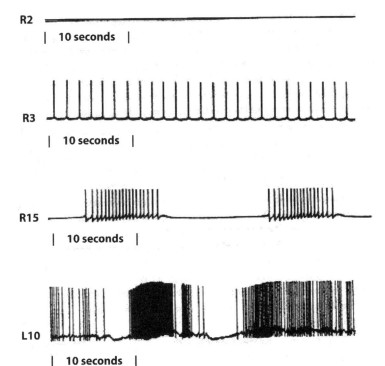

Figure 9.4. Firing patterns of different neurons. Firing patterns of identified neurons in Aplysia's abdominal ganglion are portrayed. R2 is normally silent, R3 has a regular beating rhythm, R15 has a regular bursting rhythm, and L10 has an irregular bursting rhythm. L10 is a command cell that controls cells in the system. Adapted from [57] with permission.

3. The afferent neurons terminate in contact with *interneurons*.

4. These interneurons terminate upon *efferent or motor neurons,* which pass from the central part of the cord to end in appropriate *effectors*, either muscle or glands.

An example of this mechanism is the *flexion reflex*:

> . . . the organism must contract its flexor muscles which pull the limb toward the body and simultaneously relax the antagonistic extensor muscles. This pattern of stopping an activity when its antagonist is started is the classical pattern of *reciprocal excitation* and *inhibition*. . . . At the same time that one limb is flexed, it may be necessary for the animal to extend its opposite limb for support against the ground in a *crossed-extension reflex* which calls for a contraction of extensor muscles and a relaxation of flexors. Together these two reflexes make up the basic pattern of stepping. [57]

Neurons and Nonlinear Dynamics

The rich dynamic structure of the neuron firing patterns has suggested that these patterns can be modeled by nonlinear dynamical systems. As a dynamical system evolves, it sweeps through an attractor. An attractor is a fixed geometrical structure in the space made up of the variables in the system. The trajectory goes through some regions of the attractor rather rapidly and others quite slowly, but always stays on the attractor. Whether or not the system is chaotic is determined by how two initially adjacent trajectories cover the same attractor over time. As Poincaré observed, a small change in the initial separation on any two trajectories produces an enormous change in their final separation, if the trajectories are chaotic.

Phase Space Attractors

It is possible to attain a qualitative understanding of the behavior of a trajectory without going very much into the associated mathematics. One way of doing this is to consider the limiting set (the attractor) in a phase space of a given dimension, say three. We can associate with each direction of motion in the phase space a positive number if that direction is unstable and a negative number if that direction is stable. In this way, we have a triple of numbers (α, β, γ) to determine the stability properties of every trajectory in the phase space. If one or more of these numbers is positive, the system is chaotic. What we really mean by a stable direction is that a trajectory perturbed along this direction will not change very much. On the other hand, if the trajectory is unstable in that direction, a small perturbation will produce a dramatic change in the system. Therefore, we do not need to know what these numbers are, but only what their signs are, (sign α, sign β, sign γ) .

What the Different Attractors Can Model

In Figure 9.5a, the triple $(-, -, -)$ corresponds to an attracting fixed point. In each of the three directions, there is a contraction of trajectories due to the negative number, so that no matter what the initial state of the system is, the trajectory eventually

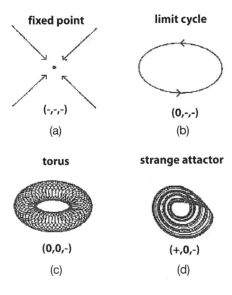

fixed point

(-,-,-)

(a)

limit cycle

(0,-,-)

(b)

torus

(0,0,-)

(c)

strange attractor

(+,0,-)

(d)

Figure 9.5. Various simple attractors in phase space. Simple attractors embedded in three dimensions are depicted. The signs of the characteristic numbers indicating the stability of the corresponding direction are also shown.

winds up at the fixed point. The arrows shown in the figure schematically represent the trajectories' approach to the fixed point. An attracting cycle is denoted by (0, –, –), in which there are two contracting directions and one that is neutrally stable. In Figure 9.5b, we see such an attractor called a limit cycle. Trajectories initiated at points off the attractor are drawn to the limit cycle as the system evolves; those initiated on the attractor stay on the attractor. The idea of using an attractor to model a physiological system was first used by van de Pol and van der Mark [60], in a medical context, to mimic the beating of the mammalian heart. They constructed an equation of motion to model what they considered to be a generic mechanism contained in a number of phenomena:

> [T]he aeolian harp, a pneumatic hammer, the scratching noise of a knife on a plate, the waving of a flag in the wind, the humming noise sometimes made by a water tap, the squeaking of a door, a neon tube, the periodic recurrence of epidemics and of economic crisis, the periodic density of an even number of species of animals living together and the one species serving as food for the other, the sleeping of flowers, the periodic recurrence of showers behind a depression, the shivering from cold, menstruation, and, finally, the beating of the heart. [60]

A Strange Attractor

The triple (0, 0, –) has two neutral directions and one that is contracting, so that the attractor is the doughnut-shaped object (torus) in Figure 9.5c. The trajectory for this dynamical system would ride along the surface of the torus. Finally, the triple (+, 0, –) corresponds to a *strange attractor,* in which the trajectories expand in one direction, are neutrally stable in another, and contracting in a third. In Figure 9.5d, such an attractor is depicted. We note here that the geometrical structure of a strange attractor is fractal and it is this fractal structure in phase space that leads to the chaotic behavior of the corresponding time series obtained by measuring any of the three dynamical variables.

Neurons and Attractors

In Figure 9.4, the normally silent neuron at the top of the figure can be viewed as a fixed point of a dynamical system. A fixed point of a dynamical system is just that. After some initial transient behavior, the system approaches the fixed point. Once the fixed point is achieved, the system no longer moves. The periodic pulse train, the second time series from the top in the figure, denotes a neuron that fires at regularly spaced time intervals. This periodicity is suggestive of a limit cycle. A limit cycle is a generalization of harmonic motion. In a simple oscillator, such as a pendulum, the amplitude of the time series increases and decreases periodically, as discussed in the previous chapter. A limit cycle repeats the values in the time series after a fixed interval of time, but the dynamics more closely resemble that of a roller coaster than the simple pendulum.

Modeling the Neurons

Finally, the erratic bursting of random wave trains is not unlike the time series generated by certain chaotic attractors. We discussed such behavior in order to distinguish between a chaotic time series and one that is strictly random. This spontaneous behavior of the individual neurons—fixed, periodic, and erratic—may be modified by driving the neurons with external excitations. It is also possible that the normal activity can be modeled through changes in internal control parameters of the isolated system. Certain dynamical aspects of neurons and their possible modeling in terms of strange attractors have been reviewed by West [8, 9], but a complete discussion requires the mathematics of nonlinear dynamics systems theory, and so we leave this intriguing subject here.

9.1.3 Electric Currents

Movements of Ions

What we are interested in here is how charged particles move and what happens to the biosystem when they do move, since charged particles carry information in biological systems. We are interested in the motion of such things as ions, that is, atoms that have either lost one or more of their electrons and therefore have a net positive charge, or that have captured one or more electrons and therefore have a net negative charge. In its original state, an atom has a given number of negatively charged electrons distributed in space around the nucleus, and an equal number of positively charged protons localized in the nucleus. Therefore, an unperturbed atom is electrically neutral, since it has as many positive as negative charges, and does not respond to an electric field. However, ions have a charge and do respond to an electric field. The electric field produced by the motion of sodium and potassium ions propagates in neurons, as we now discuss.

Free Electrons Move in Conductors

The theory describing the motion of charged particles is electrodynamics, because such motion involves the movement of the electric field associated with those charges, and these moving fields interact with other charged particles, causing them to move in their turn, and so on. For the moment, we are not concerned with the behavior of the electric field. Instead, we focus our attention on moving charged particles called currents. Materials that contain free electrons, which can move when an external electrical field is applied to the material, are called conductors. The applied field is usually in the form of a potential difference, measured in volts, between the ends of the material. In the case of a conducting wire, the free electrons move so as to reduce the electrical potential energy, just as a rock falls to the earth to reduce the gravitational potential energy. One or more electrons from each metal atom are free to move throughout the atomic lattice, and these are called conduction electrons. The conduction electrons keep flowing until the potential difference between the two ends of the wire is zero, or the total number of conduction electrons is exhausted. Thus, to maintain a flow of electrons, work must be done on the system to maintain the potential difference between the ends of the wire, even as the electrons flow through it, and to resupply the diminishing number of conduction electrons. A similar, but more complicated, situation exists in a biological cell.

Currents Across Cell Membranes

The cells that make up tissue, such as muscle, are closely packed in an *interstital fluid*. The cell plasma or *intracellular fluid* is similar, in that both fluids consist mostly of water with equal particle densities, but with different particles in them. The cell wall is approximately 7.5 nanometers (10^{-9} m) thick and is made of bimolecular lipoprotein. The cell wall severely restricts the interchange of material. Outside the cell (extracellular), the concentration of sodium ions (Na^+) and chloride ions (Cl^-) are much higher than inside the cell (intracellular). The situation is just the reverse for potassium ions (K^+), that is, their intracellular concentration is much higher than the extracellular concentration. There is also a potential difference across the surface of the cell due to the difference in ion concentrations inside and outside the cell. In skeletal muscle cells, the potential difference is about 90 mV, and negative in the interior, relative to the exterior. Of course, the interior of the cell has a great deal of structure, but it can be considered a single aqueous phase for the purposes of discussing ion exchange across the cell membrane.

Cellular Diffusion

The molecular mechanism responsible for the transport of ions across a membrane consists of diffusion and membrane permeability. The rate of diffusion depends on the difference in the concentration inside and outside the cell divided by the distance across the membrane. This ratio is the concentration gradient, as we discuss in the next chapter. Flow takes place down the concentration gradient, which is to say, from a region of high concentration density to a region of low concentration density. The relative ratios of intra- to extracellular K^+, Na^+, and Cl^- concentrations are similar in all excitable muscle and nerve tissues.

Capacitance Is Charge Buildup

Another physical mechanism that is of value in understanding the operation of a cell is capacitance. Let us assume that we have a rectangular metal plate on which we have deposited as many electrons as we can. Now we bring a second metal plate, one that is positively charged, down over the first and parallel to, but not touching it. The electrons in the first plate are attracted by the positively charged plate and are pulled to the plate's surface, thereby leaving the opposite side of the plate free to accept additional electrons. The total allowable charge on the first plate is greater than what would have been possible in the absence of the positively charged plate. The same argument holds with regard to the second plate. Because each plate lends the other a greater capacity for accumulating charge, these plates taken together are referred to as a capacitor. There is a direct relationship between the quantity of charge in coulombs, Q, and the potential difference between the plates in volts, V, which, for a vacuum between the plates yields

$$C = \frac{Q}{V} \tag{9.12}$$

The constant C is the capacitance with units of coulomb per volt. One coulomb per volt is a farad, named in honor of Michael Faraday.

Membrane Potential

Because ions carry charge and the membrane has a capacitance, charges will accumulate on the surface of the membrane, resulting in a potential difference across it. This potential difference produces an electric field within the membrane. This field exerts a force on all charged particles within the membrane. We can see from this that understanding the movement of ions requires knowledge of both diffusion and electric field forces. Further, the properties of the cell membrane are such that to retain the potential difference across the cell surface, the potential difference balances the diffusion of K^+ out of, and Na^+ into, the cell, by metabolically driven transport of Na^+ out of, and K^+ into, the cell, on the average. This active transport of ions maintains the high K^+ intracellular concentration and because the potassium permeates the membrane much more readily than does sodium, the potential difference is retained.

The Electrogenic Pump

A membrane is not the passive cover of a cell. All excitable cells are permeable to ions of metals, which is to say, certain metals, such as Na^+, K^+, Ca^{2+}, and Cl^- are passed through channels in the membrane rather easily. This is the manner in which the cell interacts with the external world, that is, by means of actively sucking ions into the cell and/or throwing them out again. In most of the physical processes we have considered, the substance that moves is transported from a region of greater stuff to a region of lower stuff. For example, heat flows from a region of higher to lower temperature, a charged particle moves from a region of greater electrical potential to a region of lower electrical potential, and so on. The active transport of ions is not from higher to lower concentrations of the solution, as one would expect, however, but is, in fact, the reverse. Because of that reverse action, the active transport is like a pump and requires a source of energy to make the pump work. The energy released during ATP splitting is used to drive active transport and is the source of energy for the pump. Na^+–K^+ ATPase catalyzes the hydrolysis of ATP to ADP and uses the energy liberated to expel three Na^+ ions from the cell and take two K^+ ions into the cell for each mole of ATP hydrolyzed. It is called an *electrogenic pump* [61].

An Analogy with Water

An analogy that is often used to help visualize the motion of electrons in a wire or ions in a neuron is the flow of water in a pipe. Consider an aqueduct coming down from the mountains, filled with the spring runoff. The difference in gravitational potential energy between the high mountains and the plains produces the pressure in the aqueduct that moves the water. As long as melting snow supplies the water, gravity provides the motive force, moving the water from regions of high to low potential energy. In the case of electricity, the flow of water is replaced with electrical current (the flow of electrons or ions) and pressure is replaced with the electrical potential difference. The rate of flow of electrons (ions) is measured in amperes, that is, the rate of flow of one coulomb of charge per second. Note that this rate is 6.25 billion billion electrons moving through a given point of the wire in each second. This is quite similar to the flow of water in a pipe measured as the quantity of water mass passing a cross section of the pipe per unit of time.

The Wire Is Neutral

One thing that is curious is the fact that positively charged protons, which are tied to the nucleus of atoms in a metal lattice, do not move under the electrical potential difference. Thus, the conduction electrons being pulled through the positively charged lattice result in the net charge of the wire remaining unchanged, which is to say, the potential of the wire remains at zero. The number of freely moving electrons and the number of relatively fixed protons in any section of the wire are essentially equal. Therefore, whether a wire carries a current or not, the net charge of the wire is normally zero during each interval of time.

Electrical Pumps

The electrical pump required to replenish the potential difference and the number of conducting electrons in a conducting wire can be of several types. The most familiar electrical pumps are the battery and the electric generator. A chemical battery works through the chemical disintegration of zinc or lead in acid; the energy that was stored in the chemical bonds is converted into electrical potential energy. To understand an electrical generator, we should really learn about magnetism, but we cannot study everything in a single course, so we will have to forego the pleasure of investigating the relationship between a moving particle, a magnetic field, and the special theory of relativity.

There Is Always Dissipation

We know from the second law of thermodynamics that there is no such thing as a free lunch—everything must be paid for. The same is true for the flow of water in a pipe, the current in a wire, and the movement of ions in neurons. In all cases, there is dissipation resulting in the loss of energy. For the water in the pipe, the dissipation arises from the drag of the fluid on the sides of the pipe and the viscosity of the water itself. For the electrons in a wire, the dissipation is somewhat more subtle, arising from the collisions of the conduction electrons with the fixed atoms of the metal lattice. In this way, part of the kinetic energy of the electrons is converted into oscillatory motion of the "fixed" atoms, manifested as an increase in the temperature of the wire. How much current there is in the wire therefore depends not only on the voltage across the wire, but on the electrical resistance the conductor offers to the electron flow as well. Finally, there is the loss of energy through the interaction of the ions with the fluid environment in the neuron, in the form of heat, which must be replenished by means of the electrogenic pump we mentioned earlier.

Resistance Measures Electrical Dissipation

The resistance of a wire to electron propagation decreases with increasing cross-sectional area, since this increase allows for the passage of more electrons per unit time. Here again, we can see the analogy to water in a pipe. A quantity called the conductivity is a property of the material that indicates the ease with which an electron moves from point to point in the metal lattice; it also strongly influences the resistance. The inverse of conductivity of the material is analogous to the viscosity of a fluid. The resistance also increases linearly with the length of the wire. Double the length of the wire and you double the resistance. Finally, the resistance is dependent on the temperature of the wire. The more the atoms of the lattice move around, the higher the temperature and the greater the impedance they present to the rectilinear motion of electrons. The greater the temperature, the greater the resistance to electron flow (for most materials).

Ohm's Law Is a Balance Between Force and Resistance

Coulomb's force law states that a charged particle will experience an electrical force when it is exposed to an electric field. The potential difference established by a battery between the ends of a wire sets up such an electric field in the wire. A free electron moves in response to the electric field, but undergoes many collisions with the fixed particles in the material, thereby giving up some of its kinetic energy to heat the material. That is why electric appliances, such as the television and computer, heat up while they are in use. Rooms with a large number of computers must be kept air conditioned. Thus, the speed of the charged particle is proportional to the applied force, which is to say the electric field, and the current through the wire is proportional to the average speed of the electrons. The average speed is a balance between the energy induced by the applied electric force and the dissipative losses due to collisions; this is manifested in the constancy of the product of the current and the resistance. The constant obtained from this product is the potential difference.

Units of Resistance

Electrical resistance is measured in units called ohms, after Georg Simon Ohm, who, in 1826, through a series of very careful experiments, discovered the empirical relationship between voltage, current, and resistance:

$$\text{current} = \frac{\text{voltage}}{\text{resistance}} \tag{9.13}$$

which, in terms of their units, is

$$[\text{ampere}] = \frac{[\text{volt}]}{[\text{ohm}]} \tag{9.14}$$

and in terms of symbols for the physical observables,

$$I = \frac{V}{R} \tag{9.15}$$

The relation in Equation (9.15) is known as Ohm's law, which states that the amount of current in a closed circuit is directly proportional to the electrical potential differences across the circuit and inversely proportional to the resistance in the circuit. Note that we have jumped from a single wire to a circuit, to a closed path for the electrical current including an electrical pump, say a battery, such as shown in Figure 9.6. The total resistance in the circuit is schematically represented by the jagged line in the figure. When the switch is closed, the circuit is completed and electrons can flow from the battery through the resistance back into the other side of the battery.

Electrical Work

We pointed out that it takes energy to keep the electrons in a circuit flowing and that the amount of energy per unit time required to maintain the current through a resistance is measured by the power. The energy can also be expressed as work. The work done, w, to move a total charge, Q, through a potential difference, V, is given by

$$\text{work} = \text{electrical potential difference} \times \text{charge} \tag{9.16}$$

or in symbols for the physical quantities,

$$w = VQ \tag{9.17}$$

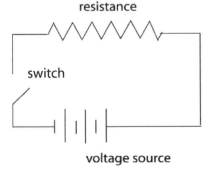

Figure 9.6. A complete electrical circuit is depicted, with a source of voltage to maintain the potential difference across the resistance and the flow of electrons. The switch is in the open position so that no electrons are presently moving in the circuit.

where the unit of potential difference is the volt and the unit of charge is the coulomb. We can test this equation in terms of its units since we know that the definition of a volt is joule/coulomb, so that in terms of units, Equation (9.17) is volt coulomb = coulomb joule/coulomb = joule, the unit of energy. Thus, we can say that a system requires one joule of energy to transport one coulomb of charge through a resistance of one ohm across one volt of potential difference. The expenditure of one joule of energy may then be converted into other forms of energy such as work, light, or heat.

The Unit of Power

It is of more practical interest to inquire as to the rate of energy expenditure, since whether a given amount of energy is expended in one second, or in one hour, can be very significant. The rate of energy expenditure is called the power and its unit is the watt, which is the utilization of one joule per second. We can use these units to devise an equation for the power, P_w. Note the string of substitutions for the units in the following equations

$$\text{watt} = \text{joule/sec} = \text{volt} \times \text{coulomb/sec} = \text{volt} \times \text{ampere} \qquad (9.18)$$

The watt is the unit of power, P_w; the volt is the unit of the potential difference, V; and the ampere is the unit of current, I, so we can replace Equation (9.18) with

$$P_w = VI \qquad (9.19)$$

See Table 9.1. Power is a much more familiar and less abstract quantity than is energy, since that is how appliances are rated, including light bulbs. In the latter case, the energy is expended to raise the temperature of the filament of the bulb and thereby increase the intensity of the light. The greater the power, the higher the temperature reached and the more intense the light radiated. This is why a 100 watt bulb is brighter than a 50 watt bulb.

Electrocutions

A common question concerns the reason why we can be electrocuted with relatively low voltage in water and not on dry land. Why is that? First of all, the electric shock that we feel is due to the passage of current through the body. The magnitude of that current can be determined using Ohm's law. For a given potential difference, the lower the resistance in the material, the greater the current that can flow through the material. When very dry, the skin has a resistance of about 500,000 ohms. When soaked in salt water, the resistance is about 1,000 ohms, or about 500 times less than very dry skin. What this implies is that the current conducted across the body may be as much as 500 times greater when the body is wet than when it is dry.

Resistance of the Skin

As we said, for normally dry skin, the resistance is approximately 10^5 ohms, so when we touch the poles of a battery with our fingers we usually cannot feel 12 volts, and 24 volts is experienced as a tingle. If we have been exercising and our skin is damp, the 24 V experience may be uncomfortable. Although usually not fatal, touching a faulty 120 V light fixture may cause death if your feet and the ground

Table 9.1. A list of basic electrical quantities is given. The notation for the dimension of each quantity along with a typical unit is also listed.

Quantity	Description	Dimension	Units
Charge	Measure of inertia	$[q] = [Q]$	coulomb
Electric field	Influence on charge	$[E] = [F/Q]$	volt
Resistance	Opposition to current	$[R]$	ohm
Capacitance	Charge buildup	$[C] = [Q/V]$	coulomb/volt
Power	Measure of heat	$[P] = [QT^{-1}V]$	energy/time

are sufficiently wet. The potential difference between your hand in contact with the voltage source and dry ground will not draw enough current to seriously harm you. When standing in water, however, the resistance may be sufficiently reduced so that the current may be greater than the body can withstand. In general, 120 V will not kill you but 200 *mA* across the heart will.

It is true that distilled water, that is, water free of impurities, is a good insulator. However, the ions in ordinary water greatly reduce the electrical resistance. The dissolved materials, especially those from a layer of salt left from perspiration on your skin, lower your skin resistance when wet to a few hundred ohms or less.

Question 9.2

If a charged particle in an external electric field is accelerated by the field, the particle travels faster and faster. What do you think might keep the charged particle from increasing its speed to infinity? Is there a braking mechanism for charged particles in nature? (Answer in 300 words or less.)

Question 9.3

Explain in your own words the meaning of Ohm's law. Do you think it is reasonable? While you are at it, explain why power is the product of current and voltage. (Answer in 300 words or less.)

9.2.1   Conduction in Neurons

A Neuron Is Not a Wire

The propagation of a nerve impulse along an axon might be thought to be analogous to an electrical pulse propagating along a wire, much like the dots and dashes sent along the telegraph lines during the last century. This analogy is further reinforced by the fact that the axoplasm (the substance filling the axon) is a solution of electrolytes that, in principle, can conduct electricity. However, the ion current in the axon cannot determine the transmission of the electrical impulse along the neuron for the following reason. The specific resistance (resistance per unit length) of the axoplasm is on the order of 10–100 ohm/cm. Therefore, the resistance per unit length of a fiber with a diameter of 1 μm is 10^9–10^{10} ohm/cm, that is, it is 10^8 or one hundred million times greater than the resistance of a copper wire of the same diameter. This high resistance results in extremely large losses and leakages of ions. However, it is experimentally observed that an axon transmits impulses over distances of more than one meter without significant attenuation or distortion. Finally, the conduction of nerve impulses is much slower than that of electricity, the latter occurring essentially at the the speed of light. Thus, we conclude that there is a mechanism to compensate for the substantial resistance of the axon and which distinguishes signal propagation along a nerve from electrical propagation of a signal along a wire.

The Biopotential

As we mentioned earlier, there is a potential difference across the membrane surface of a cell, with the internal surface being negatively charged and the external surface being positively charged. The biopotential generated at the cell surface is tremendously important to medicine: electrocardiographs and electroencephalographs are based on the measurement of the instantaneous value of the potential difference on the surface of the cell membrane. In addition, the propagation of a nervous impulse is due to changes in the biopotentials and, from the above arguments, conduction of such impulses in a neuron is an active, self-propagating process [61].

The Action Potential

The electrical events in neurons are rapid, being measured in milliseconds; and the potential changes are small, being measured in millivolts. If the potential that excites an impulse is higher than some threshold value, for example, if the potential at the membrane changes from –80 mV to +50 mV, then the cell becomes excited. During excitation, the permeabilities of the potassium and sodium channels change, which is to say, the resistance of the channels to ion transport changes, and the axon generates its own impulse, which enhances the initial external impulse. In Figure 9.7A, we can see that Na^+ enters the membrane where the pulse is rising and K^+ is ejected from the membrane where the pulse amplitude is falling. The action potential, shown in Figure 9.7B, in the spatial interval over which the membrane potential has changed, appears to be +40 mV, and propagates along the neuron with con-

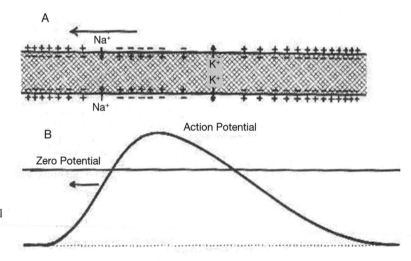

Figure 9.7. Ionic fluxes and current flow at an instant during the propagation of a nerve impulse. In (A), the ionic fluxes are depicted, and in (B), the membrane potential is shown. (Adapted from Eccles [62].)

stant amplitude and velocity. Thus, no matter how much we train, there is a physical limitation to how quickly we can react or how fast we can do things that is determined by the speed of these impulses.

All-or-Nothing Behavior of Action Potentials

The action potential of a single fiber exhibits what is known as threshold behavior. If the excitation is below a fixed value (the threshold), no propagating pulse is generated. On the other hand, if the excitation exceeds the threshold value by a great deal or by an infinitesimal amount, a stereotypical response is generated in the form of an action potential. The response is fixed in size, shape, duration, and conduction speed no matter where on the fiber the response is recorded. This is the so-called all-or-nothing nature of nervous impulses. They are either there in their characteristic form or not present at all. There is no partial, or attenuated, action potential.

Other Kinds of Thresholds

The notion of a threshold should be familiar. A match, a thermostat, and a light switch all exhibit all-or-nothing threshold behavior; the transition from one state to another occurs, or it does not. Once initiated, like the burning of the match, the process proceeds to completion. In this sense, all these phenomena are the same. One of the properties of the neuron that makes its behavior different from these familiar examples, is that once the action potential has passed a given region in space, that region returns to a resting state, ready to undergo the same transition again. This is a general property of excitable tissue—the ability to return to the original threshold state a given time after making a transition. In order to do this, the cell tissue must be supplied with energy to regain its original level of excitability, and this is done by means of electrochemical processes within the cell.

The Propagation of Action Potentials

The movement of the impulse along a neuron has to do with the induction and rejection of ions by the membrane. This process is connected with the depolarization of the membrane as it recharges at the site of the excitation. The K^+ ions leave the cell and the Na^+ ions enter the cell. As a result, the internal surface of the membrane becomes positively charged and the external surface becomes negatively charged. The impulse excites the neighboring sections of the axon, changing the membrane's permeability. After several milliseconds, directly behind the moving impulse, the directions of the flows of K^+ and Na^+ reverse, and the membrane returns to its initial polarized state. This electrochemical process is illustrated in Figure 9.7, where the propagation of the impulse is apparent. The mechanisms of the generation and propagation of the impulse were investigated by many scientists, most notably Hodgkin, Huxley, and Katz [63].

The Inappropriate Cable Analogy

We mentioned above that in the nerve fiber, the specific conductance of the core is about 10^8 times higher than that of the copper wire that the electrical engineer would use. Moreover, the sheath is about 10^6 times leakier than that of a good cable. So the cable-like performance of a nerve fiber is about $10^8 \times 10^6$ times poorer. Eccles [62] goes on to say that, in evolutionary design, this inferior performance of the biological cable was circumvented by a device also used in cable transmission over long distances to overcome the influence of diffusion. Boosters are inserted into cables at intervals to lift the attenuated signals to an acceptable level, and in the biological cable, such boosters are all along the surface of the nerve. In nerves, such boosters are chemically generated.

The Metabolism of Pulse Propagation

The dominant energy requirement of neurons is the portion used to maintain polarization of the membrane by the action of the elecrogenic pump; about 70% of the available energy is used for this purpose. The metabolic rate of neurons doubles during periods of maximal activity, whereas the metabolic rate of skeletal muscle increases by as much as a 100 times. Like muscle, a neuron has a resting heat while inactive, an initial heat during the action potential, and a recovery heat that follows activity. The recovery heat after the passage of a single impulse is about 30 times the initial heat. No wonder we sweat while we exercise; we have to dissipate all that extra heat.

Question 9.4

Why do you think nature was so careless in the design of the electrical properties of neurons? (Answer in 300 words or less.)

9.2.2 The Brain

The Electrical Signal of the Brain

It has been well over a century since it was discovered that the mammalian brain generates a small but measurable electrical signal. The electroencephalograms (EEG) of small animals were measured by Caton in 1875 and in man by Berger in 1925. It had been thought by the mathematician N. Weiner, among others, that generalized harmonic analysis would provide the mathematical tools necessary to penetrate the mysterious relations between the EEG time series and the functioning of the brain. The progress along this path has been slow, however, and the understanding and interpretation of EEGs remains quite elusive. After 125 years, one can only determine intermittent correlations between the activity of the brain and that found on EEG records. There is no taxonomy of EEG patterns that delineates the correspondence between those patterns and brain activity. The clinical interpretation of EEG records is made by a complex process of visual pattern recognition and association on the part of the clinician, and, significantly less often, through the use of the mathematics of Fourier transforms.

The Fractal Brain

The relationship between the neural physiological structure of the brain and the overall electrical signal measured at the brain's surface is not understood. Figure

9.8 shows the complex ramified structure of typical nerve cells in the cerebral cortex. Note the similarity to the fractal structure discussed earlier [8, 38]. The electrical signals originate from the interconnections of the neurons through collections of dendritic tendrils interleaving the brain mass. These collections of dendrites generate signals that are correlated in space and time near the surface of the brain, and their propagation from one region of the brain's surface to another can actually be followed in real time. This signal is attenuated by the skull and scalp before it is measured by the EEG contacts.

The Organization of the Brain

It is useful to point out that even though we do not understand the detailed relation between the EEG signal and exactly what the brain is doing, that detail has

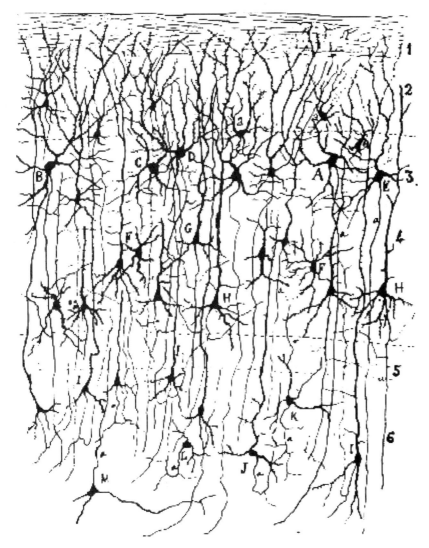

Figure 9.8. The complex ramified structure of typical nerve cells in the cerebral cortex. (Sketch made by by Cajal in 1888.)

not prevented the medical profession from mapping out the functions of the different regions of the brain. In Figure 9.9, are depicted cartoons representing functional parts of the body to show how the motor and sensor signals are organized in the human cortex. For example, as pointed out by Dethier and Stellar [57], electrical stimulation of the human cortex reveals that simple, discrete movements in localized regions of the body can be evoked by stimulation of discrete points on the cortex. It should be mentioned that these kinds of experiments were first done on monkeys and then on anthropoid apes, and knowledge of this mapping is important because, during brain surgery, it is often important to determine part of

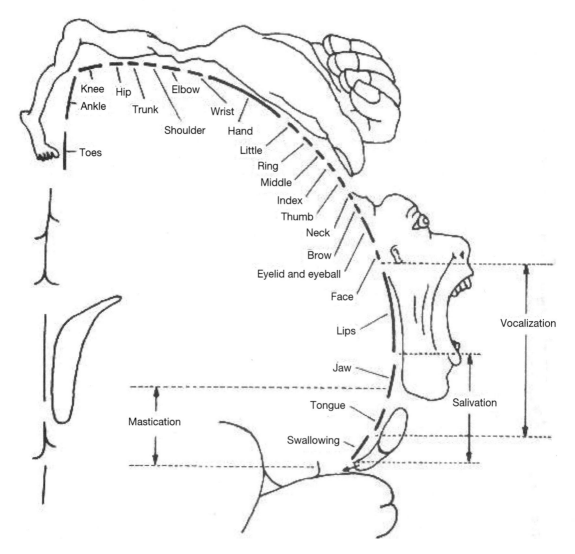

Figure 9.9. Where functions are mapped in the brain. Diagram of section cut through a human brain showing the representation of body parts as measured by evoked potentials recorded from each probing of the cortex upon stimulation of the skin at the parts of the body shown. (Redrawn from Ref. [64].)

the motor map, using for this purpose the conventional stimulating technique, in order to determine the effects of tumors or other pathological influences on the brain.

Representation Is Not Proportional

In Figures 9.9, we note that there are large cartoons of the hand, fingers, and thumb, and even larger ones for the face and tongue. This means that more surface area of the brain is devoted to one function than another function. Why this is the case is in the realm of speculation. Thus, we see that the motor cortex is not in any way spread in proportion to muscle size. Eccles [62] explains that it is skill and finesse that are reflected in the size of the representation. We see this in the disproportionately large areas for movement of the tongue, lips, and larynx, which are used to produce all the subtleties of expression in talking and singing.

Erratic Brain Waves

The long-standing use of Fourier series in the analysis of EEG time series has provided ample opportunity to attribute significance to a number of frequency intervals in the EEG power spectrum. The power associated with the EEG signal is essentially the mean-square voltage at a particular frequency. The power is distributed over the frequency interval 0.5 Hz to 100 Hz, with most of the energy concentrated in the interval 1 Hz to 30 Hz. This range is further subdivided into four subintervals, for historical rather than clinical reasons: the delta, 1–3 Hz; the theta, 4–7 Hz; the alpha, 8–14 Hz; and the beta, for frequencies above 14 Hz. Certain of these frequencies dominate in different states of awareness.

Is it Just Noise?

A typical EEG signal looks like a random time series, with contributions from all frequencies appearing with random phases; see, for example, Figure 9.10. This aperiodic signal changes throughout the day and changes clinically with sleep, that is, its high-frequency random content appears to attenuate with sleep, leaving an alpha rhythm dominating the EEG signal. The erratic behavior of the signal is so robust that it persists, as pointed out by Freeman [65], through all but the most drastic situations, including near-lethal levels of anesthesia, several minutes of asphyxia, or the complete surgical isolation of a slab of cortex. The random aspect of the signal is more than apparent; in particular, the olfactory EEG has a Gaussian amplitude histogram, a rapidly attenuating autocorrelation function, and a broad spectrum that resembles "1/f-noise" [66]. The term 1/f-noise is jargon for processes that have a preference for low frequencies over high frequencies, but which do not suppress the influence of high frequencies altogether.

Brain Wave Patterns

One of the more dramatic results that has been obtained in recent years has to do with the relative degree of order in the electrical activity of the human cortex in epileptic human patients and in normal persons engaged in various activities (Fig-

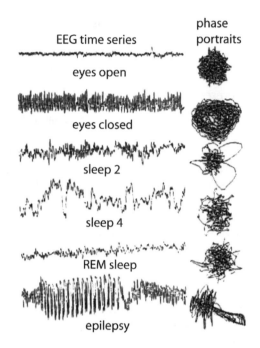

EEG time series

phase portraits

eyes open

eyes closed

sleep 2

sleep 4

REM sleep

epilepsy

Figure 9.10. Patterns in EEG signals. Typical episodes of the electrical activity of the human brain as recorded from the electroencephalogram (EEG) together with the corresponding phase portraits. These portraits are the two-dimensional projections of three-dimensional constructions. The EEG was recorded on an FM analog tape and processed off-line (signal digitized in 12 bits, 250 Hz frequency, 4th order 120 Hz low-pass filter). (From Babloyantz and Destexhe [67] with permission.)

ure 9.10). Babloyantz and Destexhe [67] used an EEG time series from a human patient undergoing a petit mal seizure to demonstrate the dramatic change in the "neural chaotic attractor" using a nonlinear data processing technique. Freeman [65] has induced a form of epileptic seizure in the prepyriform cortex of a cat, rat, and rabbit. The seizures closely resemble variants of psychomotor or petit mal epilepsy in humans. What distinguishes normal activity from that observed during epileptic seizures is a sudden drop in the degree of irregularity in the time series as measured by the fractal dimension. We shall discuss the fractal dimension and other measures of such irregular time series subsequently.

Dimensions Measure Thinking

Experiments measuring the perceptual cognitive demands of a task clearly indicate that the fractal dimension of the EEG time series increases with cognitive demand and suggest the existence of a cognitive attractor [68]. These experiments and subsequent studies establishing the fractal nature of EEG time series shall be discussed more fully once we have introduced the statistical methods that we use to analyze such biodynamical time series. It is worth pointing out that this is no less biodynamics than the time series for human gait, the beating of the heart, or breathing.

Question 9.5

Explain how you interpret the different time series for the brain depicted in Figure 9.10. (Answer in 300 words or less.)

9.2.3 Standing Is Not So Simple

The Mechanisms Are Simple

We have seen that the central nervous system (CNS) has two basic properties: excitability and conductivity. The action potential is a consequence of these two properties and is the only mode of expression available to the CNS. This simple mechanism is responsible for sensation and movement; messages traveling from the sense organ to the brain in the first case and from the brain and spinal cord to the organ in the second case. As Patton [69] observes: "Indeed, all feeling and action are reducible to orderly, sequential, neuronal exchanges of minute quantities of potassium and minute quantities of sodium."

The Pathways Are Disjoint

The CNS is not a continuous conductor but, rather, a chain of neurons linked together by the coupling of an axon of one neuron to a dendrite of another neuron. This junction of the axon and dendrite of different cells is called a *synapse*. The details of the presynaptic and postsynaptic impulses are not presented here; see, for example, Ruch [70] for a complete discussion. Neurophysiology is another of those fascinating subjects that we do not have room to pursue here, but it is important to have some clear, even if schematic, picture of how the sensors, muscles, and brain communicate within the body.

Linear Static Reactions

Let us consider the apparently simple process of standing without losing our balance or, in the extreme case, not falling down. We retain balance by means of the body's static reaction to the actual position of the head in space, the influence of the movements of one extremity on the opposite extremity, and the pull of gravity. There are a number of kinds of static reactions—local, segmental, and general—but they all serve the purpose of assuming a posture that prevents collapse of the extremities under gravity, whatever the situation—from standing quietly in Chruch to walking a tightrope across Niagara Falls.

Higher-Order Static Reactions

Normal standing and normal locomotion involve a multitude of physiological systems, starting with the brain. The cerebral and cerebellar cortex are responsible for two important postural mechanisms. The first mechanism is the hopping movement that maintains the orientation of the limbs when the body of a standing animal is pushed laterally such that the support is retained. This is appropriately called the hopping reaction. The second mechanism ensures that the foot placement on a supporting surface is proper for normal standing and/or normal locomotion. This reaction is observed when an animal is lowered to a visible surface or even when the animal is blindfolded, by allowing any part of a foot to come in contact with the surface. In cats, dogs, and primates, the limbs are extended in anticipation of contact to support the body, when the animals are lowered to a surface they can see.

The Labyrinth in the Ear

The hopping, foot placement, and other static reactions are evoked by certain sensors in the body, perhaps one of the more elaborate ones being in and around the ear. We have all heard that our equilibrium is controlled by the inner ear, and anyone who has had a severe ear infection knows the truth of that. The inner ear, also known as the labyrinth, is made up of the semicircular canals and the otolith organs (saccule and utricle), as shown in Figure 9.11. The semicircular canals are on each side of the head and perpendicular to each other, so they are oriented in three planes in space. These three planes define the three axes of the body's natural coordinate system and provide us with our sense of orientation. Loss of equilibrium is actually a loss of this orientation, as anyone who has suffered a blow to the side of the head, causing disorientation but not unconsciousness, knows.

The Semicircular Canals

The primary function of the semicircular canals in the ear is to register the change in movement of the body in space; which is to say, the semicircular canals are the body's accelerometer. We know that acceleration is a vector, so the semicircular canals must separate the vertical from horizontal accelerations. In mammals, the utricle responds to horizontal acceleration and the saccule to vertical acceleration. On the floor of the utricle there is a carpet of hair that is connected to nerve fibers in the vestibular division. A similar covering exists on the walls of the saccule in a semivertical position. Thus, when the head accelerates, the hairs in the labyrinth are distorted and generate activity in the nerve fibers. There is also a discharge due to the pull of gravity on the inner ear that is generated without movement of the head.

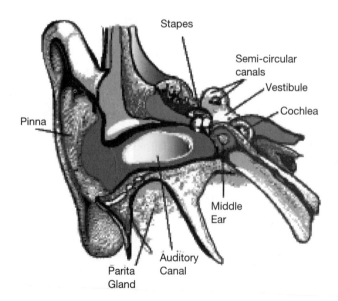

Figure 9.11. The human ear.

Angular Acceleration

We know from mechanics that there are two kinds of acceleration: linear and angular. From the discussion of the semicircular canals, it is clear that the body is well aware of the importance of linear acceleration. The same can be said for angular acceleration. As the body rotates about the vertical axis, the phenomenon of *nystagmus* occurs. Nystagmus is the fixation of the eyes on a stationary point as the body rotates, and is present in both sighted and blind individuals. At the start of the rotation, the eyes begin to move slowly in a direction opposite to that of the rotation, maintaining visual fixation in an effort to maintain visual contact with the environment. When the limit of this movement is reached, the eyes quickly snap back to a new fixation point and then again slowly track in the backward direction. As Ganong [61] explains, the slow movement is initiated by messages from the labyrinth; the quick movement is triggered by a center in the brain stem.

Caloric Stimulation of Nausea

In 1908, the Swedish neurologist, Robert Bárány, found that irrigating the external auditory canals with water cooler than body temperature produced a nystagmus that was delayed by about one minute. Subsequently, repeating the experiments of Bárány, it was determined that the semicircular canals can be stimulated by water either above or below body temperature. These experiments, in addition to producing nystagmus, can also generate vertigo and nausea. *Vertigo* is the sensation of rotation in the absence of actual rotation. Physicians soon learned that in order to avoid these symptoms when treating ear infections, it is important to use fluids at body temperature.

Awareness of Motion and Orientation

Our awareness of motion is presumably the result of impulses from the inner ear reaching the cerebral cortex, even though most of the responses of the inner ear to motion are reflexive. This awareness is also dependent on our perception of spatial orientation, which not only depends on the vestibular receptors, but on visual cues as well. At the cortical level, the brain synthesizes this information with impulses from proprioceptors in joint capsules, indicating the relative positions of various parts of the body, and impulses from cutaneous exteroceptors, especially touch and pressure receptors. The result of the cortical synthesis is a continuous impression of our orientation in space.

Signals in the Body

In this chapter, we have painted with large brush stokes how the different parts of the human body communicate with the body as a whole. The brain sends signals to the limbs through a system of interconnected neurons, the central and peripheral nervous systems, causing the limbs to move. The transmitted signals are actually action potentials that stimulate the muscles to relax or contract and thereby enable us to control how we move. This communication system also gives us the ability to maintain our sense of balance as we walk as well as when we stand quietly. It is clear from even this most cursory of examinations that we cannot understand how the body moves without an understanding of the physics of charged particles and their motion.

Electricity Moves Muscle

Electrical charge was introduced through Coulomb's force law, which looks remarkably like Newton's gravitational force law—with mass replacing charge, but having a different overall constant. The motion of electric charges leads to the concept of electrical current and thereby to the technological basis of information-age appliances. We did not pursue those applications here because it was equally interesting to observe that electricity in the body enables us to move our muscles. Membrane excitation and current in neurons enable one part of the body to communicate with another part. Thus, although mechanical analogues, such as ropes and pulleys, may be useful to visualize how muscles work, they do not explain things in a fundamental way. For such an understanding, one must have at least a cursory grasp of electricity and the properties of the electric field.

The Electric Field

The field concept does not, in and of itself, satisfy the arguments against action at a distance. For example, one might argue that the field only replaces the force conceptually, and that with a change in position of one body, the field throughout space changes instantaneously, regardless of whether we are discussing charges or masses. In the case of an electrical field, we find that this is not the case. It was the recognition that the influence of the change in the location of charged body travels at a finite velocity through space that was the crowning achievement of Maxwell's electromagnetic theory. He was able to determine that the change in the strength of the electric field produced by the motion of a charged particle propagates through space at the speed of light. Due to this measurement by Faraday, it seemed logical that light itself might be just another form of electrical disturbance, a form of disturbance produced by a charge in regular, repetitive, accelerated motion that generates a wave of energy and momentum in empty space. This, in fact, turns out to be just how electromagnetic waves, which is to say light, radio waves, radar, and radiation of all kinds, are generated. The action potential is a localized version of the same phenomenon; it is an electric pulse that propagates along the neural membrane at approximately 20 m/sec rather than 3,000,000 m/sec, the speed of light.

Nonlinear Dynamics

One of the mysteries of the human body is a detailed understanding of how the neurons do what they do. Although we do understand the dependence of the action potentials on the charged particles within the neurons and the associated elecrochemistry, we do not understand the relationship between the firing patterns of the neurons and the specific message being carried to the muscles. We have an idea that these patterns can be related to a branch of mathematical physics called nonlinear dynamics, but this relationship is an area of active research today, and what we can say about it is of a technical, rather than general, nature, and so we move on.

Power and Electrical Energy

We know from our experience that a light bulb gets hot when it is on. We now know that this heat is produced by the conversion of electrical energy into heat—heat that does not go into lighting up the room but is lost due to the resistance in the filament of the light bulb. Muscles get hot in the same way and for similar reasons. Furthermore, the heat that is generated must be dissipated or we will do injury to ourselves. We could measure the rate of generation of this heat and determine the efficiency of our metabolic system, or, for the more compulsive, the efficiency of our exercise program.

Why Is Matter Stable?

If the electrical force is so strong, what holds materials together? If we are made up of these positive and negative charges, why don't we just fly apart? Why doesn't matter with an excess of one kind of charge explode and matter with equal amounts of positive and negative charges just collapse in on itself? The answer is that this is exactly what does happen. The reason we no longer explode is, in part, that we have as many positive particles in our body as we have negative particles, so there is an exact cancellation of attraction and repulsion. If there were as little as a 1% imbalance in the number of positive and negative charges in our body, then the force of repulsion between two people separated by a distance of two feet would be sufficient to lift the weight of the earth. So what keeps the negative charges from collapsing onto the positive charges and thereby keeping the world entirely neutral? The answer to this is a quantum phenomenon that is explained by Heisenberg's *Uncertainty Principle*. We know that this is a little like saying that the explanation is magic, or that it is only for adults, but that is the best we can do. Recall that, following Galileo, we only intend to explain the how of things and not the why.

CHAPTER 10

Molecules and Metabolism— Unseen Causes

Objectives of Chapter Ten

- Understand the physical character of heat and how the body operates as a bio-thermal machine.
- Learn how to interpret empirical laws, such as Boyle's law, and how to relate them to deep theoretical truths.
- Relate work and heat to microscopic processes in order to see how the microscopic processes in the body control the macroscopic process like locomotion.
- Understand why hot things cool down and cool thing heat up, and how this process of heat exchange is related to the Second Law of Thermodynamics.
- Develop some insight into randomness, diffusion, and how we choose to describe motion in order to better understand how nutrients are distributed throughout the body.
- To understand that work is linear and additive, using simple machines.
- Understand the fractal behavior of the heart and understand the implications of such behavior.

Introduction

Energy Has Many Forms and Is Conserved

In this chapter, we are concerned with the transformation of energy in the human body. By transformation, we mean the conversion of energy from one form to another; for example, the transformation from the chemical energy contained in food to the mechanical energy of moving muscle. Energy in the body is continuously being changed from one form into another but, ultimately, it is transformed into heat and other waste products, that is, products that are not useful for running the biological engine that is our body. We focus our attention on the generation and movement of heat in the body for a number of reasons. It is the goal of this chapter to demonstrate how the body works as a biothermal machine. Bioenergetics, the application of mechanical laws to the microscopic domain, explains the macroscopic concepts of work and heat. This is where we see for the first time that the fundamental uncertainty in the molecular domain has consequences in the world of everyday experience. The reason your coffee gets cold in the morning is random molecular motion. The mechanism by which a runner cools down while resting is random molecular motion. The basic mechanism that makes possible the exchange of oxygen and carbon dioxide in the blood at the interface between the lungs and the veins is random molecular motion. So what we do not see strongly influences our lives.

Why Study Heat?

The reason to study heat is, first of all, that heat measures the waste in the body's energy conversion processes. For example, in working with our muscles, if ATP conversion of chemical energy into mechanical energy were 100% efficient, there would be no generation of heat. But, alas, that is not the case, and so we generate large amounts of heat that we need to dissipate. Second, we have introduced into social discourse the Calorie (note the capital C) as a measure of energy; the amount of energy in our breakfast cereal is measured in Calories. But calories (with a lower-case c) measure an equivalent amount of heat for a given amount of mechanical or chemical energy; specifically, in the case of cereal, the energy that would be required to burn the food in question. Note that a Calorie, the measure of energy in food, is a thousand times the calorie, the measure of the heat equivalent in a physical system. Therefore, rather than giving a direct measure of the amount of chemical energy contained in a particular food, we quote an "equivalent" number of Calories. This has always seemed a bit bizarre to me, but that is how we decided to do it. So society's Calorie is a physical kilocalorie.

Thermodynamics Is the Motion of Heat

In this chapter, we focus on the physical discipline that is concerned with the change in heat over time, that is, thermodynamics. In particular, we are interested in how what we can learn from this general physical theory can be applied to living processes. Of course, to give a presentation of thermodynamics we need to start from the beginning (the molecular theory of gases) and end with the biochemical reactions within the human body.

10.1 MOLECULAR THEORY OF GASES

Types of Molecular Motion

In living systems, nutrients are carried across membranes by means of fluids, whereas oxygen is being absorbed into the blood through gas exchange processes in the lungs. To understand these things and other processes underlying energy transfer and energy transformation in the human body, we must first understand the motion of molecules. The first kind of motion is that of simple collisions, described by Newton's laws of motion. The second kind of motion is that of diffusion, measured by averaging the statistical fluctuations in individual particle motion. In our discussion, we do not introduce quantum mechanics, which would be a more complete description of molecular motion, for the simple reason that the phenomena that interest us here do not require that level of detail.

It Is Not Easy

We emphasize at the beginning of this discussion that understanding the molecular theory of gases is very difficult. But this should not be a surprise. The molecular theory of gases, or kinetic theory as it is also known, involves the behavior of a great many particles that obey the laws of mechanics and interact at short distances by means of electrostatic forces. Since we know that we cannot even solve the mathematical equations for the three-body problem, how can we hope to understand the physical properties of a complex of particles like a gas or a solid? The three-body problem consists of three masses interacting by gravitational attraction, and was shown by Poincaré not to have any simple closed-form solutions. In fact, it has been demonstrated that if two of the masses are much greater than the third, the orbit of the light body is fractal and, therefore, its motion is chaotic; see West and Deering [38] for an extended discussion of the relevant literature.

Newton's Laws and the Motion of Gases

In mechanics, we are able to begin with a precise statement of Newton's laws, from which a host of phenomena can be analyzed and understood, and which will form the basis of our understanding of all mechanical processes thereafter. These laws also apply to gases, but the mathematical analysis becomes very complicated because the laws must be applied millions upon millions of times in each cubic centimeter of gas. It is, therefore, impossible to apply Newton's laws to gases explicitly; some averaging procedure must be used to determine the aggregate effect of the interactions of all these particles. However, the first understanding of gases was not obtained through the application of such theoretical concepts; rather, understanding was obtained through the observations of patterns in experimental data.

10.1.1 Boyle's Law

Theories We Have Not Used

Of course, we are treating microscopic particles as if they were tiny billiard balls undergoing elastic scattering. By elastic scattering we mean that both momentum and mechanical energy are conserved during the scattering process and no energy is lost due to the generation of sound and/or electromagnetic radiation (light). We know that, in fact, this is not a proper description of how molecules and atoms behave. The scattering should be described by quantum mechanics, a subject that we have not introduced, nor do we intend to discuss it here, except when it is needed. It might be annoying that we intermittently mention the fact that we do not discuss quantum mechanics, but we do this to emphasize that certain phenomena may have important quantum effects. So, given that we have neither developed probability theory to give a proper classical description of a gas, nor have we discussed quantum mechanics, what can we do to give a proper microscopic description of a gas? We can do what any good scientist does; we can go back to experimentation.

Pressure and Volume

The experimental situation was summarized by Boyle in 1661 in his statement that the pressure and volume of air are "in reciprocal proportions." Today, we would say either that pressure was inversely proportional to volume, or that the product of pressure and volume is a constant:

$$\text{pressure} \times \text{volume} = \text{constant} \tag{10.1}$$

It is remarkable that Boyle's empirical law is so simple, and yet so important in establishing the atomic nature of matter, as we shall see.

Constant Temperature

In addition to Boyle's law, at that time it was also known experimentally that air expands when heated, so Equation (10.1) is only valid subject to the caveat that the temperature is held constant. If we plotted Boyle's law on graph paper, with pressure and volume as coordinates, the law would yield a set of hyperbolae, each one differing from the other by the constant on the right-hand side of the equation. It would be difficult to unambiguously determine the form of a hyperbola using experimental data, so instead of plotting pressure versus volume, one often sees a plot of pressure versus 1/volume. In such a plot, Boyle's law yields a straight line from

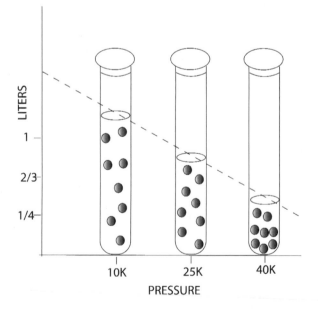

Figure 10.1. Schematic representation of Boyle's law. The volume occupied by the gas in the tubes decreases as the pressure in the tubes increases. At a constant temperature, gas compressed to ¼ of its original volume will be at quadruple its original pressure, as indicated by Equation (10.1). (Suggested by the *Dictionary of Science* [18].)

data. The test tubes in Figure 10.1 graphically indicate how the volume of the gas decreases with increasing pressure.

Initial and Final States

A second way of writing Boyle's law is to note that if we have a sample of gas initially at pressure P_i and volume V_i, and we change the volume to a final value V_f, we find that the gas adjusts to a pressure P_f. This adjustment of pressure is made in such a way that the product [Equation (10.1)] remains constant so we can write

$$P_i V_i = P_f V_f \qquad (10.2)$$

which is also an expression of Boyle's law for constant temperature. How does Boyle's law imply the atomic theory of matter? We shall explain how, but we caution the student that it took almost 200 years of thinking and experiment to come up with this argument, so do not be too surprised if the argument requires some concentration to follow. Who knows, you may want to read it twice.

A Bellows

The behavior of a bellows is determined by Boyle's law. One often sees a bellows in old western films, next to the blacksmith's hearth. The bellows is a device for blowing a stream of air in a given direction. Typically, a bellows consists of two flat pieces of wood, each in the shape of a heart, with a handle lying along the sides and extending back from the round side of the wood. A flexible bladder joins the two pieces of wood along the outside edges, creating an accordian-like structure with a small hole at the pointed end of the heart for the air to pass in and out of the accordion box. The handles can be moved towards or away from one another, expanding or collapsing the bladder, thereby decreasing or increasing the volume of the box. Then, according to Boyle's law, with a decrease in volume comes an increase in pressure and air is expelled from the bellows, since the internal pressure exceeds the external pressure; with an increase in volume comes a decrease in pressure and air is taken into the bellows, since the internal pressure is less than the external pressure.

Question 10.1

A SCUBA diver acquires a depth such that the experienced water pressure is twice that at the surface of the water. How does the diver's breathing have to change in order to maintain the same averge oxygen level in his blood. Use Boyle's law to explain your answer. (Answer in 300 words or less.)

The Lungs Are Similar to a Bellows

The bellows action sketched out above is very much like breathing. During inhalation, we expand the chest cavity, thereby increasing the volume of the lungs and reducing internal pressure relative to that outside the body. The pressure differential between the air inside and outside the lung forces the outside air into the lungs. This fresh air is necessary for the gas exchange at the interface of the bronchial tubes and the venous system. A pressure differential is a change in pressure over a given distance and is also known as a gradient. After a brief relaxation, we contract the chest cavity, reducing the volume of the lungs and increasing the internal pressure. When the pressure inside the lungs exceeds the external air pressure, we exhale, thereby passing the waste from our blood that has been released into the air in our lungs out into the atmosphere. Measurements of the air expelled by the lungs enable us to determine the amount of oxygen taken up by the blood and provides an indirect measure of the metabolic rate.

A Simple Machine

Let us proceed from the bellows to the piston. Consider the piston depicted in Figure 10.2. We assume that the gas consists of a large number of individual particles racing randomly around within the volume of the cylinderical sleeve of the piston. These particles collide with the face of the piston on a regular basis and as they do, they impart momentum to the piston. We assume the collisions to be elastic, so the particles do not change their energy, only their direction of motion. Therefore, the gas transfers momentum to the piston, but not energy, unless the piston moves. If the piston moves then the gas is doing work on the piston and so the gas loses energy. Thus, we must apply a force to the piston to keep it from moving. Let us write F as the force required to just balance the momentum being imparted to the piston by the gas. It is more convenient to express things in terms of the force per unit area, the pressure. Therefore, if P is the pressure and A is the area of the surface of the piston in contact with the gas, then the force required to just balance the influence of the gas on the piston is

$$PA = F \qquad (10.3)$$

We write things this way because you will recall that area is a vector and force is a vector, but we have not yet learned to divide two vectors.

Newton's Laws and Collisions

Thus, because of the continual bombardment of the piston by the gas, if we release the piston, it will move and begin to increase in speed. With each collision the pis-

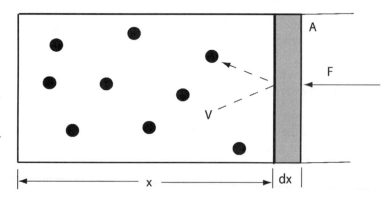

Figure 10.2. Atoms of gas in a cylinder with a frictionless piston. The piston is a distance x from the bottom of the cylinder, the gas has moved the piston a distance dx, A is the cross-sectional area of the piston, and F is the force applied to balance the influence of the gas on the piston.

ton receives, it increases its speed a little more, so the piston accelerates. From Newton's second law of motion, we know that the rate at which the piston accelerates is proportional to the force acting on it. In this way, the force, as given by Equation (10.3), is equal to the change in the momentum per unit time delivered to the piston by the colliding particles.

Work Is Due to Change in Volume of the Gas

If the external force overcomes the effect of pressure and moves the piston a distance dx, work is done because the force is exerted over a distance parallel to the force. We denote the work done by dw, so that inserting Equation (10.3) into the definition of work, and introducing the change in volume as $dV = A\ dx$, we obtain for the work done on the colliding particles

$$dw = F(-dx) = -PA\ dx = -P\ dV \qquad (10.4)$$

The minus sign is an indication that the force is compressing the gas and therefore decreasing its volume. The work done on the gas is produced by the change in volume of the gas at a constant pressure due to the external force. On the other hand, if no external force is applied and the piston is moved by the colliding particles, then the work is done by the gas to expand the volume at constant pressure, and $dw > 0$.

Cleanse the Blood and Air

We have many examples of living systems that use Equation (10.4), that is, the principles of mechanical laws, to relate work to changes in volume. Consider the simple act of breathing. We change the volume of the lung, while pretty much keeping the pressure constant, using our chest muscles and diaphram. The work being done is to bring fresh air into the lungs and expel waste-laden air from the lungs. The heart does the same thing through the pumping of blood. It is the change in volume of the heart chambers, produced by the contraction of the heart muscles, that forces the blood through our bodies. These simple mechanical examples provide fairly good models to keep in mind in order to understand how some parts of our body perform their function. So given this brief detour, let us return to Boyle's law.

Conservation of Momentum

We want to use the force expressed by Equation (10.3), along with the molecular hypothesis, to obtain Boyle's law. We begin by considering how much momentum is delivered to the piston by a single particle. Assume that p_x is the x component of the momentum, just before the collision with the piston. Assuming that the piston is a perfect reflector, the momentum of the particle just after the collision is $-p_x$, as in a tennis ball bouncing back from a concrete wall. Therefore, conservation of momentum requires that the piston absorbs the momentum $2p_x$. Remember, we are dealing with vectors, so the fact that the original momentum of the particle is in the positive x direction, and the final momentum of the particle is in the negative x direction, implies that the wall must have changed the particle's direction of motion.

To change the particle's direction of motion, the wall had to impart a momentum to the particle sufficient to turn it around.

Recall a Pain

Perhaps you could better visualize what the particle and wall are doing if you recall a time when you were hit by a baseball or by a spitball. The pain you experienced was due to the object depositing its momentum into your muscle and your muscle becoming deformed and ultimately transforming that deformation into heat. Typically, the ball just falls to the ground after delivering its momentum to you. For the ball to rebound a significant distance from where it strikes you requires that your body impart additional momentum back to the ball, rather than just absorbing the shock. This might happen, in part, if you are hit in the head, since the cranium is relatively rigid, or if the ball hits another bone. Unfortunately, what often occurs in that case is that the bone cannot absorb the rapid change in momentum and is fractured, leaving you quivering on the ground. Now back to the piston.

Another Dimensional Argument

We could continue with the above mechanical argument, but this would break our word to the reader about limiting the mathematics. So let us consider a dimensional argument to determine the force on the piston in Figure 10.2. The important variables and parameters for the argument should involve the properties of the gas, such as the number density, the number of particles per unit volume ($\rho = N/V$), the mass of the individual gas particles (m), the x component of the momentum (p_x) (since this is the component of the momentum along the axis of the cylinder), and the cross-sectional area of the piston (A). The equation for the force is, in terms of these quantities,

$$F = C\rho^{\alpha}m^{\beta}p_x^{\gamma}A^{\delta} \tag{10.5}$$

The corresponding dimensional equation is, given that we know the units of each of these quantities from previous chapters,

$$[ML_xT^{-2}] = [L_xL_yL_z]^{-\alpha}[M]^{\beta}[ML_xT^{-1}]^{\gamma}[L_y L_z]^{\delta} \tag{10.6}$$

where we have taken into account the vector nature of the process we are considering by keeping the directional subscripts for the various lengths. Equating the exponents in Equation (10.6) for dimensional homogeneity yields the following five equations:

$$\text{exponent of } M: 1 = \beta + \gamma$$
$$\text{exponent of } L_x: 1 = -\alpha + \gamma$$
$$\text{exponent of } L_y: 0 = -\alpha + \delta \tag{10.7}$$
$$\text{exponent of } L_z: 0 = -\alpha + \delta$$
$$\text{exponent of } T: -2 = -\gamma$$

so that $\gamma = 2$, $\alpha = 1$, $\delta = 1$, and $\beta = -1$. Therefore, using the values of these parameters, the equation for the force acting on the piston is

$$F = C\rho m^{-1} p_x^2 A \qquad (10.8)$$

so that inserting the definition of the number density into this equation and dividing by the area of the face of the piston yields

$$\frac{F}{A} = C\frac{N}{V}\frac{p_x^2}{m}$$

Replacing the ratio of force to area with the pressure and cross multiplying by the volume we obtain

$$PV = CN\frac{p_x^2}{m} \qquad (10.9)$$

This is Boyle's law, albeit in a form that might not be immediately recognizable. Note that we still have this annoying constant C that we cannot determine by this method.

The Average Kinetic Energy

Now we have something that looks almost right. The product of a pressure and a volume on the left-hand side of Equation (10.9) is part of the perfect gas law, but on the right-hand side $(p_x^2)/2m$ is the instantaneous kinetic energy associated with the x component of the motion of a single particle, and N is the total number of particles. The kinetic energy can vary significantly from particle to particle and from one time interval to the next; therefore, we replace the single particle kinetic energy by an average kinetic energy. This is like making free throws with a basketball. You are not interested in whether a particular basket is made or not. What is of interest is the average number of baskets made. The same is true with the motion of these microscopic particles. We do not care if the instantaneous value of any one particle is large or small. It is the overall average of the aggregate of particles that is of significance and this is given by the product of the total number of particles and the average kinetic energy of a single particle.

Thermodynamic Energy

Thus, the right-hand side of Equation (10.9) becomes the product of the total number of particles in the gas and the average kinetic energy for each particle. This energy quantity is often called the total internal energy, U, so that Equation (10.9) is written as

$$PV = \frac{2}{3}U \qquad (10.10)$$

where the numerical coefficient 2/3 is dependent on the details of the averaging procedure and comes from the definition of the overall constant C. In real gases, the structure of the gas molecules may contribute to the internal energy so that U is no longer just the average kinetic energy of all the gas particles, but may have other contributions as well. In any event, energy conservation requires that the right-hand side of Equation (10.10) is constant and remains a good approximation for gases over a significant range of pressures. Boyle's law was referred to as "the spring of the air"—a spring of variable strength compared with Hooke's law of solid springs.

Change in Phase

The density of water is one gram per cubic centimeter, but when water is boiled, gaseous steam is formed. The change in density from water to steam is drastic and dramatic. Water is transfigured from a density of 1 gram per cubic centimeter to steam with a density of 6×10^{-4} grams per cubic centimeter. Steam is 1/1700 as dense as water. This can only be reasonably explained by adopting a molecular point of view, in which the water molecules move far apart in the conversion of liquid water to gaseous steam, and the steam is as low in density as it is because it consists mostly of empty space between molecules. This spreading out of molecular particles would account not only for the extremely low density of gases, but also for their low pressure, their small frictional forces, as well as their other physical properties.

Temperature and Internal Energy

Note that the content of Boyle's law given by Equation (10.10) is that, as long as the total internal energy of a system is left unchanged, we can change the pressure and volume in any way we want and we will still get the same result. The product of the new pressure and the new volume is the total internal energy, just like it was for the product of the old pressure and old volume. Further, since Boyle's law is only valid for constant temperature, a change in temperature must be associated with a change in the internal energy of the system. We show this to be the case below. We have not as yet introduced the concept of physical temperature and the associated idea of heat into the discussion, except to say that temperature is a constant in the experiments used to verify Boyle's law. Let us now see what happens if we do not require the temperature to remain constant during experiments of this kind.

So What Is Temperature?

We are all familiar with heat and temperature. The stove gets hot when dinner is being cooked. As babies, we learned that it is imprudent to touch the stove while your mother is cooking. Heat is generated by fire and conducted through the metal of cooking utensils to the food that is then heated to the point where we say it is "cooked." Some metals conduct heat better than others do and so chefs often have their favorite pots and pans for cooking certain dishes. However, we do not set the oven for the amount of heat we want to use to cook the roast but, rather, we set it for a certain temperature. The idea is then formed that the temperature of an object is a measure of how much heat is contained within the object. The higher the temperature, the more heat the body contains; the lower the temperature, the less heat the body contains, or so it would seem.

So What Is Heat?

This idea relating heat and temperature cannot be completely right, however. Consider a glass of water with a temperature of 80 °F and a swimming pool with water also at 80 °F. It is obvious that there is more heat contained in the water in the pool than there is in the water in the glass, if for no other reason than because the pool is so much bigger. Therefore, temperature cannot depend on the total amount of heat in an object. Temperature must be an intensive rather than an extensive property of the material. An extensive quantity depends on the amount of material you have; for example, double the mass of the material and you double the heat energy, but you do not change the temperature. On the other hand, both the glass of water and the pool have the same temperature, so what is the physical process that is measured by the temperature, which we interpret as heat?

Newton's Law of Cooling

This kind of question always drives us back to our experience and to the results of experiments. If we leave a hot cup of coffee freely standing in a room, its temperature will fall until the coffee is the temperature of the ambient air. A cold soda freely standing in a room will become warmer until the soda's temperature is that of the ambient air in the room. Newton was even able to quantify this process by noting that the rate of change in the temperature of an object is directly proportional to the difference in temperature of the object and its surroundings:

$$\text{rate of temperature change} \propto \text{temperature difference} \qquad (10.11)$$

Thus, the greater the difference in temperature between an object and the ambient air, the more rapidly the object will warm up (cool off), eventually coming to equilibrium with the air. We discussed this process somewhat earlier.

Temperture Measures Energy

The above molecular argument establishing Boyle's law can be made more rigorous by taking the exact motion of the millions and millions of gas molecules into account, but the final result will remain: *the mean kinetic energy of the particles is dependent only on the temperature of the gas.* In fact, we can use this relationship to define the scale for temperature. For this purpose, we relate the energy of a molecule to a degree of absolute temperature called a degree Kelvin. The conversion factor, first given by the nineteenth century physicist Boltzmann, is $k_B = 1.38 \times 10^{-23}$ joules per °K. Notice from the units of this constant that if we multiply k_B by a temperature, the degrees cancel out and one is left with energy. The centigrade scale is just the Kelvin scale with a zero chosen at 273.16 °K, so to convert from degrees centigrade to degrees Kelvin, simply add 273.16 °K to the number of degrees centigrade to obtain the number of degrees Kelvin. Thus, if T is the absolute temperature in degrees Kelvin, the definition states that the mean molecular kinetic energy is

$$\frac{\text{mean molecular kinetic energy}}{\text{particle}} = \frac{3}{2}k_B T \qquad (10.12)$$

The numerical factor of 3/2 is included for convenience; it comes about in part because we associate $\frac{1}{2}k_BT$ with the average kinetic energy in each spatial direction of the gas. This association of a specific quantity of energy with each spatial direction is true whether there are forces acting on the gas particles or the gas is in complete isolation. It is always the same. Equation (10.12) is often called the thermal energy of the gas. In the absence of interactions among the particles, we obtain for the internal energy of the gas $U = Nk_BT$ in Equation (10.9).

The Ideal Gas Law Replaces Boyle's Law

We can combine our definition of temperature from Equation (10.12) with Boyle's law in the form of Equation (10.10), where we assume all the internal energy is thermal, to obtain the *ideal gas law,*

$$PV = Nk_BT \tag{10.13}$$

The product of pressure and volume of a gas equals the total number of molecules times the temperature of the gas and Boltzmann's universal constant k_B. Amazingly, if we now take nearly any gas at a fixed pressure, fixed volume, and fixed temperature, we can obtain the number of molecules (atoms) and another universal constant. This universality in the number of particles is a direct consequence of Newton's laws, independent of the kinds of particles in the gas.

Question 10.2

If the gas in a balloon increases from 100 cm³ to 141 cm³ and the corresponding temperatures increases from 50 to 71 °F, what happens to the pressure (approximately)?

Partial Pressure for a Mixed Gas

Suppose we have two gases in a cylinder. Both gases impact on the walls, creating partial pressures P_1 and P_2, respectively. These pressures are independent of one another, since the two gases do not interact with one another, except perhaps for the occasional collision, which we ignore. Therefore, the total pressure on the walls is the sum of the partial pressures:

$$P = P_1 + P_2 \tag{10.14}$$

Lest you think that this is one of those idealized laws from physics that does not have real-world applications, we point out that it is used in one of the standard procedures for determining metabolic rates. The pressure of the oxygen in our breath can be determined by subtracting the partial pressure of the water vapor from the ambient pressure in our exhalation. We shall show that the ideal gas law is used to determine the volume of oxygen that is used by the body in metabolizing food and from which the metabolic rate is inferred.

Heat Transfer Due to Temperature Difference

Now we know that matter does not contain heat, at least not in the sense of a caloric fluid flowing within the material. Gas, like other matter, contains molecules that are whizzing about at high velocities, and the gas exchanges energy with everything with which it comes in contact, through collisions of the molecules. When this transfer of energy is due to a temperature difference, that being a difference in the mean kinetic energy of the gas and the mean kinetic energy of the body with which it is in contact, then we say that heat is transferred. Even in a solid, the heat is actually the thermal activity of particles tied to their local sites and oscillating randomly. These random oscillations determine the kinetic energy of the bound particles, and the average value of the kinetic energy of these bound particles determines the temperature of the solid. In fact, when the temperature gets high enough, the particles have sufficient energy to tear themselves away from their fixed sites and move around freely. We call this process melting.

Units of Heat

The unit of heat is defined as the energy necessary to produce some agreed-to change in temperature of a given mass. The most commonly used unit for heat is the calorie, the amount of heat required to change the temperature of one gram of water by one degree centigrade. Another common unit is the British Thermal Unit (BTU), the amount of heat required to change the temperature of one pound of water by one degree Fahrenheit. The calorie is an important unit, in part, because outside of the physics and engineering community, it is the only unit of energy with which most people are familiar.

Question 10.3

Explain the difference between heat and temperature. (Answer in 300 words or less.)

10.1.2 Diffusion

Fluctuations and Dissipation

The slow spreading of a smoke plume from a chimney, the tendrils of color from ink dropped in a glass of water, and the odor of a fellow traveler sitting next to you on the train are all examples of diffusion. This mechanism of diffusion also explains the transport of oxygen molecules in the blood moving across membranes. Diffusion is a process of transport, whereby large particles (aromatic molecules, oxygen, ink, and so on) are moved from here to there using the random behavior of the environment, typically a fluid, to generate the motion. We have learned that the particles that make up a medium are in a continuous state of agitation and this average motion is measured by the temperature. Transport is accomplished by the collisions between the particles of the medium and the object being transported. But we know that, unless there is a bias in the medium, the object does not move very far from its initial position, on average. The object (large particle) merely jiggles around some central location, unless there is a temperature difference or a mass density difference in the ambient medium.

Diffusion and Speculation

The concept of diffusion was first successfully described, not in the physical context of these examples, but in a social context. In his doctoral dissertation, Bachelier, a student of Poincaré, addressed the effect of speculation on stocks in the French stock market, and in so doing invented the mathematical description of the process of diffusion that is used today. His concept of temperature was the general state of excitation of the background stocks, and the collisions of the ambient particles with the particle of interest were seen as the interactions among the various stocks in the market with that of the stock of interest. This particular view of the stock market has been generalized and extended in recent years to include the average evolution of the stock as well as the fluctuations.

A Most Peculiar Kind of Motion

The process of diffusion is clearly seen by placing a heavy particle in a fluid of lighter particles, as in the physical examples above. The lighter particles collide with the heavier particle and, via Newton's laws, move the heavier particle erratically in space. It is not unlike a huge push ball on a beach being pushed around by a crowd of people surrounding the ball. Viewed from a sufficient distance, only the movement of the ball is seen, not the people doing the pushing. The ball's motion is produced by the random imbalance in the forces being applied to the ball's surface, whether it is the push ball on the beach, or the heavy particle in the fluid. This particle motion was seen by the Scottish botanist Robert Brown [71], who in the late 1820s observed through his microscope pollen motes suspended in water undergoing a most peculiar motion, such as that depicted in Figure 10.3. The lines in the fig-

Figure 10.3. Brownian motion. The motion of a heavy pollen particle in a fluid of lighter water molecules was sketched by Perrin [72] using the observed position of the heavy particle through his microscope at equally spaced time intervals.

ure denote the trajectory of the heavy particle, just like the trajectory of the cannon-ball we studied earlier. Notice that unlike the trajectory of the cannon ball, the path of the particle's motion is disjoint, moving first this way and then that way, with no apparent rhyme or reason. Brown could not explain what he was seeing, and eighty years of incorrect theories were developed to explain this behavior. Then along came Einstein.

The Motion Is Two Hundred Years Old

The actual effect Brown viewed through his microscope was first seen by the Dutch physician Jan Ingen Housz [73] in 1785, who observed that finely powdered char-coal floating on an alcohol surface exhibited a highly erratic random motion. Here again, it was the lighter molecules of alcohol colliding with the charcoal powder that produced the motion. Of course, today we do not refer to either Bachelier or to Ingen Housz. The reason these two scientists are relatively obscure is because five years after Bachelier's thesis, a clerk in a patent office published a paper in a physics journal deriving the same mathematical description of the phenomenon. Bachelier had the misfortune to write a seminal paper in an area of research that was of interest to the young Albert Einstein and, consequently, Bachelier's work was eclipsed by the work of Einstein and went unrecognized until well after his death. Of course, due to the context of Bachelier's research, speculation in the French stock market, there is no reason why a young physicist working in isolation in the Swiss Patent Office should have known about it.

Brownian Motion

Einstein gave the first correct description of physical diffusion in terms of molecu-lar collisions and in his second paper on this subject referred to the possibility that the phenomenon he had described might be the same as that observed by the botanist Robert Brown. It was this random confluence of events that lead to Brown's immortality, through our adoption of the name Brownian motion for this phenomenon, and to Bachelier's and Ingen Housz's relative obscurity, at least in the domains of diffusion and random phenomena in the physical sciences. Ein-stein's theory of the random motion of "large" particles in fluids was used as a proof of the existence of molecules, and Perrin [71] and Svedberg's quantitative measurements of particle displacement as a function of time yielded the numerical value of Avogadro's number and lead to Nobel prizes [74].

Fluctuation–Dissipation Relation

We also know from our discussion of the kinetic theory of gases that Brown's pollen motes absorb energy from the background medium via collisions and in turn give up energy to the fluid, on the average, through dissipation. The average rate of energy gain by the pollen mote is determined by the temperature of the fluid, and the rate of energy loss is determined by the coefficient of friction. This balance be-tween the energy imparted to the heavy particle due to erratic forces and its average energy loss due to friction leads to a dynamical equilibrium situation in which the average energy per degree of freedom of the heavy particle is $\frac{1}{2}k_BT$. The mass of the pollen mote is much greater than that of the fluid particles, so its speed is very much reduced from the average speed of the ambient fluid particles. We estimate the speed of the Brownian particle below.

Random Fluctuations

A particle having a diameter of a micron or two will move, by the process of diffusion, a distance on the order of a millimeter or two in a second. This motion is still very difficult to observe in a microscope because of the vast number of reversals in direction. Even the erratic trajectory depicted in Figure 10.3 has been smoothed, in that the reversal points are only aids to the eye. If the neighborhood of a turning point were magnified, we would see that a large number of reversals have been averaged over to obtain the single reversal depicted. Then again, if one of these turning points at the latter level of magnification is in turn further magnified, we again would see multiple reversals. It is clear that the trajectory of the heavy particle cannot be described by a smooth analytic function. It is this unpredictable structure within structure within structure that physically defines a random fractal function. We shall find, in later chapters, that many of the measured "signals" we obtain in experiments have this erratic form.

Coarse Graining

It is evident from Figure 10.3 that it will take a great deal of data in order to properly characterize the erratic process of particle motion. Historically, the measures that have been used to represent variable data are the average and the variance, whether we are discussing the variations in height in a population, the extremes of income in that same population, or the variation in the intervals between beats of the heart. The average is taken in order to smooth out the fluctuations and provide a one-number representation of the data. The variance is calculated to provide a measure of the degree of variability in the data and give an estimate of how reliable the average is as a one-parameter characterization of the data. But for many processes of interest, the mean and the variance are very coarse instruments with which to characterize the phenomenon. Let us imagine that we divide a second into one hundred equally spaced intervals. In terms of the rate of collision, a hundredth of a second is a very long time. Experimentally, we find that in each second there are 10^{14} collisions of the ambient fluid particles with the pollen mote at room temperature. Therefore, in a hundredth of a second there are approximately 10^{12} collisions, so each reversal in Figure 10.3 represents a coarse graining over an incomprehensibly vast number of events.

Langevin Equation

Suppose we want to write down Newton's law of motion for a heavy particle, such as Brown's pollen mote, being pushed around by water molecules. The force would consist of two parts. One part is the random force produced by all the collisions of the ambient fluid particles with the Brownian particle. The other term represents a dissipation mechanism, a consequence of fluid drag on the motion of the heavy particle. Fluid drag was shown by Stokes to be proportional to the particle velocity. In addition, the specific form of the dissipation parameter is determined by the geometry of the heavy particle and the viscosity of the ambient fluid. This is true in general; for example, these considerations are equally valid for a person walking in a swimming pool as they are for a Brownian particle. Thus, for a person with a given shape, the resistance to their walking in a pool increases with their speed in the water. You can actually feel the increased amount of water you drag along with you as you increase your speed of walking and, therefore, there is an increase in the work you do in order to maintain a higher speed. This is discussed more fully where we investigate the dynamical properties of fluids.

Brownian Motion Using Dimensional Analysis

In order to determine the properties of Brownian motion, the statistics of the random force would have to be known. However, in keeping with our philosophy of introducing as few equations as possible, we wish to determine the important statistical measures of diffusion using only dimensional analysis. One such measure is the mean-square separation of a pollen mote in a fluid at two different times, or ink particles in a diffusing ink droplet, or oxygen molecules in blood. Because the motion of the particle is determined by forces, we can write the mean-square position of the diffusing particle at a given time $\sigma^2(t)$ as a dimensional relation using the dissipation parameter, λ, the mass of the heavy particle, m, the temperature of the ambient fluid, T, Boltzmann's constant k_B, and the time t to obtain

$$\sigma^2(t) = C\lambda^\alpha m^\beta T^\gamma k_B^\delta t^\mu \tag{10.15}$$

Here, the dissipation parameter has the dimensions of inverse time, the temperature is in degrees Kelvin (Θ), and k_B has the dimensions $[ML^2T^{-2}/\Theta]$ so that the dimensional equation corresponding to Equation (10.15) is

$$[L^2] = [T^{-1}]^\alpha [M]^\beta [\Theta]^\gamma [ML^2T^{-2}/\Theta]^\delta [T]^\mu \tag{10.16}$$

The algebraic equations for the exponents in Equation (10.16), making the homogeneity assumption, are given by

$$
\begin{aligned}
&\text{exponent of } L: 2 = 2\delta \\
&\text{exponent of } T: 0 = -\alpha - 2\delta + \mu \\
&\text{exponent of } M: 0 = \delta + \beta \\
&\text{exponent of } \Theta: 0 = \gamma - \delta
\end{aligned} \tag{10.17}
$$

which yields $\delta = 1$, $\beta = -1$, $\gamma = 1$, and $\alpha = -2 + \mu$, where μ is undetermined. Inserting these values for the exponents into Equation (10.15), the variance of the displacement of the particle of interest is

$$\sigma^2(t) = C\frac{k_B T}{m\lambda^2}(\lambda t)^\mu \tag{10.18}$$

and the exponent of the time, μ, is a number that must be determined by experiment.

Ordinary Diffusion and Scaling

Experimentally, we learn that the exponent of the time in Equation (10.18) is $\mu = 1$ for ordinary or classical diffusion. In fact, using the experiments of Brown, that is, measuring the position of the pollen mote for different times, yields the exponent. Thus, we obtain for the mean-square separation of the diffusing particles,

$$\sigma^2(t) = \frac{k_B T}{m\lambda} t \tag{10.19}$$

where we have set C equal to 1. We see that the mean-square separation increases linearly with time and with the ratio $k_B T/(2m\lambda)$, which occurs so often it is given a special name, the diffusion coefficient, D:

$$D = \frac{k_B T}{2m\lambda} \tag{10.20}$$

The mean-square displacement [Equation (10.19)] can be written as

$$\sigma^2(t) = 2Dt \tag{10.21}$$

Equation (10.20) was first derived by Einstein [75] in 1905 and is called the fluctuation–dissipation relation. But why is this particular combination of parameters important? The numerator $\frac{1}{2}k_B T$ is the average energy per degree of freedom delivered to the Brownian particle by the random force characterizing the ambient fluid. The denominator λ is the rate of energy dissipation. Thus, the ratio of these two quantities determines the relative efficiency of the pollen mote in balancing the energy being supplied to it and the energy being extracted from it, and the diffusion coefficient determines the average area covered per unit time by the Brownian particle in maintaining this balance. Note that all the quantities in Equation (10.19) can be determined by experiment: the mean-square separation of the Brownian particles after a given time $\sigma^2(t)$, the temperature of the ambient fluid T, the rate of energy dissipation λ, and the time t. Thus, Equation (10.19) can be used to directly determine Boltzmann's constant and thereby another universal constant.

Transport of Molecules by Diffusion

Diffusion is one of the primary mechanisms by which nutrients are transported within the body. For example, the heart pumps blood to the various regions of the body, but it is diffusion that enables oxygen to move through the blood and into muscle. The same is true for the gas exchange process that occurs at the interface between the lung and the body's blood supply system. The diffusion coefficient, D, characterizes the migration of particles of a given kind, in a specific medium, at a particular temperature. The diffusion coefficient depends on the size of the particle, through its mass; the structure of the medium, through the dissipation parameter; and on the absolute temperature of the medium. The diffusion coefficient for small molecules in water at room temperature is $D \cong 10^{-5} \text{ cm}^2/\text{sec}$.

The Long and the Short of It

Berg [76] points out that a particle with a diffusion coefficient of the order of magnitude 10^{-5} cm^2/sec will diffuse a distance of 10^{-4}cm (about the width of a bacterium) in a time determined by the scaling equation [Equation (10.21)] to be 5×10^{-4} sec, or about half a millisecond. The same particle diffuses a distance of one centimeter, about the width of a test tube, in a time 5×10^{4} sec, or in about 14 hours. Thus, diffusion can transport molecules across microscopic distances in fairly short times, but it takes dramatically longer to cover macroscopic distances. This change in efficiency from the microscopic to macroscopic times is due to the linear time dependence of the square of the distance. For a particle to travel twice the distance takes four times as long; for a particle to travel ten times the distance requires a time that is one hundred times as long. Thus, we have a nonlinear relation between space and time, and this nonlinearity leads to all kinds of interesting phenomena involving diffusion.

10.1.3 *Microscopic Blood Flow*

Capillary Beds

We mention in passing, in order to fill out our sketch of the human body, that there is a great deal of anatomical structure associated with blood flow and blood–tissue interaction. Most of this structure we cannot discuss here, but we do want to examine how diffusion enters our understanding of the gas exchange process in blood–tissue interaction. The cardiovascular system has as its lowest level of subdivision the capillaries, which are 10 to 20 microns or less in diameter. The capillaries are where the exchange of O_2, H_2O, CO_2, ions, nutrients, metabolites, and humoral products takes place between the blood and tissue; see Catchpole [77] for a complete discussion of the physiology. The capillaries form a dense network between arterioles (small arteries leading into capillaries) and venules (small veins leading up from capillaries), and form the basis of "microcirculation." The greater part of the total circulatory blood within the body at any given time is at the microscopic level (see Figure 10.4). The speed of blood flow in microcirculation is actually quite low, between 0.8 and 2.0 mm/sec.

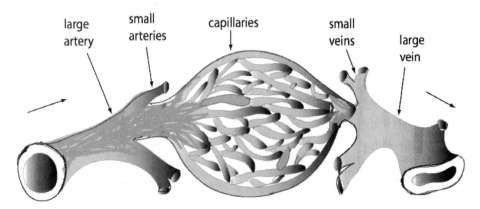

Figure 10.4. The capillary bed between arteries and veins.

The Capillary Bed Is Actually Huge

Estimates of the extent of the capillary bed in human muscle has been given as 6000 square meters for an average person, and all the this area is available for gaseous exchange [77]. In the alveoli of the lung, for example, the capillary network is quite dense and the capillary bed is only a few times the diameter of the capillary itself. Futhermore, these capillary beds are not homogeneous, but vary markedly in various organs and tissues. The red blood cells are seen to deform as they squeeze their way through the smallest of these elastic tubes, taking on a characteristic parachute-like appearance.

Diffusion and Pressure in the Capillary Tube

As blood moves through the capillaries, oxygen and nutrients are delivered to tissue and metabolic products are received in return. This exchange is never ending and consists of two processes: filtration and diffusion. Filtration is driven by a hydrostatic pressure difference between the inside and outside of the capillary tube. The hydrostatic mechanism will be discussed subsequently. Here, we note that it forces the passage of water and dissolved substances through the capillary wall. Diffusion, on the other hand, is produced by a mass concentration difference across the capillary wall, forcing material from a region of low to high concentration. Note that diffusion and hydrostatic pressure are distinct mechanisms and may, therefore, act together or oppose to one another in different situations, depending on the electrochemical environment.

10.2 WORK AND HEAT

Energy as Fuel

The idea of work in a physical system was introduced earlier in connection with our discussion on the nature of mechanical energy. Energy was defined, in general, as a system's capacity for doing work. You may recall that in mechanics, work is the force multiplied by the distance over which the force acts, which we now know is the scalar product of two vectors. We did not give many examples of work in our earlier discussion because we were then more interested in the forms of energy. One category of energy that we did not discuss is *fuel*. We can think of fuel as stored usable energy in much the same way as we think of money in stocks, bonds, or in the bank as stored wealth. The human body stores energy as fat. But this brings up an interesting question. Why do humans have to eat in order to remain alive?

Energy Loss

It is clear that the human body is not designed to consume or produce energy; rather, it is designed to convert energy from one form to another. We describe this as burning fuel, that is, metabolizing food. But we know that energy is neither created nor destroyed, so why can't the human body, once it has stopped growing, reach some maximal size and then stay that way forever? One reason lies in the fact that the processes that constitute life lose energy over time. The body is not really very efficient. For example, blood is quite viscous and, therefore, it continuously loses mechanical energy as it is passed through the circulatory system. The rate of pressure drop from the vicinity of the heart to the peripheral regions of circulation is a measure of this viscous energy loss. The energy lost in this and other dissipative phenomena must be replentished in order to sustain blood flow and, therefore, support life.

Staying Alive

We should also note a number of other energy-associated properties of the human body. First, our energy only comes in the particular form of molecular configurations found in food that we ingest and digest. We have no way to shortcut the process, say by eating sunlight. The radiant energy produced by nuclear fusion in the sun reaches the earth, photosynthesis transforms the sunlight into the energy stored in plants, and animals eat the plants and store the energy in muscles and fat. We, being at the top of the food chain, eat both plants and animals. Second, in the energy utilization process, we generate heat that cannot be used to do work. We observed earlier that mechanical, chemical, and electrical energy can be readily converted into heat in the body. However, the reverse process cannot easily be done; that is, we cannot use body heat to, say, generate mechanical energy by flexing a muscle. Heat is generally waste energy that the body must shed in order to regulate body temperature and maintain that temperature in the narrow range necessary to sustain life. But this is not the entire story, since we determined above that heat is the engine that drives diffusion and, therefore, molecular transport, such as oxygen in the blood.

10.2.1 Work Is Linear and Additive

Double One and You Double the Other

One property of doing work that we read directly from the force multiplied by the distance relation is that the energy required to do a job is a linear additive process. For example, if we are lifting a load $F = W$ through a distance s, we do work $E = Ws$. Then, if we double the load so that $F = 2W$, we also double the work. But, equivalently, if the load remains the same, and we double the distance, $2s$, through which we move the load, W, then the work is also doubled. When the output is directly proportional to the input, the system's behavior is said to be linear. Although most phenomena, when examined in detail, are found to be nonlinear, for the purposes here, we restrict most of our discussion on energy to linear processes. There is typically some domain of parameter values over which this is an adequate approximation.

Friction and Dissipation

It may be evident from our previous discussions that the process of utilizing fuel to do work, like hauling a load, is merely a way of transforming energy from one form to another. The mechanical processes we have been discussing are all reversible, so that lifting a load with a steam engine, for example, transforms energy from steam pressure to gravitational potential energy. In the absence of energy loss through such processes as friction, the gravitational potential energy could, in principle, be returned to the steam in a piston. We have not discussed friction in any detail, however, because it is a very difficult concept and not easily understood in a mechanical context. When we drag a wooden box over a concrete floor, the resistance we encounter is called friction. The rough wood of the box "catches" on the porous surface of the concrete. To move the box, we must literally break off small pieces of the wood to overcome static friction. To continue to move the box once it has been set in motion is easier, as long as we do not allow the wood to "reattach" itself to the concrete. One might label this the Velcro theory of friction.

Friction Generates Heat

However, no matter how much we polish two surfaces, when they come in contact with one another under relative motion there is always friction. Even when we have polished a piece of sculpture to a high gloss, if we rub it rapidly back and forth, we feel our hand getting warm. This increase in temperature is due to friction and suggests that friction directly converts the mechanical energy of motion into molecular motion and, therefore, heat. From this "experimental result," we conclude that friction is actually a molecular process that transforms useful energy into useless energy in the form of heat. This may be a rather harsh indictment of heat, since we know that rubbing our hands together on a cold night in the winter warms them, and may help prevent frostbite. Therefore, keeping an organism, such as our body, at a constant temperature, is life affirming. On the other hand, heat is manifest on a macroscopic scale, even though the origins of friction reside in the microscopic nature of matter and quantum mechanics is necessary to truly understand it.

Archimedes Invents the Machine

So far, we have failed to mention the mechanisms that humanity has used to convert fuel to energy to do work in a maximally efficient way. To accomplish this, we invented machines such as levers and pulley systems to produce big forces from small ones. As we remarked earlier, Archimedes, in the third century BC, in response to a challenge from King Hiero, developed a series of compound pulleys, by which he was able to move a ship that had been lifted by many men out of the harbor onto dry land [48].

Machines Increase Efficiency

Machines can magnify the effect of our fuel resources and produce more work from less fuel. One very simple machine is the child's seesaw, which we will call a lever when we want to use it to do work. In equilibrium, the torque produced by two weights (forces) about the pivot point are the same, and the seesaw does not move. In our discussion of muscle, we introduced the three classes of levers and discussed the mechanical advantage realized by each.

The Seesaw

In Figure 10.5, we again depict a class one lever, where we apply a torque T_1 to the plank on the left of the fulcrum, rotating the plank in a counterclockwise direction.

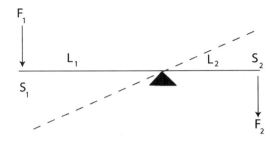

Figure 10.5. The torque, $T_1 = F_1 L_1$, applied at the left and acting counterclockwise, is just sufficient to overcome the torque of the load, $T_2 = F_2 L_2$, which acts clockwise at the right. The applied force moves through a distance s_1 and the load moves through a distance s_2.

This torque consists of a vertical force F_1 acting over the lever arm L_1, producing a rotation about the pivot point. The load has a vertical force F_2 that acts over the lever arm L_2, producing a clockwise rotation about the pivot point, thereby producing a torque T_2. Suppose that T_1 is just sufficient for the applied torque to overcome the torque produced by the load so that they are equal. Then we have

$$F_1 L_1 = F_2 L_2 \tag{10.22}$$

We have dropped the vector notation, since the angle between the direction of the force and the lever arm is 90° and we use sin 90° = 1. By geometry, since the plank is a continuous rigid object, the angle from the horizontal must be the same on both sides of the pivot point. Using this fact we have, using the definition of the sine function,

$$\sin \theta_1 = \frac{s_1}{L_1}$$
$$\sin \theta_2 = \frac{s_2}{L_2} \tag{10.23}$$

and since $\theta_1 = \theta_2$,

$$\frac{s_1}{L_1} = \frac{s_1}{L_1} \tag{10.24}$$

so that the ratio of the displacements equals the ratio of the lever arms, which when substituted into Equation (10.22) yields

$$F_1 s_1 = F_2 s_2 \tag{10.25}$$

Thus, since $F_1 s_1$ is the work done by an engine, say a human body generating the force F_1, and $F_2 s_2$ is the work done on the load, and they are equal, we can see that the lever transfers energy from the engine to the load without loss. Therefore, the input energy and the output energy of the lever are equal in the absence of dissipation.

Physical Exercise

We can use our arms to raise and lower weights during exercise. The mechanical work done on the environment in raising and lowering the weights is generated by chemical energy in muscle. In addition, there is heat generated in the energy conversion process, since the conversion process is not 100% efficient. Recall the relation between the application of force through the use of muscles and the interpretation of these muscles in terms of levers.

Conversion of Energy into Work

Another example of an extremely efficient machine, that is, a machine whose energy output is nearly equal to its energy input, is the hydraulic press. If all the energy supplied to a system goes into the work the system does (there are no energy loses), then the efficiency is 100%. According to Pascal's principle, which we shall elucidate in the next chapter, fluids at rest transmit pressure unchanged in all directions and to all horizontal distances. This principle is used to do work. Consider an oil-filled container with two pistons of unequal cross-sectional area. We know that the force acting on the face of a piston is the pressure times its cross-sectional area. Therefore, from the sketch in Figure 10.6, on the larger piston with area A_l we have the force

$$F_l = PA_l \tag{10.26}$$

and on the smaller piston with area A_s we have the force

$$F_s = PA_s \tag{10.27}$$

where the oil acts to transmit the pressure from one piston to the other. Both pistons experience the same pressure P, so that the force ratio is just the ratio of the areas:

$$\frac{F_l}{F_s} = \frac{A_l}{A_s} \tag{10.28}$$

The larger piston can, therefore, support a weight much greater than the applied force F_s. Suppose that the ratio of areas is a factor of nine, then the applied force F_s

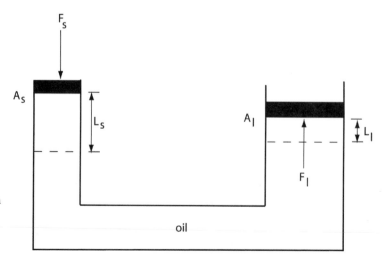

Figure 10.6. Two pistons are shown connected by a closed hydraulic tube. The forces acting on the separate and distinct surface areas of the two pistons are discussed in the text.

can support or lift a weight nine times as large, $F_l = 9F_s$. This is sketched in Figure 10.6, where the diameters of the two pistons are in the ratio of 3 to 1.

Work In Equals Work Out

Here again, we find that the side of the machine with the load moves a far shorter distance than the side to which the force is applied. The volume of oil pushed in the small cylinder by its piston is the cross-sectional area times the length of the stroke, $A_s L_s$. This volume is shifted over to the big piston and pushes it up. If the big piston rises a distance L_l, then, assuming the oil cannot be compressed, both the volume generated by the movement of the small piston and that experienced by the large piston are the same, that is, $A_s L_s = A_l L_l$. Equating the volumes implies that the ratio of the areas equals the ratio of the stroke lengths:

$$\frac{L_l}{L_s} = \frac{A_s}{A_l} \tag{10.29}$$

But, according to Equation (10.18), this also equals the ratio of the forces. Substituting Equation (10.29) into Equation (10.28) then yields

$$F_s L_s = F_l L_l \tag{10.30}$$

or equality between work in and work out; the input work on the left-hand side of Equation (10.30) equals the output work on the right-hand side. Further, what is gained in the forces due to the ratio of areas is lost in the distances traveled, such that mechanical energy is conserved.

Heat Is an Energy Tax

Of course, any real machine has at least a small energy tax in the form of friction and a corresponding loss of energy in the form of heat. This loss of useful energy can be measured through an increase in the temperature of the oil for a hydraulic press, and a similar increase in temperature of the pivot point of the lever. But let us now turn our attention from these idealized machines and consider one of the most remarkable machines in the human body, the heart.

10.2.2 Cardiac Pump

An Electromechanical Pump

You have undoubtedly heard that the heart is a mechanical pump, but now you are in a position to better understand the meaning of that remark, based on the simple principles regarding work that we have just considered. Using a combination of Boyle's law (the product of pressure and volume being a constant at a given temperature) and the hydraulic lift, we can see how the cardiac pump works. First of all, we note that the beating of the heart is a rhythmic, but not periodic, process. Heartbeats are a sequence of contractions and relaxations, resulting from the cycle of electrical depolarization and repolarization of the cardiac muscle. It does not matter where in the cycle we start our discussion.

Waves of Excitation

We mentioned earlier that cardiac muscle is striated like skeletal muscle but, like smooth muscle, it is involuntary, which is to say that the heart's contraction is not under conscious control. Therefore, we cannot activate cardiac muscle as an act of volition, like we can the skeletal muscles, claims regarding the slowing of the heartbeat made by fakirs, yogis, and others notwithstanding. Further, the separate elements of the cardiac muscle are interconnected, so that once electrical depolarization is initiated in any location, it propagates throughout the muscle. The wave of excitation propagates radially from the sinoatrial node at a speed of nearly one meter per second, so that the spread of depolarization takes on the order of 80 ms to cover the atrium. This is the P-wave in the electrocardiogram depicted in Figure 10.7. The electrical wave then travels through the ventricular muscle and produces an excitation referred to as the QRS complex in the electrocardiogram. Finally, there is the T-wave signaling the repolarization, that is, the returning to a resting state, of the ventricular muscle.

The Fractal Heart

In Figure 10.7, the cardiac cycle appears regular, and this apparent regularity contributes to the reason why these time series are called normal sinus rhythm. Measure the distance between peaks and verify for yourself that the interbeat interval changes, ever so slightly, from beat to beat [78]. Subsequently, we shall see how to interpret this variability; in particular, we shall show that this variability implies that the cardiac control system has certain fractal random characteristics.

Systolic and Diastolic

The general physical structure of the heart (see Figure 10.8 for a schematic) consists of the atria and ventricles, each of which is divided into a right and a left chamber, making four chambers in all. Systolic pressure is the highest pressure produced by contraction, and refers to the peak pressure of the aorta. Diastolic pressure is the lowest pressure reached during ventricular relaxation. As Scher [79] describes, a portion of the blood ejected into the aorta and pulmonary artery during systole distends the elastic walls of these vessels, storing potential energy. This stored potential energy is released into the blood flow during diastole. The aortic and pulmonary

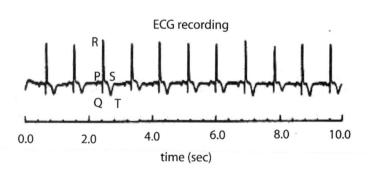

Figure 10.7. The cardiac cycle. The basic cycle (P-QRS-T) repeats itself again and again.

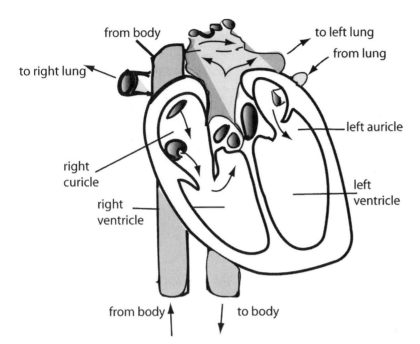

Figure 10.8. A cross-sectional schematic of the human heart.

arterial pressures rise to the systolic value during the contraction of the ventricle, but the pressures in either the greater or lesser circulation do not fall to zero between beats. Peak systolic pressure is normally about 120 mmHg in the aorta and 25 mmHg in the pulmonary artery. The diastolic pressure in the aorta is about 80 mmHg; that in the pulmonary artery is about 7 mmHg. From these values it is clear why the physicians talk about the ratio 120/80 as being an indicator of health. However, the proper interpretation of deviations from these values and the possible causes of these changes is relegated to the domain of medicine and we shall not pursue this discussion further.

Right and Left Ventricles

The right ventricle of the heart fills with freshly oxygenated blood, returned from the interface with the lungs. The pacemaker cells in the sinus node would spontaneously fire, but in the heart they are entrained by impulses from the autonomic nervous system and these impulses initiate muscle contractions. This ventricle is a pressure pump, which, upon muscle contraction, ejects a constant stroke volume of blood into the aorta, against widely varying pressures. Scher [79] explains that the right heart can easily adapt to changes in stroke volume; the left heart can adapt to demands in increased pressure. The valves that separate the atrium and ventricle, on each side, open and close in response to pressure-induced flow changes. The valves are designed to allow for the one-way flow of blood from the atria to the ventricles, via the atrioventricular valves, and from the ventricles to the aorta or pulmonary artery, via the aortic and pulmonary valves, respectively.

The Lawful Heart

In this discussion it would be quite easy to become heavily involved in an examination of the physiology of the cardiovascular system. However, our purpose here is to highlight the heart's electromechanical nature and to indicate the complexity of the system, but not to pursue a detailed examination of the anatomy and physiology of its workings. Rather, we hope to reveal to the reader the number and kinds of principles involved in the operation of the cardiovascular system. For example, the application of Boyle's law to breathing; the application of Pascal's law to the difference in hydrostatic blood pressure from our heads to our feet; the application of diffusion to the transfer of oxygen from inhaled air, across the membrane tissue of the lungs, to the blood in the capillary system interfacing the alveoli, and so on.

10.3 SUMMARY

Theory Versus Experiment

We may have exceeded your level of comfort concerning the detail presented in the discussion of the molecular theory of gasses, but let us consider the alternative. We could have introduced Boyle's law as an empirical relation between volume and pressure in a gas and left it at that. But then we would also have had to introduce the perfect gas law as a second empirical equation. A second relation is appropriate for the case when temperature is allowed to change as the pressure and volume of a gas change. Of course, the deep connection between these two empirical relations might have been suspected, but not proven. It is the role of theory, like the molecular theory of gases, to provide the connections among experimentally determined equations.

Temperature and Kinetic Energy

Another thing that we learned through the attention we paid to the molecular theory of gases is that temperature is a measure of the kinetic energy of the gas molecules; that is, temperature characterizes how rapidly the gas particles are moving, on average. We established the relation between temperature and kinetic energy using a couple of conservations laws and dimensional arguments. This is not too bad, since there are volumes written about this relationship.

Diffusion and Microscopic Transport

One thing that temperature does is to provide a measure of how effectively a medium can transport material. The discovery of Brownian motion established that the phenomenon of diffusion provides a mechanism by which the lighter particles of a fluid can move around a heavier particle suspended in that fluid. How quickly the heavier particle is moved from point to point is determined by the diffusion coefficient, which in turn is dependent on the temperature of the ambient fluid, the dissipation rate, and the mass of the particle. Diffusion is the mechanism by which oxygen is transferred from the air in the lungs to the blood in the arteries, at the interface of the arterial and respiratory systems. Diffusion is also the mechanism by which most molecules come in contact to initiate a chemical reaction. Here again, this mechanism could not be understood without the kinetic theory of gases and was in fact used to prove the existence of molecules.

Heat, Work, and the Cardiac Pump

A picture should be forming as to how the concepts from mechanics, momentum, force, energy and so on, not only work in our everyday world of the visible, but also work at a level that we only access indirectly, the microscopic. We examine the implied relations between these two worlds in the next chapter, but here we comment on the fact that mechanical laws are applicable even to systems like the cardiac pump, where we cannot make direct measurements.

CHAPTER 11

Bioenergetics—
The Dynamics of Heat

Objectives of Chapter Eleven

- Learn that the motion of heat, whether in the body or elsewhere, is described by the four laws of thermodynamics.
- Understand how a thermometer works and in so doing learn how temperature is related to the physical properties of a system.
- Look a little deeper into energy conservation and its implications, in particular, the fact that heat is energy and why energy that is not conserved winds up as heat.
- Find that the second law means there is no such thing as a free lunch, which means that you always pay, in one form or another.
- Understand that entropy measures the useable energy in a physical system as well as in a biosystem, so that entropy can be used to measure the order in biological systems.
- Learn how the body's metabolism can be understood through an application of thermodynamics to biology.

Introduction

The Importance of Thermodynamics

In this chapter, we give a fundamental, if phenomenological, explanation of how the energy exchange processes in the body works, using the laws of heat motion. Of course, thermodynamics is the motion of heat, and one of the things we investigate

Biodynamics: Why the Wirewalker Doesn't Fall. By Bruce J. West and Lori A. Griffin
ISBN 0-471-34619-5 © 2004 John Wiley & Sons, Inc.

is how heat can be used to do work and when it cannot be used to do work. We know that a fire generates heat, heat can change water to steam, steam can be used to operate a steam engine, and a steam engine can be used to do work. So that is a case where heat can do work. But heat can only be used to do work when there is a difference in temperature. It is the temperature difference that determines the amount of energy that is available to do work. The heat generated by the human body cannot do work under most circumstances. It is the waste product of chemical reactions and serves no purpose other than to keep us alive by maintaining the body temperature at a constant value.

We Make Laws

The laws of thermodynamics have been discussed in other forms in earlier chapters, particularly the first law—conservation of energy. The four laws of thermodynamics are a way of understanding the organized behavior of large, many-faceted systems, such as the human body, that are open to the environment. An open system is one that takes a quantity in, like food, uses that quantity to perform a particular task, like jumping a hurdle, and expelling what is not used as waste. This continual flux of material (food) through the body is what keeps us alive and, more generally, is what produces organized patterns in nature.

Entropy Is a Measure of Order

Entropy, defined by the second law of thermodynamics, is, in general, a measure of order in the universe and, in particular, a measure of order in the human body. Entropy measures how we transform the energy the body takes in through food into the energy the body uses to do work and the energy that is lost in waste products, such as heat. In short, entropy is what is measured through the metabolic rate.

Temperature Is a Measure of Heat

11.1 HEAT AND TEMPERATURE

We have used the idea of heat quite a bit in our discussions but, historically, this has been a difficult concept to nail down. Even today, the concepts of heat and temperature are often confused. We have attempted to make the distinction between heat and temperature apparent in our previous discussions, but it is an uphill battle since the two terms are so often interchanged in everyday usage. For example, the poetic phrase ". . . in the heat of the day" does not have any scientific meaning. The correct phrase might be ". . . in the temperature of the day," which, although scientifically meaningful, lacks a certain texture that the original phrase contained. Thus, there are some modes of expression that must be abandoned for scientific accuracy, if not precision. The term heat actually refers to the stuff necessary to increase the temperature of an object, and temperature is a measure of hotness.

Temperature Units

To make the temperature scale correspond to our experience of the world, we select two familiar temperature-dependent phenomena and assign numerical values to the temperature at which these phenomena occur. Two such phenomena are the melting of ice (and freezing of water) and the boiling of water (turning water into steam).

These two-phase transitions occur at definite temperatures. The freezing of water (and the melting of ice) occurs at a temperature assigned the value of zero degrees centigrade (0 °C) and the turning of water into steam (boiling water) is assigned the value of one hundred degrees centigrade (100 °C). The scale is then divided into one hundred equal intervals, each one of which is called one degree centigrade. These are the two standards of hotness: melting ice and scalding steam. Of course, these numbers are arbitrary, as is apparent from the fact that in the United States we stubbornly cling to a different scale from that used by the rest of the world, the one due to the German physicist, Gabriel Fahrenheit. Water freezes at 32 °F and boils at 212 °F in this latter scale, and there are 180 equally spaced intervals between these two extremes, not one hundred as on the centigrade scale. Each of these scales measures the hotness of an object and in both cases we call this number the temperature. The two scales can be related by the equation $F = 9/5\ C + 32$, where F is the temperature in degrees Fahrenheit and C is the temperature in degrees centigrade.

Absolute Temperature Scale

Finally, we examine the temperature scale of Lord Kelvin. Using thought experiments and ideal heat engines, Kelvin devised a temperature scale on which the zero point corresponds to the situation in which the average kinetic energy of an object is zero. Furthermore, since we now know that heat is the motion of microscopic particles, in the absence of heat, when the kinetic energy is zero, the particles in an object cease all movement. It turns out that the absolute temperature scale of Kelvin can be put in one-to-one correspondence with the centigrade scale of temperature and with the Fahrenheit temperature scale as well. Therefore, we see again that the scale one chooses is determined by convenience, no one scale being favored by the phenomenon of interest. It is, of course, possible to transform temperatures between scales. For example, given the temperature in °F we can convert to °C by subtracting the zero-point of 32 °F and multiply the resulting number by 5/9. Consider the boiling point of water, 212 °F, from which we subtract 32 °F to get 180 °F and multiplying by 5/9 we obtain 100 °C.

So What Is Temperature Really?

Temperature is that property of a material that is measured by a thermometer. Of course, we know a deeper truth. We learned earlier that the temperature of a gas, for example, was proportional to the average kinetic energy of the molecules of the gas. Thus, a thermometer actually measures the average kinetic energy of the particles making up the medium in which it is in contact. Recall that if two objects are brought into contact, they will come into equilibrium with one another, that is, they will eventually come to a temperature that is between the beginning temperatures of the two objects. If the thermometer is very small and the object whose temperature we require is very large, then the thermometer will come into equilibrium with the object without disturbing the heat content of the object.

Particle Collisions

We know that when two objects are put in contact with one another, the two types of particles that make up the different objects will collide at the interface of the objects and exchange momentum and energy. The object with the greater average kinetic energy will give up energy and momentum to the object with the lesser average kinetic energy. As the average kinetic energy of the colder object increases, so too does its temperature. Thus, at equilibrium, the average kinetic energies of the two objects are equal and their corresponding temperatures are, therefore, the same. Using the perfect gas law–the product of pressure and volume is proportional to the temperature–we can see that as the temperature increases, then so too does the volume of, say, the mercury in the glass tube of a thermometer. For constant pressure, the mercury rises in the evacuated glass column of a thermometer. The physical mechanism producing the expansion of the mercury is the increase in the average separation between mercury atoms as the average kinetic energy increases.

And Heat Is . . .

So what is heat? Heat is actually a form of energy. When we put a pan of water over a flame, we are transforming the radiant energy of the fire into the kinetic energy of the molecules of the pan. As these molecules absorb the radiant energy they jiggle faster and faster about their fixed positions in the metal. This rapid motion of the molecules of the pan is transformed by means of collisions to the water molecules in the pan adjacent to the inside surface of the pan. The water molecules with increased kinetic energy due to collisions move away from the pan's surface, being replaced by "colder" molecules. Thus, as the energy is transferred to the water, the water is heated and its temperature increases. In fact, we know from experiment that:

$$\text{amount of heat} \propto \text{mass of water} \times \text{change in temperature} \qquad (11.1)$$

Therefore, if the mass of water is in kilograms and the change in temperature is in degrees centigrade, the heat is measured in kilocalories or, for food lovers, in Calories. One kilocalorie is needed to raise the temperature of one kilogram of water one degree centigrade. Since heat is a form of energy, it can be expressed in terms of joules (J) so that we can convert between the units of energy:

$$1 \text{ cal} = 4.187 \text{ J} \qquad (11.2)$$

Note again that there are two kinds of calories, one with a capital C that denotes the calories found in food and one with a lower-case c that is a factor of one thousand smaller and is the scientific measure; it is the latter calorie that is determined by Equation (11.2).

Power of Humans

The rate of energy conversion per unit time is called power and in human metabolism it is expressed in kilocalories per hour or Calories per hour. These units can be converted into joules per second or watts in the form

$$1 \text{ Cal/hr} = 1.16 \text{ watts} \qquad (11.3)$$

Brown and Brengelmann [26] describe the physiological "average" man as having a mass of 70 kg (154 pounds) and a surface area of 1.73 m². This man, while sitting quietly, converts energy at the approximate rate of 80 Cal/hr or has the same power as a 100 watt light bulb. They go on to say that one hundred of these average men sitting in a room will generate 10 kilowatts of heat, an amount sufficient to cause extreme discomfort in the absence of adequate ventilation. This might account for the discomfort many students experience in classrooms.

Question 11.1

Suppose a person exercising generates 20% more heat than when at rest. How much heat must be dissipated in a gym with ten people exercising strenuously for an hour?

11.1.1   Heat Capacity

Heat is Linear and Additive

Let us denote the quantity of heat by the symbol Q. Therefore, in Equation (11.1) the quantity of heat is given by the product of the mass m and the change in temperature ΔT, and can be expressed by the equation

$$Q = C_H m \Delta T \qquad (11.4)$$

Here C_H is an empirical parameter called the heat capacity, a quantity that characterizes how readily material transports heat from one point in space to another point in space. Equation (11.4) tells us that the quantity of heat in a substance is *additive* and *linear*. Double the mass of the object on the right-hand side of the equation and the amount of heat contained in the object represented by the left-hand side of the equation is doubled. The same is true if the change in temperature is doubled. This linear relation can be used to quantify the efficiency with which heat is absorbed by masses of different kinds. Different materials transport heat at different rates so that certain things, such as the apples in a hot apple pie, retain their heat much longer than do others, such as the pie crust. You bite into the piece of pie thinking it is cool because the crust is cool, and then burn the roof of your mouth on the apples themselves. This is because the C_H for apples is much less than the C_H for pie crust.

Water Is the Standard

The constant C_H in Equation (11.4) is a measure of the heat transport property and is also called the *specific heat*. The greater the heat capacity or specific heat, the greater is the ability of a substance to store internal energy. The standard or reference specific heat capacity is that of water. For water, C_H is defined to be unity. This is a useful standard since water has a much higher heat capacity than most other substances. For example, $C_H = 0.2$ for aluminum, so that for equal masses of water and aluminum, and equal temperature differences, water stores five times the internal energy in the vibrational degrees of freedom of its atoms and molecules than does aluminum. It is the oscillatory motion in these molecules that increases the internal energy of the system. Of course, energy can also be stored in the internal vibrations or rotations that do not increase the temperature. In general, a molecule absorbs energy in all three forms—translational, rotational, and vibrational—but only the average kinetic energy of translation is related to the temperature. This is the reason the beach is always hotter than the water, it takes much more energy to raise the temperature of the water by a given amount than it does to raise the temperature of the sand by the same amount, due to the difference in molecular structure.

Latent Heat

Another useful concept is that of latent heat, which means heat that lies hidden. Take, for example, the heat supplied to ice in order to melt it. As the ice is being transformed to water, the temperature does not rise at first. The specific heat of ice is $C_H = 0.5$ and the heat first goes into tearing apart the molecules in the ice crystals and not into the translational or vibrational motion of the molecules. Therefore, as the ice melts, the temperature remains the same, until all the crystal bonds are broken, or at least this is the case for ice that is being homogeneously heated. By the same token, when boiling water is converted into steam, even the weak attraction between the water molecules is overcome and the individual molecules escape from the fluid.

Phase Changes

From experiment, we know that one kilogram of ice requires 80 calories to melt at 0 °C, that is, to change phase from solid to liquid with no change in temperature. So it is possible for water to be either solid or liquid at a temperature of 0 °C. Similarly we can have either water or steam at 100 °C, but it requires an additional 540 calories to convert one kilogram of water into steam at this temperature. The reverse process, condensing steam to water, delivers a lot more heat to the environment than does cooling hot water. In both these processes, heat is extracted from the fluid—steam and hot water in the two examples, respectively—and delivered to the environment. Live steam gives up more heat than does hot water, as is evidenced by the fact that live steam burns the skin more severely than does boiling water. By the same token, live steam also heats apartment buildings more efficiently than does hot water.

Question 11.2

Explain why water is a better cooling agent than sand. Also discuss how the resistance of water to cooling, 1 calorie per degree for each gram of water that cools, can mitigate climate and weather. (Answer in 300 words or less.)

11.2 THE LAWS OF THERMO-DYNAMICS

Averages Yield Properties of Complex Phenomena

The properties of materials, whether gas, liquid, or solid, cannot be completely understood without examining the properties of the individual particles constituting the material. This is not to say that certain general properties that are independent of individual characteristics do not exist, for they indeed do. Some properties, like temperature, for example, are characteristic of large aggregates of particles and only emerge as average properties. It does not make sense to talk about the temperature of a single particle. When we say the temperature of the room is 72 °F, what we mean is that the temperature of the air in the room has this temperature, which, as we now know, is a measure of the average kinetic energy of the air molecules. We might express the kinetic energy of a single particle in terms of the ambient temperature, such as when we say a Brownian particle has an energy $k_B T$ in each degree of freedom. But the temperature T is a property of the ambient fluid, not of the Brownian particle.

Dynamics of the World Seen Large

We can understand the nature and behavior of aggregate properties of complex systems, such as pressure and temperature, phenomenologically without a detailed mechanistic understanding of the underlying particle dynamics. The discipline of thermodynamics was developed to understand the nature of matter and its influence on the world of our senses, without recourse to a microscopic model. It is fortunate for us that this is the case, since we still have not established the fundamental connections between the microscopic and macroscopic worlds. We understand some of it, but certainly not all.

Simple Empirical Laws

An example of a thermodynamic relationship is the ideal gas law, $PV = Nk_B T$, from which we can understand the inverse relationship between pressure and volume, and the direct relationship between pressure and temperature. We discussed the justification for this law using the kinetic theory of gases, in which we related the temperature to the average kinetic energy of the particles. Of course, we can also take the ideal gas law as a description of the general behavior of a certain class of gases and not concern ourselves with how the equation is derived. However, there are certain insights about the process that come from theory or a model of the process of interest. We know from the microscopic model that the ideal gas law is not dependent on the details of the collision process among the gas molecules. This is both the strength and weakness of thermodynamics. It encapsulates, within its four laws, the empirical knowledge of science for all of recorded history, but this generally does not answer the question of why.

The Four Laws of Thermodynamics

In bioenergetics, where we study energy transformation in living systems, we do not address the question of why, but only that of how. We shall, where appropriate, borrow from Brody's extensive, if somewhat dated, discussion of the generalization of thermodynamics to bioenergetics [22], but to start we need to emphasize that thermodynamics has four laws, not only the usual two that Brody, as well as others, quote and discuss. In shortened form these laws are:

Zeroth law: If two objects are in thermal equilibrium with a third object, they must be in thermal equilibrium with each other.

First law: Energy can be neither created nor destroyed; it can only be changed in form.

Second law: Heat cannot be spontaneously transferred from a colder body to a hotter body.

Third law: An isolated system becomes completely ordered as its temperature approaches absolute zero.

11.2.1 The Zeroth Law of Thermodynamics

Definition of Temperature

The zeroth law of thermodynamics is actually what allows us to define the temperature of a system, that is, it allows us to define a thermometer. What this means is that we can define an instrument that measures a single property of a system that behaves in the way we would expect temperature to behave. In a pedantic way, we consider a system Λ and focus on one property of the system, say β. The parameter β changes when we put Λ in thermal contact with other systems and all other properties of Λ are held constant. The parameter β, which is allowed to vary, we call the "thermometric parameter" of Λ. For a mercury bulb thermometer, for example, β is the height of the mercury in the glass tube. There are many other such devices.

Let Us Define Temperature More Carefully

Suppose we have two bodies A and B. We put the thermometer Λ in contact with body A and allow Λ to come to equilibrium, that is, the mercury in the bulb stops rising. We then do the same thing with object B. If Λ registers the same height in both cases, the two systems can be put in thermal contact with one another, without changing their state of equilibrium. Contrarily, if the readings on the thermometer Λ are not the same for bodies A and B, then when the two systems are put in thermal contact, neither one remains in equilibrium; they both change.

Anything Can Be Used to Measure Temperature

According to this general definition, that which we call a temperature can be a length, a pressure, or any other physical quantity that responds to the thermal state of a system. There is a more fundamental definition of an absolute temperature, having to do with the statistics of the microscopic degrees of freedom of a system, but we shall spare the reader this discussion.

11.2.2 The First Law of Thermodynamics

Conservation of Energy

The first law of thermodynamics is easy to articulate since we have encountered it already; it is the conservation of energy. It is not just the conservation of mechanical energy, however, but the conservation of all forms of energy, including heat. In this way, it is clear that if energy disappears from one location, it will, like the blocks in the child's room, reappear somewhere else, perhaps in another form. Note, however, that this law does not specify the rate at which things occur, but only that when everything is sorted out, nothing will be lost. This is true for both living and nonliving systems and as put by Brody [22],

. . . the energy equivalent of work performed by an animal, plus the maintenance energy of the animal, plus the heat increment of feeding must equal the energy generated from the oxidation of nutrients. This definition gives the first law a sense of universal finality, and a firm basis for bioenergetic investigations even if the mechanisms of the reactions are unknown.

Heat and Work

Thus, if one puts heat into a system, say an amount ΔQ, and, simultaneously, the system does work on the environment $(-\Delta w)$, then the internal energy of the system is increased by the quantity of heat injected and reduced by the amount of work done. If we denote the internal energy of the system by U, then the change in the internal energy, ΔU, made by the addition of heat and the doing of work is

$$\Delta U = \Delta Q - \Delta w \tag{11.5}$$

One way heat is introduced into a body is through food. The body's metabolism releases the heat content of the food by burning it. The body uses the energy to do work and stores what remains in the form of fat.

What this Energy Means

In the form of the first law given by Equation (11.5), we need not specify how the work is done by the system. It could be done by pushing against a piston and compressing the gas in a cylinder so that the applied force is $F = PA$, as we obtained before, where P is the pressure and A is the cross-sectional area of the piston. If Δx is how far the piston has moved, the work done by the gas is $\Delta w = F\Delta x$, and using ΔV for the change in volume $A\Delta x$, we obtain

$$\Delta w = P\Delta V \tag{11.6}$$

Equation (11.6) could also refer to the rhythmic motions of the heart pumping blood or the lungs forcing air in and out as the organs go about their business of keeping the body alive. In any event, this is called PV-work of the body on the environment.

Chemistry Can Also Do Work

The work in Equation (11.5) could also be done *on* the system rather than *by* the system, and in this way increase the internal energy rather than decrease it. This contribution could be made by changing the number of molecules in the system against a chemical potential μ, so that if ΔN is the change in the number of particles, the work done on the system is

$$\Delta w = -\mu \Delta N \tag{11.7}$$

The minus sign is used in Equation (11.7) because Δw is defined as the work done by the system; the work done on the system has the opposite sign. This process represents, for example, the oxidation of nutrients by the body, thereby releasing energy into the body. Thus, incorporating these two examples into Equation (11.5), we can write for the change in the internal energy of the system

$$\Delta U = \Delta Q + \mu \Delta N - P \Delta V \qquad (11.8)$$

In fact, the universality of Equation (11.5) means that we can introduce any and all other means of changing the internal energy of the system into the first law of thermodynamics.

Not All Energy Is Useful

One of the things we should note is that not all the energy obtained from food is available to do work. There is always some of this energy that is converted into heat, and we take this heat into account with the content of the second law of thermodynamics.

Question 11.3

The last term in Equation (11.8) is the energy lost by the system doing work on the environment, for example, in breathing or pumping blood. Discuss how one might include walking in the energy balance equation and suggest a functional form. (Answer in 300 words or less.)

11.2.3   The Second Law of Thermodynamics

Entropy and the Flow of Heat

The second law of thermodynamics is probably the most widely known law, after the conservation of energy, and is probably also the least well understood of the laws of physics. This law involves the much maligned and misused concept, *entropy*. Before discussing the concept of entropy, however, we can express the second law in a relatively simple form based on everyone's experience of the flow of heat. Primitive man must have noticed that the air around a fire is warmer than the air outside the cave and that cooked meat could not be eaten immediately from the fire, but must be allowed to cool. It could have taken more than a few tens of thousands of years for these mental giants to realize that in any closed system heat will spontaneously flow from a hot region to a cold region. A hot body can be cooled by emersing it in a stream of cold water and a cold body warmed by placing it over a camp fire. It is in this form that we express the second law.

Heat Flow Is Driven by Temperature Differences

Now we explore the implications of this form of the second law using some rather subtle reasoning. It is important to follow the train of argument for it is here that science lies, not just in the identification of a phenomenological truth such as heat always flowing from a region of higher temperature to a region of lower temperature. Note that this behavior is not unlike a fluid flowing from regions of high pressure to regions of low pressure. The transport of heat is due to the heterogeneity of temperature within the system; heat flows in the direction of the temperature gradient, just as particles move in diffusion due to concentration gradients.

Diffusion and Heat

We mentioned a number of times that the effect of friction is to convert usable energy into nonusable energy or heat. We have already discussed the fluctuation–dissipation relation of Einstein, an example of the conversion of kinetic energy into heat during the process of diffusion. The conversion occurs due to the imbalance of the collisions of the lighter fluid particles against the surface of the heavier Brownian particle. This imbalance of collisions may be visualized as millions of small impulses on the surface of the large particle, each one acting in a different direction, the sum of all the impulses producing a net force of random size and pointing in a random direction. The net effect of the collisions is to impart a kinetic energy to the Brownian particle, that is, to make the heavy particle move. In moving, the heavy particle pushes through the fluid, experiencing a resistance due to the fluid drag. This resistance is proportional to the speed of the heavy particle. This drag on the particle is the force of friction. Thus, the thermal fluctuations of the ambient fluid move the Brownian particle about in a random way and the average interaction of the heavy particle with the fluid particles attenuates this motion.

Covert Heat to Work?

Thus, if we inject the heavy particle with additional energy, the phenomenon of diffusion will convert this energy into heat at a constant temperature, due to the resistance of the fluid. Since there is so much more energy in the ambient fluid than in the diffusing particle, the conversion of the heavy particle's kinetic energy into heat does not significantly affect the temperature of the fluid. The question is whether the inverse process can be accomplished as easily, that is, the conversion of heat into work, at a constant temperature. The second law of thermodynamics, as it is stated above, implies that this reverse process cannot occur. Since there is no temperature difference in simple diffusion, heat cannot be made to do work. An equivalent statement of the second law is: Heat *cannot* be taken in at a certain temperature and converted into work with no change in the system or surroundings.

There Is No Free Lunch

This is the form of the second law put forth by the 19th century scientist/engineer Sadi Carnot. It is worth mentioning that the conservation of energy had not been established in its full generality at this time (1824) so the second law of thermodynamics was articulated before the first law. Carnot, being an engineer, formulated his arguments in terms of the workings of a steam engine. In a steam engine, heat flows from a hot region, the steam cylinder, to a cold region, the condenser. The heat flows from the high temperature, T_H, to the low temperature, T_L, at a rate that is determined by the temperature difference, $\Delta T = T_H - T_L$. Thus, the rate at which a steam engine can do work is determined by the same mechanism as the cooling of a cup of coffee; recall Newton's law of cooling. Note that this is a different mechanism from that of diffusion. It is fair, then, to represent the energy available to do work in terms of the temperature difference within the steam engine, ΔT.

Maximum Efficiency

The maximum efficiency of a steam engine, ε, can be expressed as the ratio of the temperature difference to the higher of the two temperatures:

$$\varepsilon = \frac{\Delta T}{T_{\mathrm{H}}} \tag{11.9}$$

which is the ratio of available energy to total energy when the temperature is given in degrees Kelvin. Recall that at absolute zero ($T_{\mathrm{L}} = 0°$) the average kinetic energy is also zero and there is no energy available to do work. Thus, if all the energy of a system could be made available to do work in the steam engine, then the efficiency would be unity, $\varepsilon = 1$. If half the total energy were converted, then ε would be 1/2, and so on. But, of course, this notion of efficiency assumes no losses due to mechanical friction, radiation, and other dissipative mechanisms. In the physical world, the efficiency of any process is much less than that stated by Equation (11.9), but this equation gives the ideal limit that cannot be surpassed in the absence of all losses.

Summary

In summary then, the first law of thermodynamics states "You cannot win," whereas the second law states "You cannot even break even."

11.2.4   Entropy

Order and Disorder

Entropy is one of those scientific terms that, from time to time, has captured the imagination of the general public. It has been used to describe the eventual "heat-death" of the universe, the devastation that exists at the periphery of civilized society, the organizing principle for biological and sociological evolution, the degree of memory in random time series, and on and on. Our task here is a rather less ambitious than these attempts to model general physical, social, and biological phenomena. We propose to give a broad outline of the general physical meaning of entropy and suggest the possible reason why it should be of interest to those studying biodynamics, that is to say, locomotion.

Available Energy

The concept of entropy grew out of thermodynamics, the motion of heat, and so its first interpretation was in terms of heat engines and the notion of a thermodynamic temperature. However, there is no reason to restrict the argument to heat engines. The argument works just as well with chemical, wind, human, or any other type of engine. If the idea of entropy did not work in a general setting, it would be a relatively impoverished physical concept.

Importance of Inhomogeneities

It is true that in heat engines, it is only possible to convert energy into work by means of a temperature difference. However, we know that batteries can do work when there is no temperature difference involved. Rather, there is a difference in electrical potential. Like the difference in gravitational potential, the difference in electrical potential allows us to do work, since the electrical potential is available for such work. However, we know that the efficiency of such processes is not 100%; in physical systems, some heat is always generated. We mentioned doing work by humans. What we actually mean is doing work by biochemical reactions, without changing the temperature of the system. The difference in chemical potential represents the available energy in this latter case. It is clear that we can continue to expand this idea of available energy to any and all physical systems, both living and nonliving.

Unavailable Energy

It is becoming apparent that the second law of thermodynamics must apply to electrical energy, chemical energy, and, indeed, energy of all forms, to be completely general, and not just to heat alone. It follows that regardless of the form in which energy exists, it must be nonuniformly distributed in the system of interest, which is to say, the energy must be more concentrated in one region of space than in another. In the case of heat, this difference in energy concentration is measured by the temperature, so one portion of the system has a higher temperature than another. For other forms of energy, there are other measures of energy intensity rather than temperature. It is the overall difference in energy intensity that determines the energy available to do work. When the energy density is uniform, there is no energy available for doing work. This uniform distribution of energy is referred to as the *unavailable energy,* since it is nonzero but cannot be used to do work. What drives a system is, therefore, not homogeneity but heterogeneity.

Temperature Difference Gives Available Energy

Let us now try to understand the second law of thermodynamics based on this more general perspective. We know that the temperature difference in a heat engine is a measure of the energy available to do work. Further, as time unfolds, heat in the closed system flows from the higher- to the lower-temperature regions, cooling the hot and heating the cold. Thus, the temperature difference decreases with time and so too does the available energy. From the first law of thermodynamics, we know that the total energy in a closed system is constant, so that as the available energy decreases, the percentage of unavailable energy must increase. So at very long times, all the energy of the closed system must become unavailable for work and the energy is therefore uniformly distributed in the system.

Open System

What happens when we open the system up to the environment and allow heat to enter the hot region to keep it from decreasing in temperature. We can also pump heat out of the cooler region to keep it from increasing in temperature. Of course, this cannot be done in an energy-free way; it requires energy to maintain the temperature of the hot and cold regions at a constant difference. Thus, it requires the input of energy to maintain a system at a constant available energy. We see that the total energy of the system increases in order to keep the available energy constant, so that the unavailable energy increases even while the available energy is being held constant.

Waste Enegy Increases

Regardless of how we argue the situation, in the case of a heat engine, the energy that is unavailable to do work increases with time. Further, what applies to heat engines ought to apply to all devices that convert energy to work, alive or dead. One can repeat the above arguments, replacing temperature differences with potential differences for a battery or a hydroelectric dam, and thereby come to the same conclusions for very different kinds of systems. In this way, we can say that the unavailable energy in any real system increases with time.

Back to the Steam Engine

In 1850, the German physicist Rudolf Julius Emanuel Clausius coined the word entropy to identify the measure of the unavailable energy in a system, that is, the energy that is unavailable to do work. The argument he used was, of course, based on steam engines and the absolute thermodynamic temperature. The idea is that in a single reversible engine running between two temperatures T_1 and T_2, where Q_1 is the heat absorbed at the temperature T_1 and Q_2 is the heat delivered at the temperature T_2, the ratio of the heats to the temperatures is the same for both and is denoted by S:

$$S = \frac{Q_1}{T_1} = \frac{Q_2}{T_2} \tag{11.10}$$

S is the entropy of the system, and if the engine is reversible, this relation between the heats and temperatures must follow. However, the entropy of the system can be defined even when Equation (11.10) is no longer valid, that is, when the engine has some dissipation and is therefore no longer reversible. The units of entropy are calories (units of heat) per degree centigrade (units of temperature). We can say that the entropy of a system may remain constant under ideal conditions, such as perfect insulation of the system, but it always increases in time under actual physical conditions. An equivalent expression of the second law of thermodynamics is that entropy either increases or remains the same in a physical system, so that

$$S \geq \frac{Q}{T} \tag{11.11}$$

Free Energy Number One

Carnot, again using a loss-free, reversible, ideal heat engine, was able to show that the theoretical maximum energy available to do work is

$$\Delta w = Q \frac{\Delta T}{T} \qquad (11.12)$$

where the ratio of the temperature difference, to the temperature, is the efficiency, ε, just discussed. Further, using the definition of entropy [Equation (11.10)], we can write

$$\Delta w = S \Delta T \qquad (11.13)$$

thereby expressing the work done by the system in terms of the entropy. However, living systems do not do work in this way, that is, by means of temperature differences. Human bodies work isothermally, or without a change in temperature. The heat produced by the body is an end product, rather than the motive power for the work the body does.

Free Energy Number Two

Thus, rather than Equation (11.13), the theoretical maximum amount of work available from an isothermal (constant temperature) chemical reaction is

$$\Delta F = \Delta U - T \Delta S \qquad (11.14)$$

where again ΔU is the change in internal energy, ΔS is the change in entropy, and ΔF is the Helmholtz free energy of the reaction. Brody [22] points out that the Gibbs' free energy in this case can be written as

$$\Delta F = \Delta H - T \Delta S \qquad (11.15)$$

where ΔH is the change in heat content, that is, the *enthalpy*, or heat reaction, determined by direct calorimetry. In both free-energy equations, $T \Delta S$ is the energy in the system that is not convertible to work.

A Little More Chemistry

The free energy is a measure of the driving force of the given reaction, that is, a measure of the chemical affinity. If $\Delta F < 0$, the reaction is exothermic, meaning that energy is given off in the reaction, which, when initiated, takes place vigorously and completely. If $\Delta F > 0$, the chemical reaction is endothermic and will not take place unless supplied with energy from external sources. If $\Delta F = 0$, the system is in equilibrium and no reaction occurs. You must bear in mind that the body can only obtain energy by which to sustain itself by means of chemical reactions. Even on the sunniest day, the body cannot use radiant energy to fuel itself.

Entropy-Driven Reactions

Physical and chemical reactions proceed spontaneously only from states of high organization to states of low organization, which is to say, from states of lower entropy to states of higher entropy. The change in the free energy defined by Equation

(11.15) for any chemical system is inversely related to the change in the entropy. A system's free energy is generally equated with its ability to do work. A highly organized system is capable of performing more work than a disorganized, or minimally organized, system, because a highly organized system is capable of greater changes in entropy than a less organized system. Thus, the driving force in the chemcial reaction is toward an increase in entropy.

The Cost of Free Energy

Brown and Brengelmann [26] explain that even if the maximum amount of free energy is extracted from the food we eat, when it is metabolized, some heat must be produced. However, the fraction of energy that goes into heat is typically not very large; for example, in the metabolizing of glucose, the $T\Delta S$ term in Equation (11.14) accounts for about 5% of the total energy. This would be the case if the conversion of food energy into high-energy biochemical compounds like adenosine triphosphate (ATP) were efficient, but it is not. It is true that a large percentage of the ingested energy is potentially available as free energy, perhaps as high as 95%. It is the intermediate stage, where the food energy is converted into the chemical form in which the energy can be used by the body, that is inefficient. This process can be, for example, the synthesizing of the high-energy phosphate bonds of ATP using the free energy of food. Recall that ATP, by releasing the energy from its bonds, can produce muscle contraction. It can also activate sodium transport. The complete oxidative metabolism of 1 mol of glucose whose free energy is 686 Cal/mol produces only 38 mols of ATP, in which the free energy of hydrolysis is about 9 Cal/mol. Thus, nearly 50% of the potential free energy goes into the generation of heat, which the body must eventually transmit to the environment. This is a far cry from the 5% in the basic chemical reaction.

Question 11.4

Consider the reaction converting glucose into ATP, as discussed above. Using the numbers supplied discuss the reaction using the second law. (Answer in 300 words or less.)

Entropy Increase Means More Disorder

It should be noted that the arguments presented up to this point mostly concern closed systems. These kinds of arguments have lead to some confusion over the uses of entropy. Take, for example, the seminal work of Schrödinger on the physical basis of life [20]. In this classic work, Schrödinger, one of the founders of quantum mechanics, introduced the concept of negentropy, or negative entropy, for biological systems. This was in recognition of the fact that entropy actually decreases, rather than increases, for biological systems. Another way to say this is that a biological system is ordered (it is alive) and remains ordered for a long time (until it dies). The fight to retain order, in the face of the world's attempts to destroy that order, is a metaphor for life itself. So the question arises: How do the ordered biological systems, with their lowering of entropy, remain consistent with the second law of thermodynamics and the need for entropy to increase?

Open Systems

The key to biology's apparent violations of the second law of thermodynamics is the fact that a biological system is not closed. A system is thermally closed when it is thermally insulated from the environment, so that its temperature does not change. Thermal insulation means the system does not exchange heat with the environment. A completely closed system does not exchange mass or energy with the environment either. On the other hand, in an open system, there is an exchange of stuff with the environment. There is an energy flux through the biological system—food and air coming in; heat, breath, and excrement going out. The waste products that humans give up to the environment increase the entropy of the environment while working to lower or keep constant the body's entropy.

Violations of the Second Law

The traditional way of resolving biology's apparent violation of the second law is to expand the system of interest to include its immediate environment and to then treat the enlarged system as if it were closed. A physical example of this is the refrigerator. Heat is continually being pumped from the cold objects within the refrigerator to the warm atmosphere outside the refrigerator, in apparent disregard of the second law. A warm object, placed within the refrigerator, cools down; therefore, the available energy, as measured by the temperature difference between the ambient air in the room and the object within the refrigerator, increases. However, objects within the refrigerator do not spontaneously cool down. The cooling takes place only because a motor is working within the refrigerator. Although the entropy of the refrigerator's interior is decreasing, the entropy of the motor is increasing. It can be shown that the increase in the motor's entropy exceeds the decrease in the entropy of the refrigerator's interior, so that the net entropy—that of the refrigerator interior, plus the motor, plus the air in the room—increases, in full accord with the second law.

Ever-larger Systems

In this way of expanding the notion of closed systems to large-scale, planet-wide, processes, we force a decrease in what we define to be the entropy. Examples of such entropy-decreasing phenomena are the uneven heating of the atmosphere, which gives rise to wind and weather; the lifting of uncounted tons of water miles high against the pull of gravity, which gives rise to rain and rivers; and the conversion by green plants of carbon dioxide in the atmosphere to complicated organic compounds, which is the basis of the earth's food generating processes, and of its coal and oil as well. However, none of these phenomena can be considered in isolation, for they are initiated by the solar energy reaching the earth. It is the solar energy that heats the atmosphere unevenly, evaporates water, and is the driving force for the photosynthetic activity of green plants. When the entropy increase of the fusion reactions within the sun is compared with the entropy decrease produced here on earth by solar radiation, the latter increase is almost negligible. Thus, the enlarged system, earth plus sun, obeys the second law.

Entropy Always Increases

Therefore, if we continue to extend our definition of a system to include all the processes that affect the evolution of the system, it turns out that the net entropy always increases. If we detect an entropy decrease, such as in a biological system, it is invariably because we are focusing on a part of the system rather than on the system as a whole. In fact, due to the interactive nature and interdependence of phenomena, we can only be sure that the system is isolated if we consider the entire universe. In terms of the universe, we can, along with Clausius, express the laws of thermodynamics with utmost generality as:

1. The total energy of the universe is constant.
2. The total entropy of the universe is continuously increasing.

Available energy decreases with time

From these two laws, we may make a prediction. If the universe is finite in size, then it must contain a finite amount of energy. This energy is constant and only a fraction of it is available to do work. As time unfolds, more and more of this energy becomes unavailable for work. Eventually, all the energy in the universe is unavailable for work, entropy is at a maximum, and the universe has "run down." As T. S. Elliot put it:

> . . . this is the way the world ends,
> this is the way the world ends,
> this is the way the world ends,
> not with a bang,
> but a whimper.

11.2.5   The Third Law of Thermodynamics

Order Varies Inversely with Temperature

So now that we have determined what will happen to the universe at the end of time, let us consider another small detail and that is what happens to all matter at absolute zero. The zeroth law establishes that we can define a thermodynamic temperature. The second law defines the absolute temperature and entropy. From the point of view of statistics, the entropy of a physical system is a measure of that system's disorder. The content of the third law is the trend of all physical systems to increase their order as the temperature is lowered. This increase in order actually means that fewer and fewer microscopic degrees of freedom are available for excitation as the temperature is lowered. The expectation is that all systems will be dominated by quantum mechanics as the temperature approaches absolute zero.

A Frozen Body

Examples of the macroscopic coherent motion that arises near absolute zero are superconductivity and superfluidity. Of course, none of this has anything to do with the human body, but it might be of interest for biodynamics in a broad sense. People interested in having their bodies frozen and revived at some future date, say after cancer is cured, have more than a passing interest in this topic.

Dependence of Metabolism on Weight

Mature animals of different species show a remarkable consistency in their basal metabolic rate, that being a dependence on weight given by $W^{0.73}$. This was depicted in the mouse-to-elephant curve, Figure 2.1, covering the range of body weight from 0.02 to 4,000 kg. Based on experiments involving cows, horses, humans, goats, and rats, Brody [22] argues that one would not expect this type of consistency during rapid growth of the animal for a number of reasons. First of all, thermodynamically, there is a work-energy cost of growth that is not separable from the basal metabolic rate. This cost increases with advancing age and increasing growth rate. Second, in the prenatal stage, the individual does not use thermoregulatory mechanisms; subsequently, it requires some time to develop their function after birth. Finally, the neuroendocrine system, which has the function of thermoregulatory control of the metabolic rate, does not operate optimally until relatively late in life. One example is the sex endocrines, adrenals, and thyroid, all of which influence energy metabolism, and do not mature until after puberty.

11.3.1   What Is Metabolism?

Transformation of Energy

Metabolism is the sum total of chemical activity occurring within a living organism. The metabolic rate is the amount of energy per unit time that is used to accomplish the transformation generated by this chemical activity. The energy a person uses per unit time is considered to be their metabolic rate, that is, the rate at which the body transforms free energy into waste products, and, in so doing, prolongs life, which is the work of living systems.

Life and Equilibrium

The living state referred to here is difficult, if not impossible, to define in any definitive way. We typically define life by what it is not; for example, it is not an equilibrium state. Then we list a number of properties that distinguish life from a state of equilibrium. As we mentioned earlier, the human body is open to the environment and operates through a number of imbalances with the environment, including chemical concentrations, temperature, and pressure. It is by means of these differences that the body expels mass and energy in the form of heat and other waste products, all of which must be replenished through the metabolizing of food.

Hess's Law

The energy available to the body from food is completely independent of the particular chemical pathways by which the chemical reactions proceed. This is known as Hess's law, formulated in 1840, which states that the heat delivered or absorbed in a reaction is independent of the manner in which the reaction takes place. Therefore, it does not matter if the reaction takes place in the human body or in a test tube, the energy change is the same. Thus, we can determine the caloric content of the food we eat by burning it, much as Huxley did with his larval mealworms in Figure 2.7.

Suppose the oxidation of glucose is given by a certain stoichiometric relation:

$$C_6H_{12}O_6 + 6O_2 = 6CO_2 + 6H_2O + 686 \text{ Cal} \qquad (11.16)$$

$$\underset{180 \text{ gm}}{} \quad \underset{134.4 \text{ L}}{} \quad \underset{134.4 \text{L}}{} \quad \underset{108 \text{ mL}}{}$$

From such a relation, it may be calculated that ingesting and absorbing 1 gm of glucose increases the body's free energy content by 3.81 Cal (686 Cal/180 gm = 3.81 Cal/gm). We should always keep in mind that these calculations provide the total energy that would be available if the processes within the body were 100% efficient, which, of course, they are not.

Photosynthesis and Respiration

The two most important chemical processes in biology are, arguably, photosynthesis and respiration.* Photosynthesis is the process by which energy is transformed from sunlight into the chemical energy in green plants. Respiration is the process by which the energy stored in food is released in the body to maintain life. The photosynthesis reaction is an endergonic one in which green plants combine solar energy, carbon dioxide, and water to produce, for example, the food molecule glucose. The energy level of glucose is higher than those of the reactants, which is to say, it is a highly organized molecule. Respiration is an exergonic reaction and involves the breakdown of glucose or other fuel molecules to release the stored potential energy and produce the end products of carbon dioxide and water [80]. The breakdown of glucose is indicated in the overly simplified expression in Equation (11.16); it is simplified since there are a large number of reversible biochemical reactions that take place between the left-hand and right-hand sides of the equation that are not indicated here. In respiring cells, however, the carbon dioxide diffuses through the cell membrane and is removed. The removal of CO_2 from the reaction site makes the overall reaction proceed inexorably to the right, even though the individual biochemical reactions not shown are reversible.

Biochemical Mechanisms

From another point of view, we perceive the human body to be a reaction vessel and ignore the biochemical mechanisms employed. For example, it is possible to determine the oxygen consumption necessary to burn 1 mol of glucose, the corresponding volume of carbon dioxide, and so on, using only the simple reaction relations of chemistry. Brown and Brengelmann [26] explain that the biochemical mechanisms are very important in determining how the energy is used and the fraction of that free energy that goes into heat, but not in determining the useful energy content of food. Recall, from our discussion of thermodynamics, that useful energy is the energy available for work. We know from thermodynamics that the energy content of food

---

*Respiration is not to be confused with breathing. Breathing is the process by which we take oxygen from the environment into our lungs to that it can be carried to the individual cells. Respiration is the chemical process that occurs within these cells and is responsible for releasing energy to carry out metabolic processes [22].

is entirely determined by the initial and final products of energy intake. Therefore, for our purposes here we do not need to discuss the biochemical mechanisms in detail.

Typical Calories

The energy contents of different carbohydrates are sufficiently similar to permit one number to be used for all such foods. The carbohydrates in our diet typically have 4 Cal/gm. Recall that this is 4 kilocal/gm using the physical unit of heat. Proteins have essentially the same energy content on average as carbohydrates. The amino acids of the proteins are not all oxidized, however; some are incorporated into cellular constituents, hormones, and so on. Lipids are the most "energy dense" of common foods. These fats and oils have a net dietary mean value of 9 Cal/gm.

Equation for the Metabolic Rate

These values can be used to determine the energy required to simply maintain the body in a supine state. Calder [25] mentions that one method is to measure the intake of food and the output of feces and urine. Samples of the food and waste are then analyzed in a bomb colorimeter to determine their respective energy contents. In this way, the simple equation for the metabolic rate, the energy required per day, is

$$\frac{\text{gm food}}{\text{day}} \frac{\text{Cal}}{\text{gm food}} - \frac{\text{gm waste}}{\text{day}} \frac{\text{Cal}}{\text{gm waste}} = \frac{\text{energy required}}{\text{day}} \quad (11.17)$$

where we essentially use the above energy values for carbohydrates and proteins. This equation is only accurate if the animal is not gaining or losing fat. This is not the technique for measuring the metabolic rate, which we shall discuss in detail.

Different Rates for Different People

The basal metabolic rate is a theoretical quantity that is supposed to capture the rate at which the body uses energy while at rest. Therefore, any amount of activity increases the metabolic rate above this basal level. The further implication is that the greater the level of activity, the greater the expenditure of energy and the higher the metabolic rate. Figure 11.1 shows the calories expended by people in various vocations. What is evident from the figure is that as the level of activity increases, so does the metabolic rate, as one would expect.

11.3.2 Measuring the Metabolic Rate

Measuring the Energy Content of Food

The energy value of different foods given above is determined in vitro (outside the body) using what is called a bomb calorimeter. This is a sealed container filled with pure oxygen to ensure complete combustion. A known quantity of food is placed inside the chamber, oxygen is pumped in, and the food is ignited electrically. The heat released during oxidation is then used to determine the caloric value of the food. Note that the efficiency of the bomb calorimeter is greater than that of the human body, so not all the energy measured using this instrument is used by the body.

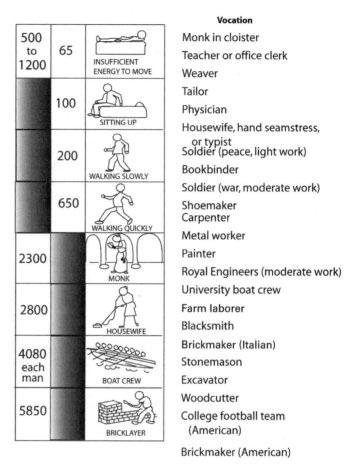

			Vocation
500 to 1200	65	INSUFFICIENT ENERGY TO MOVE	Monk in cloister
			Teacher or office clerk
			Weaver
	100	SITTING UP	Tailor
			Physician
			Housewife, hand seamstress, or typist
	200	WALKING SLOWLY	Soldier (peace, light work)
			Bookbinder
	650	WALKING QUICKLY	Soldier (war, moderate work)
			Shoemaker
			Carpenter
2300		MONK	Metal worker
			Painter
			Royal Engineers (moderate work)
2800		HOUSEWIFE	University boat crew
			Farm laborer
			Blacksmith
4080 each man		BOAT CREW	Brickmaker (Italian)
			Stonemason
			Excavator
5850		BRICKLAYER	Woodcutter
			College football team (American)
			Brickmaker (American)

Figure 11.1. Various ways to burn calories. The more work done per unit time the greater the metabolic rate. (Redrawn from Bourne [81].)

Some of the energy is lost in the form of waste products, such as uric acid, ammonia, and so on, all of which are passed through the body. These are the waste products being subtracted in Equation (11.17). Other substances, such as cellulose, contribute to the energy measured by the calorimeter, but since it cannot be digested, it has no energy value for the body.

Direct Measurement of Heat

Direct measurement of the metabolic rate is achieved by means of a human calorimeter. This does not mean that a human being is burned in a chamber, but the heat generated by a person in an insulated chamber is measured very precisely. The energy balance is determined so that the total energy loss from the chamber is equal to the subject's rate of energy utilization. However, we shall not explore the details of this technique here.

Only One Technique

There are a number of techniques for measuring the metabolic rate of humans, but since this is not a book on techniques, we shall only describe the most popular one.

It turns out that this technique does not measure the metabolic rate directly but, instead, measures the difference between the input and output energies to a person's body in a given time interval. Knowing this energy difference and the time, we use the law of conservation of energy to deduce how much energy has been metabolized and thereby retained by the body.

Indirect Measurement of Metabolism

The indirect method of determining the metabolic rate requires the measurement of oxygen utilization and/or the production of carbon dioxide. This gas measurement approach takes advantage of the fact that in the steady state the metabolizing of food, involving the release of energy, is associated with the consumption of oxygen and the corresponding production of carbon dioxide; see Equation (11.16). According to Brown and Brenglemann, this method uses the fact that [26] "... a subject at rest, in a steady state, post-absorptive, oxidizing an "average" diet generates about 4.83 Kcal for every liter of oxygen he consumes."

Use Oxygen Consumption

Brown and Brenglemann [26] go on to say that the Calories per liter of O_2 are relatively insensitive to the type of food being digested. Thus, the metabolic rate can be determined from the rate of oxygen consumption using the formula

$$\text{metabolic rate} = 4.83 \, \dot{V}_{O_2} \qquad (11.18)$$

Therefore, the metabolic rate is directly proportional to the O_2 flow volume. How do we measure the O_2 from volume $\dot{V}_{O_2}$? The decrease in oxygen concentration between the inspired and expired gas, the oxygen extraction, at a given gas flow rate is a measure of the oxygen consumption rate. In this procedure, a person breathes in the ambient air and exhales into a large bag. The expired air is collected in this bag, and the composition of the air in the bag and the air in the room are determined. From these values, the oxygen consumption is calculated. Using the total time of collection, an average consumption rate can also be determined.

Gas is Perfect

The details of this procedure, as well as other methods of determining the rate of oxygen consumption, are given by Brown and Brengelmann [26]. We merely point out that in the metabolic investigations the measured gases are referred to a standard state, which is the gas at 0 °C temperature, 760 mmHg pressure, and dry (standard temperature, pressure, and density, or STPD), in which state one mole of gas occupies a volume of 22.4 liters. This relation is important because it is the number of moles of the gas that is used in chemical reaction equations and not the volume of the gas. The perfect gas law is used to transform the data from the physiological state in the lungs to STPD. The perfect gas law can be used because, in the physiologic range, gases behave effectively as ideal gases. The moles and volumes of gases are directly proportional to one another, and the proportionality constant depends on the state of the gas, which can be determined from the pressure–volume–temperature relation of the perfect gas law.

Thermodynamics Has No Theoretical Basis

A handful of scientists in the nineteenth century, concerned with making the principles of mechanics compatible with thermodynamics, attempted, in vain, to derive the latter from the former. Every attempt to use the atomic theory in conjunction with the laws of mechanics to derive the first and second laws of thermodynamics failed. The thermodynamics laws could only be derived through the introduction of some form of randomness into the physical models.

Knowledge Is Incomplete

The resolution of the paradox associated with introducing randomness into the deterministic equations of physics turned on the notion that man's knowledge is of an imperfect kind. It was believed, following Laplace, that if a person had knowledge of all the conditions for all the particles in the universe at one instant of time, then that person, using Newton's laws, would be able to predict the future of the universe forever. Of course, given the limited intellect of mortals, this cannot be done. Therefore, we make predictions under conditions of imperfect, that is, incomplete, knowledge, and these predictions are flawed. Scientists, therefore, abandoned their willingness to know everything in principle for the view that the universe is deterministic, very complicated, and therefore, in principle, unknowable. Only God can know everything, but the scientists took succor in the belief that the universe could be known by some entity.

The Four Laws of Thermodynamics

The zeroth law states that two bodies at different temperature will, when put in contact with one another, come to equilibrium at a temperature between the two initial temperatures. This is the thermodynamic version of the observation that when you work out and get heated up, the surrounding air will absorb the heat from your body and cool you down. The first law states that energy cannot be created or destroyed; it can only change its form. This is realized by anyone that has had to diet. The energy from the food you eat will be stored in fat, unless that energy is used to generate heat during exercise. The second law states that entropy either increases or remains the same in a thermally isolated system. The fact is, things run down and, eventually, dissipation wins out over organization. The third law states that all thermal motion ceases at absolute zero.

Metabolic Rate

The energy per unit time used by a person is considered to be his or her metabolic rate, the rate at which the body transforms free energy into useful work and waste products. Respiration is the process by which the energy stored in food, for example, the highly organized molecule of glucose, is released into the body to maintain life. The breakdown of glucose and other fuel molecules releases the stored potential energy and produces the end products of carbon dioxide and water. The CO_2 diffuses from the reaction site through the cell membrane and is removed by the blood.

CHAPTER 12

Fluids at Rest

Objectives of Chapter Twelve

- Learn the physical properties of fluids at rest, such as the fact that fluids can resist motion in a predictable way.
- Learn that fluids have different laws of motion than do particles and apply the principles of Archimedes (fluid displacement), Bernoulli (energy density in a fluid), and Pascal (fluid pressure).
- Understand how the motion of a body relates to the fluid environment, for example, the dependence of energy loss on the speed of a swimmer in water.
- Learn about hydrostatic pressure and what it means for physiology in such things as high and low blood pressure.
- See what a soap bubble and an artery have in common.

Introduction

New Laws Are Required for Fluids

We leave behind the particle picture of mechanics and biomechanics in this chapter and plunge into the world of the continuum. We examine the behavior of vast aggregates of particles acting as an unbroken fluid, in which the identity of individual particles is lost. The laws of fluid motion are, of course, consistent with the laws of particle motion, but they are not the same laws. One of the goals of this chapter is to use the static properties of fluids in the physical world to better understand the workings of the biological world.

Biodynamics: Why the Wirewalker Doesn't Fall. By Bruce J. West and Lori A. Griffin
ISBN 0-471-34619-5 © 2004 John Wiley & Sons, Inc.

Elusive Phenomena

Most of what we understand about the world is based on our interactions with the solid stuff around us. It is probably safe to say that we understand the properties of a piece of wood more completely than we do those of a wisp of smoke. Wood is tangible. We can see the soft, colored fibers that make up the wood's grain, we smell the acrid odor of its sap, taste its sweetness as we chew it, and feel its firm resistance to our touch. Our five senses make solid objects accessible to us and then, on top of all that, we have the direct application of the laws of mechanics and thermodynamics that seem to give these properties an objective reality, independent of our senses. But what about smoke, water, and other things that flow, change shape, and never appear to stay the same? How do we understand these more elusive phenomena?

The Third State of Matter

It is not that smoke has less of a reality than a piece of wood, or that smoke has significantly fewer dynamical variables for its description than does a piece of wood. The difference between the two is that the particles in the wood, by and large, are held in fixed positions relative to one another, whereas smoke consists mostly of empty space and particles that freely move around. Therefore, wood holds together and smoke thins out, eventually dissipating altogether. Liquids, the third state of matter, not being solid and not being gas, contain particles that neither retain their position relative to one another, nor move around freely. It is primarily the dynamics of liquids, for example, blood and water, that we address in this chapter.

Fluid Flows

In this chapter, we discuss the stationary properties of fluids; we investigate their motion in the next chapter. One of the first things we have to understand is that fluids are not just liquids, but are, in fact, anything that flows. Therefore, liquids, gases, and polymers are all examples of fluids. Our primary concern here is liquids, even though much of what we say about fluids and their motion is also valid for more complex media such as glasses. We should mention that, contrary to common belief, glasses are fluids and not solids; they just flow very slowly. Another contrast is the difference in the rate of flow of blood and water. As everyone knows, blood is thicker than water; therefore, it flows more slowly.

What Is a Fluid?

A fluid differs from a solid in that the individual particles are not tied to a particular spatial location. In a solid like table salt, for example, the two kinds of atoms— sodium (Na) and chlorine (Cl)—are arranged in a repeating spatial pattern in which each atom is enclosed by six atoms of the opposite kind. The structure of table salt (NaCl) has just as many Na atoms as it does Cl atoms, but there is no salt molecule. It is the entire structure that constitutes table salt. We introduce this example to contrast its rigid structure with that of water. Unlike table salt, water is made up of molecules consisting of two hydrogen atoms and one oxygen atom (H_2O). The hydrogen and oxygen atoms are strongly bound together to form the water molecule, but water molecules only weakly interact with one another. The strength of the interactions between water molecules is, of course, temperature dependent. If the temperature is above 100 °C, the water is a gas. If the temperature is below 0 °C, the water is a solid. For the moment we are only interested in water in the temperature interval 0 °C < T < 100 °C, at which it is a liquid at sea level.

Transfer of Momentum

The clear liquid in a cylindrical glass of water consists of many millions of H_2O molecules flying around inside the glass, occasionally colliding with one another, but much more often colliding with the sides of the glass. Molecular collisions with the sides of the glass are elastic, so, using the conservation of momentum, we find that the component of a molecule's momentum perpendicular to the glass during a collision merely changes sign, that is, its direction changes but not its magnitude. Therefore, elastic collisions do not change the energy content of the water. There is no change of energy content when the water, glass, and ambient air are all at the same temperature. You might recall the argument from the analysis of the gas in a cylinder. If the water in the glass has a higher temperature than the ambient air, it will transfer momentum to the sides of the glass, thereby heating it up. The molecules of ambient air will collide with the hotter glass, absorb momentum, and take that added heat away, thereby cooling the glass and, indirectly, the water as well.

Evaporation Is the Loss of Water

But what happens at the top of the glass where the water is open to the air? The water molecules actually pop out of the surface if they are going sufficiently fast. But they must overcome surface tension in order to escape. Surface tension is a consequence of the cohesive nature of the water molecules and those molecules on the surface have fewer water molecules with which to interact. We shall discuss this phenomenon subsequently, but for the moment we note that when a water molecule escapes, the water in the glass loses energy. This is the process of evaporation.

Condensation Is the Gain of Water

Evaporation is not all that happens to the water in the glass. There will be many water molecules diving from the air into the water as others leave. When the number of molecules escaping from the surface exceeds the number entering the surface, we have evaporation. On the other hand, when the number of water molecules exiting the surface is less than the number entering the surface, we have condensation. Condensation is most commonly seen on the sides of the glass, where the water vapor from the air imparts its momentum to the glass and the water molecules from the air remain stuck there, build up, and form water droplets. Finally, when, on average, the two numbers, those particles leaving and those particles entering the water in the glass, are the same, we have equilibrium and the temperature of the water in the glass no longer changes.

Inside and Outside the Body

We want to understand fluids from the perspective of a swimmer and how the properties of water determine the forces acting on the locomotion of a person in water.

That is one aspect of biodynamics. We also want to understand how blood, a fluid different from water, flows in the elastic pipes we call veins and arteries. Further, we seek to understand how the characteristics of blood affects the exchange of O_2 and CO_2 at the interface of the venous and arterial systems. Finally, we investigate how the properties of gases influence the exchange processes that occur in the lungs. These physiological processes constitute biofluidics.

12.1 WHEN MECHANICS IS NOT ENOUGH

Newton's Laws Must be Modified

Fluid mechanics or the mechanics of fluids is the study of the physical properties of fluids, using the same physical laws that we used in studying the mechanics of material bodies: the application of Newton's laws to the billions of particles that make up the motion of a mountain stream, the water from the faucet, and the chlorinated contents of the swimming pool. This may seem hard to accept at first, but the perspective is that, given the manner in which particles interact with one another in a fluid, Newton's laws, and the appropriate averages, we can obtain equations of motion for the physical observables of the fluid. Typically, the physical observable is the fluid velocity or fluid momentum. However, there are other physical observables, such as the fluid pressure, in which we are interested. There are one or two problems with this picture, such as the fact that we still do not understand turbulent fluid flow, but we shall not concern ourselves with these limitations here.

12.1.1 Pascal's Principle

Pressure Is Force per Unit Area

Remember that we have already defined pressure using the momentum imparted by the particles of a gas to the walls of its container. We were able to show that pressure is the ratio of force F perpendicular to the area A,

$$\text{pressure} = \frac{\text{force}}{\text{area}} = \frac{F}{A} \tag{12.1}$$

We now extend this definition to include fluids of all kinds and not just gases. Pascal was the first to enunciate one of the general properties of pressure in fluids. He observed that in a closed container of fluid, the pressure at a given depth is the same everywhere in the fluid. More generally, the water pressure one foot under the surface of the ocean and the pressure one foot under the surface in a bucket of water at the beach are exactly equal. Further, the pressure exerted at a point in the fluid is the same in all directions; consequently, the pressure on the walls of the container is perpendicular to the walls.

Why Is Only Depth Important?

Let us examine why the pressure depends only on the depth of the water. The force acting at a depth s over an area A is determined by the weight of the water above, say, the location of a diver at that depth:

$$\text{force} = \text{mass} \times \text{acceleration} = \rho V g \tag{12.2}$$

where ρ is the mass density of the water and V is the volume of the water column above the diver, so that the total mass above the diver is ρV. The volume in Equation (12.2) is the product of the area at the bottom of a column of water of depth, s, that is, $V = As$, and g, of course, is the acceleration of gravity. We note that we have explicitly referred to ρ as a mass density and not just a density. The reason for this is that we shall encounter a number of densities in our discussion and we do not want to confuse them with one another. For example, we define the weight density of the water as

$$\text{weight density} = \rho g \tag{12.3}$$

where we have removed the dependence on any particular volume by using the density of the water rather than the mass. The weight density can also be referred to as a force density, since weight is a force. If we substitute Equation (12.2) into Equation (12.1), we obtain for the pressure

$$\text{pressure} = \frac{\rho V g}{A} = \text{weight} - \text{density} \times \text{distance} \tag{12.4}$$

using Equation (12.3). Therefore, the pressure is the same at all points of a given depth, depending only on the density of the water and its depth. Increase the depth by a factor of two and pressure is doubled; increase the density by a factor of three for a given depth and the pressure is tripled. But increase the horizontal dimension (increase the area) by any amount and the pressure on the diver remains the same. Thus, the pressure one foot below the surface in your bathtub is the same as that one foot below the surface of the Atlantic Ocean; only the depth of the fluid matters, not the horizontal area. This statement, of course, assumes that the water density in your bathtub and in the Atlantic ocean is the same. A note of caution, however: for very great depths, the density of the fluid may change, so the pressure in Equation (12.4) would also change due to the change in density.

Air Pressure Also Depends on Depth

The observation of Pascal regarding the dependence of pressure on depth is also true of the biosphere of the earth. We live in an ocean of air that is miles deep, with the weight of all that air pressing down on us. The pressure of that air at sea level is 14.7 pounds per square inch and is equivalent, in terms of pressure, to living under a column of water ten meters high. Thus, we have this pressure on every square inch of our body and the only reason that it goes unnoticed is that the pressure on the inside and the outside of our body is the same. The air in our lungs, the blood in our veins, the living tissue of our bodies that is essentially a thick, viscous fluid, is all at one atmosphere pressure. Since the pressure on the inside of the human body is the same as that on the outside of the human body, the net pressure, that being the difference in the pressure inside and outside of the body, is zero, and we are, therefore, unaware of the weight of the air during normal activity.

The Barometer Measures Air Pressure

The Italian physicist Evangelista Torricelli in 1644 invented an instrument to measure air pressure. He took a long tube closed at one end and open at the other and filled it with mercury (Hg). He then inverted the tube in a dish of mercury. Of course, the mercury in the tube began to run out into the dish creating a vacuum at the closed top of the tube. Not all the mercury ran out, however. At a certain height in the column, the weight of the mercury in the column exerted a pressure at the dish that was exactly balanced by the pressure on the surface of the mercury in the dish due to the weight of the atmosphere. That is to say, air pressure was transmitted in all directions within the body of the mercury, including a pressure upward into the tube to balance the mercury in the column at a height of 760 mm. Torricelli had invented the barometer.

One Atmosphere of Pressure

The height of 760 mm of mercury (760 mmHg) is often referred to as one atmosphere, or as 760 Torr, after Torricelli. If one takes a barometer up a mountain, the height of the mercury in the column should decrease. This is due to the fact as one goes up the mountain, more and more of the earth's atmosphere is left below and therefore does not contribute to the air pressure at the location of the barometer. There is less air above the barometer to push down on the mercury. This was actually verified experimentally by Pascal in 1648 on the Puy de Dome in what was called "The Great Experiment." As Ashcroft [82] mentions in a figure caption, Pascal did not actually carry out the experiment himself. He persuaded his brother-in-law and various local dignitaries to climb the Puy de Dome carrying a barometer. As a control, a second barometer was left under the watchful eye of the Reverend Father Chastin in the town of Clermont. Only the baraometer that took the trip was observed to change during the time period of the experiment.

Breathing Depends on Pressure

Air is not only important because of the pressure it provides, but also because it is what we breathe. We know that air consists of 21% oxygen, 0.04% carbon dioxide, and that the remainder is mostly nitrogen. From the kinetic theory of gases, we also know that the total pressure of a mixture of gases is the sum of the partial pressures of the separate gases. For example, the 760 Torr of the atmosphere consists of 159 Torr of oxygen, 0.03 Torr of carbon dioxide, and approximately 600.7 Torr of nitrogen. Note that this small percentage of carbon dioxide is extremely important in maintaining the stability of the earth's biosphere. The partial pressures of oxygen and carbon dioxide in the blood determine the rate of breathing. The primary mechanism for increasing respiration is a reduction in the partial pressure of oxygen in the air as detected by chemoreceptors located in the carotid arteries. The reduced partial pressure triggers a signal to the respiratory center in the brain to increase breathing. Carbon dioxide is also a significant regulator of breathing, acting on a different set of chemoreceptors in the brain than does oxygen. If the concentration of carbon dioxide in the blood falls, breathing is inhibited. As pointed out by Ashcroft [82], the switching between oxygen control of breathing and carbon dioxide control of breathing is not always smooth and may result in the phenomenon of "hunting" found in poorly regulated central heating systems of buildings. Engineers would call this a periodic instability, in which the feedback mechanism consistently overshoots the desired result.

Question 12.1

Explain why pressure is only a function of depth of the fluid. (Answer in 300 words or less.)

12.1.2 Hydrostatics

Motionless Mercury

Hydrostatics as a discipline consists of the laws of fluids at rest, and relies on Pascal's observations regarding the importance of fluid pressure. In medicine, such things as blood pressure are measured in millimeters of mercury (mmHg), based on Equation (12.4) for the hydrostatic pressure:

$$P = \rho g s \tag{12.5}$$

The pressure of 1 mm of mercury is obtained using this equation from the density of mercury given by $\rho_{Hg} = 13.6$ gm/cm^3, the acceleration of gravity $g = 980$ cm/sec^2, and a height of one millimeter, $s = 0.1$ cm:

$$P = (13.6 \text{ gm/cm}^3)(980 \text{ cm/sec}^2)(0.1 \text{ cm}) = 1{,}333 \text{ dynes/cm}^2 \tag{12.6}$$

a remarkably high pressure. Let us compare the pressure determined in Equation (12.6) with the pressure generated by the blood in your body.

Motionless Blood

The density of blood is approximately 1.055 gm/cm^3, so that a pressure of 1 Torr, given by Equation (12.6) to be 1,333 dynes/cm^2, corresponds to an equivalent height of 1.29 cm of blood. The pressure in the arteries at heart level is pulsatile, with a mean value determined by experiment to be about 100 Torr. This pressure will support a column of blood determined by the product of the height of mercury and the density of mercury divided by the density of blood [(100 mm × 1.36 gm/cm^3)/1.055 gm/cm^3] or 129 cm high, as the Reverend Stephen Hales found in 1733 [83]. Do not be overly concerned. The good reverend did not do the experiment with a person; instead he stuck a tube into the vein of a horse and measured the height of the blood in the tube (see Figure 12.1). This early experiment showed that the hydrostatic equation [Equation (12.5)] applies to the vascular system, as it should.

Blood Pressure for Humans

The rather benign looking Equation (12.5) has some rather interesting implications for medicine. First of all, if a person is lying down, the mean arterial pressure in the brain and feet are approximately the same as in the heart. However, when a person is standing, the hydrostatic pressure in the head is reduced, while that in the feet is increased. Burton [83] uses the number 50 cm for the distance of the artery in the head above the level of the heart; the mean pressure at the head is then calculated to be (100 − 500/13) Torr or approximately 62 Torr. Recall that 100 Torr is the aver-

Figure 12.1. Blood pressure of a horse is given by height of the blood in the column. Here is a rendition of how Reverend Stephen Hales must have looked in 1733 measuring the blood pressure of a horse.

age value of the pressure in the arteries at heart level, 500 mm is the distance of the head above the heart, and dividing by 13 converts mm of mercury to mm of blood. The pressure in the blood is reduced by 38 Torr over the distance from the heart to the head. Similarly, Burton calculates the arterial pressure in the feet, assumed to be 130 cm below the level of the heart, to be (100 + 1300/13) Torr or 200 Torr, an increase of a factor of two above the hydrostatic pressure at the heart. This factor of three change in pressure from your head to your feet requires a very active cardiovascular system.

Water Pressure for Divers

According to Equation (12.5), the pressure produced by a column of water increases faster than that of a corresponding column of air due to the ratio of the density of water to that of air. This ratio is approximately 1300. Thus, the pressure under water increases by one atmosphere for every 10 m (33 ft) of depth. Ocean divers denote an increase of one atmosphere as a bar, so the pressure at a depth of 20 m is 3 bars, one for each of the ten meters plus the pressure of one bar at the sea surface. Ashcroft [82] points out that using Boyle's law, the volume of a dissolved gas in a diver's bloodstream at 30 m below the ocean's surface is compressed by a factor of four below that at sea level. The expansion of the gas volume with the ascent of the diver to the surface will be subsequently shown to have profound effects on the diver.

Question 12.2

Discuss the experiment done by the Reverend Stephen Hales and explain the significance of his results. (Answer in 300 words or less).

12.1.3 Archimedes' Principle

Pressure Is the Same in All Directions

We know from Pascal's principle that the pressure of a fluid is transmitted not only to the walls of a container but also perpendicularly to the surfaces of any solid object within the fluid. Imagine a swimmer submerged in water near the bottom of a pool. The pressures on the sides of the swimmer's body at any given distance from the top of the water are all the same and, therefore, yield a zero net force in the horizontal direction. On the other hand, the height of water above the upper part of the swimmer's body is less than that above the lower part of the swimmer's body. Therefore, there is a relatively greater upward pressure on the lower part of the body and a relatively lesser downward pressure on the upper part of the body. Consequently, there is a net upward force exerted by the water upon the submerged swimmer. This upward force of fluids against submerged bodies is called *buoyancy*.

How Strong Is the Force of Buoyancy?

To determine the strength of buoyancy, we can use Newton's third law, the equality of action and reaction. As a body is submerged in a fluid it displaces the fluid until, when it is completely submerged, it has displaced a quantity of fluid equal in volume to its own. The body exerts a downward force on the fluid that is greater than the weight of the fluid being displaced. That is how it pushes the fluid out of the way. The reaction of the fluid to the body is to push back with a force equal in magnitude to the weight of the fluid that has been displaced. It is this reaction force that constitutes buoyancy.

Buoyancy Force

We may express the law of buoyancy in equation form by writing the volume of the body as V and the mass density of the body as ρ_B, so that the weight of the body is the product of the two, $V\rho_B$. If the mass density of the fluid is ρ_F, then since the volume of the displaced fluid is the same as the volume of the body, we may write the weight of the displaced fluid again as the product of the volume and the fluid density $V\rho_F$. Therefore, the measured weight of the submerged body is the difference between the actual weight and the weight of the displaced fluid (buoyancy force)

$$\text{measured weight} = \text{body weight} - \text{fluid weight}$$

or in terms of symbols,

$$W = V\rho_B - V\rho_F \qquad (12.7)$$

We can use this equation to determine the mass density of a person. We obtain the weight of the person, W, from a scale positioned in the fluid. This scale measures the weight of the body in fluid. The fluid of choice is water, so the density is 1 gm/cm^3. We can obtain the volume of the body by multiplying how much the water rises when the person is submerged by the cross-sectional area of the pool. Thus, we obtain for the mass density of the person,

$$\rho_B = 1 \text{ gm/cm}^3 + \frac{W}{V} \qquad (12.8)$$

Note that W is not the true weight of the person, but the weight when submerged, so that $W/V < 1$ gm/cm^3. It is clear from Equation (12.8) that the mass density of a person differs from that of water by less than a factor of two.

Archimedes Finds a Thief

The method of measuring the density of solid bodies, discussed above, is due to the Greek mathematician Archimedes and is nearly 2300 years old. He did not apply the method to humans, however, but to the task of determining whether the crown of King Hiero of Syracuse was pure gold or not. The story goes that the King instructed Archimedes to determine the gold content of his crown without damaging it. Since he could not change the crown into a shape whose volume he could calculate, such as a sphere, Archimedes was at a loss.

"I've Got It," Shouts a Naked Greek

Later the same evening, Archimedes lowered himself into his bath, and watching as the water rose, the principle of buoyancy occurred to him. He then supposedly ran out of his house naked and through the streets of Syracuse shouting "Eureka! Eureka!" ("I've got it! I've got it!"). He subsequently weighed the crown immersed in water, and measured the displacement of the water. He did the same thing with an equal volume of pure gold and was able to determine that the King had in fact been the victim of graft on the part of the goldsmith. We are fortunate that such things no longer occur. The principle of buoyancy is also known as Archimedes' principle.

That's Why Things Float

However, what happens when the density of the body is less than that of the fluid? In that case, the weight of the fluid displaced, in a volume equal to that of the body, is greater than the weight of the body and Equation (12.7) yields a negative number. This negative number means that the body will float. In fact, the body will rise in the fluid until the volume of fluid being displaced by the body floating on the fluid has a weight that is exactly equal to the weight of the body, in which case the effective weight of the body on the fluid is zero.

Torque in the Water

A swimmer attempting to float on the surface of a pool experiences two vertical forces. The pull of gravity acts at his/her center of gravity and the push of the force of buoyancy acts on what might be called the center of buoyancy. If the line of action of the two forces, the one downward and the other upward, do not coincide, they exert a torque on the body, that is, a rotational force. The center of gravity and center of buoyancy do not overlap because the distribution of water in the profile of the displaced fluid is not the same as the distribution of mass in the swimmer's body.

Center of Buoyancy

As Hay [2] explains, if the two force centers do not coincide when the swimmer is lying prone on the surface, as shown in Figure 12.2a, then stability might be restored by swinging the legs down so the two centers do lie along the same line of action. When the centers lie on a line, there is no resultant torque and the swimmer will retain his/her horizontal position. Experimentally, the center of buoyancy is found to lie between the swimmer's center of gravity and head, thereby producing an unstable situation in both males and females [84].

Question 12.3

Explain what a buoyancy force is and give two examples of how it might be used that are not discussed in the text. (Answer in 300 words or less.)

Question 12.4

A piece of floating wood has a density of 0.8 gm/cm^3 with the shape of cube and a volume of 10 m^3. How much of the wood will be above water?

12.1.4   Bernoulli's Principle

The Bernoulli Clan

It is worth pointing out that the name Bernoulli, found in mathematics and physics texts, does not refer to a single person. There were at least five generations of

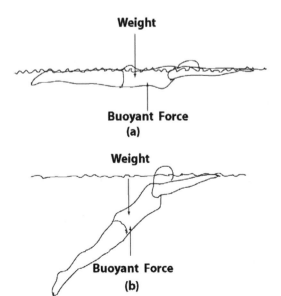

Figure 12.2. In the upper figure, the weight and the buoyancy force are seen to exert a torque on the swimmer. In the lower figure, the swimmer has adjusted the distribution of water mass so that the weight and buoyancy force act along the same line.

Bernoullis, originally from Switzerland, who held academic positions, primarily in mathematics and physics, at various universities in Europe. In the first three generations, starting with the brothers Jacques I (Jakob) and Jean I (Johann) in the last half of the seventeenth century, and ending with Jean III in the last half of the eighteenth century, there were seven members of the family that made fundamental contributions in the fields of the calculus, probability theory, and hydrodynamics, and there are various laws that bear the family name.

Force On and In a Fluid

Colliding fluid particles impart a force to the container wall that is perpendicular to the wall. This is easily observed experimentally by poking a hole in a styrofoam cup filled with water. The water jet escaping from the side is observed to exit at right angles to the side of the cup, prior to its curving downward under the force of gravity. Suppose we wanted to know how fast the water was flowing when it exited the side of the cup. What do we know about the physical situation? First of all, we know that the pressure at the point at we wish to make our hole is given by Equation (12.5), where the weight density has been written out explicitly and s is the distance from the center of the hole to the top of the water in the cup. You might recall from the form of Equation (12.5) that this quantity is a potential energy density, that is, the internal pressure, P, is a potential energy per unit volume. The right-hand side of this equation would be a potential energy if the mass density were replaced by the mass. Using this idea of a potential energy density and a corresponding kinetic energy density, we write for the sum of the potential (PE) and kinetic (KE) energy densities,

pressure + KE density + PE density = total energy density

or in terms of symbols

$$P + \frac{\rho u^2}{2} + g\rho s = \text{constant} \tag{12.9}$$

so that the total energy density of the fluid is a constant, at a fixed depth. This is an extension of the conservation of energy to a per-unit volume basis. The pressure is an energy density due to a mechanical force, $\rho u^2/2$ is a kinetic energy density due to the motion of parcels of fluid, and $g\rho s$ is the potential energy density due to gravity pulling down on each particle in the fluid, referred to the top of the cup.

Fluid Exits the Hole in a Cup

Equation (12.9) is called Bernoulli's principle and it may be generalized to include the internal energy of the fluid. Here u is the speed of the fluid at the depth s. Thus, if we measure the zero of potential energy to be at the top of the fluid in the cup, and denote the pressure of the air at the top of the cup by P_0, then at the location of the center of the hole in the side of the cup, the pressure can be written

$$P_0 = P + \frac{\rho u^2}{2} \tag{12.10}$$

Now replace the pressure P with the pressure at the specific location a distance s below the top of the cup. At the point of the hole, the pressure is reduced by the potential energy density, so that $P = P_0 - g\rho s$. When this value of the pressure is substituted into Equation (12.10) we have a factor of P_0 on both sides of the equation, so they will cancel one another. From the equation resulting from Equation (12.10) after this substitution, we obtain the initial speed of the fluid leaving the hole in the cup to be

$$u = \sqrt{2gs} \tag{12.11}$$

the self-same speed an object would have if it had fallen a distance s under the influence of gravity.

Question 12.5

What is the similarity between the energy of a pendulum and Bernoulli's principle? Use equations if you must. (Answer in 300 words or less.)

Question 12.6

If the side of a water tower is punctured 10 ft above the bottom and there is 20 ft of water in the tower, what is the speed at which the water will exit the hole? What is the pressure at the bottom of the water tower?

What Does Energy Density Mean?

Consider how we may use Equation (12.9). Suppose we have a pipe filled with flowing water, but having a constriction such as shown in Figure 12.3. When a fluid flows steadily in a pipe of uniform cross section, the fluid parcels follow smooth paths of steady flow shaped by the solid boundaries of the pipe, and the pressure is uniform. Because the pressure is uniform in all directions, the open pipes along the top of the main pipe act as barometers, and the water rises in them to a level determined by the pressure in the main pipe. Since the flow in the main pipe is constant, so too is the pressure in the main pipe, as indicated by the constant levels of fluid shown. However, we find that when a pipe narrows, it crowds the fluid parcels closer together. The fluid must move faster in the constricted region, because the same mass of fluid has to flow each second, but through a narrower opening. This is familiar to anyone that has ever partially blocked the end of a garden hose with their thumb to make the water squirt farther. The smaller the opening, the greater the distance the water is projected. The height of the fluid in the "barometers" in the lower figure indicates a reduction in pressure in the constricted region. In terms of Bernoulli's principle, we have in the three regions of the pipe,

energy density$_{\text{region 1}}$ = energy density$_{\text{region 2}}$ = energy density$_{\text{region 3}}$

or in terms of symbols,

$$P_1 + \frac{\rho u_1^2}{2} = P_2 + \frac{\rho u_2^2}{2} = P_3 + \frac{\rho u_3^2}{2} \tag{12.12}$$

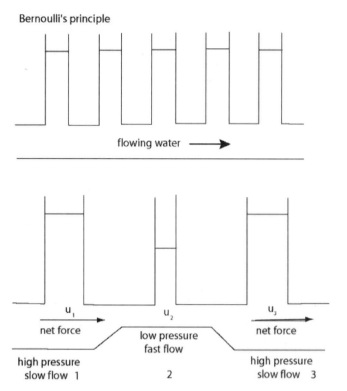

Figure 12.3. Height measures pressure. The height of the water in the vertical columns gives the pressure of the fluid in the pipe below. In the constricted region, the pressure is seen to decrease, P_1, $P_3 >$ P_2, along with the increase in the fluid flow, u_1, $u_3 < u_2$, due to the conservation of energy density.

where the separate regions have been denoted by subscripts. Note that this relation is only valid in the parameter range in which the fluid flow is smooth, which is called laminar flow.

Hydrodynamic Stream Lines

The concept of an electric field was facilitated by the introduction of field lines. These lines show the strength and direction of the electric field. The same geometrical construction was introduced into hydrodynamics, where the lines are called streamlines. Here again, the density of the lines in a region of space indicate the magnitude of the velocity field, with the fluid moving in the direction of the field lines. In Figure 12.4, the fluid flow is shown in a pipe containing a region where the diameter is reduced. The crowding of streamlines is a pictoral indication of the increase in velocity in the constrained region of the pipe. It is evident that the streamlines crowd together as the pipe narrows and spread out again as the pipe opens.

Different Regions Have Different Flows

If Equation (12.12) is to be valid, then the speed of the fluid in region 2 must exceed that in region 1, $u_2 > u_1$, due to the constriction in the pipe. It must also be true that the pressure in region 1 is greater than that in region 2, $P_1 > P_2$, which is to say that the pressure drops in going from region 1 to region 2 of the pipe. Note the height of the fluid in Figure 12.3, indicating the reduction of the pressure in

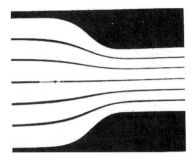

Figure 12.4. Streamlines at a two-dimensional boundary contraction, made visible by the injection of dye into water flowing between closely spaced glass plates (from Rouse [85] with permission).

region 2. The pressure is less when the flow is faster and this is true in general for laminar flow. Of course, in going from region 2 to region 3, when the diameter of the pipe increases back to its original size, the fluid slows down and the pressure increases. In fact, if the diameter returns to the value it had in region 1, then $P_1 = P_3$ and $u_1 = u_3$.

Different Regions Have Different Pressures

We can understand this behavior of the flow in a general way by considering a parcel of fluid at the border of regions 1 and 2. Because the flow rate changes, $u_2 > u_1$, there must be a region over which the fluid parcel is accelerated to account for its change in momentum. This external force must be produced by the difference in the pressure in regions 1 and 2. In order for the force to be of the proper sign, that is, to accelerate the fluid parcel, the pressure in region 1 must exceed the pressure in region 2, $P_1 > P_2$, which is what is observed experimentally.

Pressure and Speed Are Reciprocally Related

Bernoulli's principle is equivalent to the observation that the pressure in the fluid decreases as the speed of the fluid increases. This should not be surprising when you consider that pressure is a potential energy density. Thus, the exchange between pressure and fluid velocity is another example of energy conservation. One possible application of this idea is to obstructions in blood vessels. When there is a buildup on the walls of a blood vessel that constricts the flow of blood, the blood flows faster in the region of the restriction and the pressure is reduced, as we can see from Figure 12.3. We shall have more to say about this mechanism when we examine the flow in a pipe and discuss the phenomenon of drag.

Heart Sounds Reveal Disruption of the Flow

If the velocity becomes sufficiently large in the constrained region of the pipe, the conditions for laminar flow are violated and a new type of flow is generated. This new kind of flow is turbulence. The white water in mountain streams is turbulent, as is the swirling water behind a diver after she enters the water and the breaking of waves on the ocean surface. Turbulence can be oscillatory, pulsate, and/or random, thereby creating noise. This noise is the sound heard by the physician with his stethoscope, the so-called *heart sounds*. The sound is either produced directly by turbulence of the blood flow, or by the vibrations of the walls of the vessels, or a combination of the two. The medical student might remember:

Streamline flow is silent,
Remember that, my boys,
But when the flow is turbulent
There's sure to be a noise.
So when your stethoscope picks up
A bruit, murmur, sigh,
Remember that it's turbulence
And you must figure why.

An Ideal Fluid

It is worth pointing out that Bernoulli's equation only applies to ideal fluids. Of course, such fluids do not exist in nature, that is, fluids without viscosity. No fluid moves without energy losses due to friction, either internally, where the energy is converted into heat, or externally due to contact with the walls of the container. To be pedantic about the whole thing, a moving ideal fluid is isothermal (does not change its temperature), incompressible (does not changes its volume), continuous (does produce holes spontaneously), and inviscid (has no internal friction). Although ideal fluids do not exist in nature, they are often good approximations to real fluids.

12.2 THE LAW OF LAPLACE

Relations Between Pressure and Tension

We know the rules for pressure when the walls of the container are rigid, but how does fluid react to the responsive heart, the supple arteries, the pliant veins, and all the other flexible membranes in the cardiovascular system. Here we must account for the tension in the muscles and membranes that act to oppose the pressure of the fluid (blood) that is being pumped through them. The first person to recognize the balance of forces in such a general configuration was Laplace in his study of soap bubbles around 1820. Laplace sought to relate the physical quantities, the pressure P of the fluid, which we know to be a force per unit area, and the tension T in the membrane containing the fluid, which is a force per unit length. These two quantities are proportional to one another in a soap bubble, and the proportionality constant is the radius of curvature R of the surface, the only other physical quantity available with the proper units. Thus, Laplace's law can be written in the dimensionally consistent form

$$P = T/R \qquad (12.13)$$

up to an overall dimensionless constant. For the case of a cylinder, a primitive model of a blood vessel, this overall constant is unity, so that Equation (12.13) is exactly the law of Laplace for balancing the pressure within a blood vessel by the tension in the walls of the vessel.

Application to Physiology

In physiology, we can use Equation (12.13) to determine the total tension in the walls of blood vessels of various kinds, using the mean values of the pressure in them and the radii determined from histological analysis; see Burton [83]. Burton points out that tension in the blood vessels varies from 2×10^5 dynes/cm for the aorta to only 14 dynes/cm for the capillaries. This reduction of nearly ten thousand in tension, from the thick-walled aorta (2 mm) to the thin-walled (1 μm) capillaries, is accomplished by a decrease in radius of curvature so that the aorta and capillaries withstand the same pressure. Vogel [86] discusses this effect and mentions that the tensile stress obtained by dividing the tension by the wall thickness, yields 10,000/2,000—a factor of five. Consequently, the stress on the walls of the aorta and a capillary are nearly the same, even through the pressure in their respective vessels differs by 10^4.

Soap Bubbles

Let us now consider a more general two-dimensional surface, that of a soap bubble. At a point on the surface there are two principle radii of curvature acting at right angles to one another, call them R_1 and R_2. The relation between pressure at the surface and tension in the surface is

$$P = \frac{T}{R_1} + \frac{T}{R_2} \tag{12.14}$$

Notice that as the radius of curvature in a given direction becomes larger, the surface bends more slowly in that direction. In the case of a cylinder, the curvature along its length is infinite, that is, it does not bend at all in that direction, thereby reducing Equation (12.14) to Equation (12.13). In the case of a soap bubble, however, the curvature away from a point is the same in all directions, so $R_1 = R_2 = R$ and Equation (12.14) reduces to

$$P = 2T/R \tag{12.15}$$

where R is the radius of the soap bubble. Thus, the smaller the bubble, the greater the pressure. Consider the consequences of Equation (12.15) on a bubble of gas trapped in the blood.

Henry's Law

The T for a gas bubble in a fluid is the surface tension of that fluid. The surface tension compresses the gas, creating an increasingly greater internal pressure with decreasing bubble size. A tiny bubble of, say, 0.1 μm (10^{-7} m) diameter works out to 30 bars, for a viscosity of 0.073 N/m. This dramatic hydrostatic pressure enhances the solubility of gas in the fluid—Henry's law. As the gas goes into solution in the fluid, the radius of the bubble decreases, causing the pressure to increase, thereby increasing the rate at which the gas goes into solution. Thus, as poetically stated by Vogel [86], ". . . quick as a wink the bubble ceases to exist."

Nonlinear Response Curve

Laplace's law is like the other simple laws in physics: it is a linear relation among physical observables. The tension in the surface increases in direct proportion to the pressure as well as in direct proportion to the radius of curvature. As desirable as

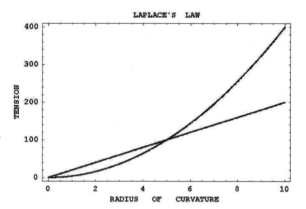

Figure 12.5. Here we depict Laplace's law with the linear curve for a constant value of the pressure. The concave curve depicts the relation between the elastic tension and the radius of a vessel under a given transmural pressure.

Laplace's law might be, it does not adequately describe the interrelation among the three variables for blood vessels. This failure to capture the nonlinear character of real physiological systems is a consequence of not taking into account the elastic tissue in the walls of the various blood vessels, and the difference between the idealized relation and an experimental one is depicted in Figure 12.5. We ran across this same nonlinear behavior in our earlier discussion of Hooke's law in biological systems, where we found that the system response was not directly proportional to the excitation.

What Does the Figure Suggest

The two curves in Figure 12.5 have very different interpretations. Laplace's law (the straight line), like Hooke's law, indicates that the vessel wall resists in direct proportion to how much it is stretched. The concave curve, on the other hand, indicates that the wall does not resist in direct proportion to the stretch, but increases its resistance to each successive stretch. Burton [83] explains that this peculiar property of blood vessels is the result of the combination of elastin and collagen fibers in the vessel wall. The elastin fibers respond to the slightest stretch, whereas the collagen fibers respond to much stronger distensions of the wall and do not reach their maximum extension until the wall is very far from equilibrium.

12.3
SUMMARY

Pascal's Principle

We have introduced a number of laws concerning the properties of fluids in this chapter, summarizing a great deal of experimental observation. The hydrostatic pressure in a fluid is determined by the weight of the fluid pressing down from above. In a solid, the strong molecular bonds transmit the force as a vector in the direction the force is acting in, for a gravitational field this is preferentially downward. In a liquid, the weak bonds have no orientational preference, so the fluid pushes equally in all directions at a given depth, and the force of the fluid weight is not just downward, but acts in all directions. Therefore, the blood pressure in my feet is greater than the blood pressure in my head and I can calculate that difference using Pacal's law for hydrostatic pressure. That is not the complete story, of course, because some pressure is lost due to the viscous nature of blood and the pumping of the heart increases the pressure above the hydrostatic level. But these are the properties of fluids in motion, which we take up in the next chapter.

Archimedes' Principle

A solid object, denser than a fluid, sinks to the bottom when placed in water and displaces a volume of water equal to that of the body. A scale at the bottom of the water does not measure the true weight of the object, but indicates the weight of the object minus the weight of the fluid that has been displaced. This is the principle of Archimedes and accounts for the fact that objects less dense than water can float. Along with this principle comes a new force, that of buoyancy. The force of buoyancy is a consequence of Newton's law of action and reaction and its magnitude is determined by the volume of fluid displaced by the object being submerged. Once the diver has been submerged, the bouyancy force is the same regardless of the depth of the diver, since the volume displaced by the diver remains the same independent of depth.

Bernoulli's Principle

The transition from hydrostatic to hydrodynamic properties in a fluid is made by Bernoulli's principle. This principle is a generalization of the conservation of energy to the continuum, extending the conservation law for point particles and solid objects. Because a fluid consists of point particles, it can have both kinetic and potential energy, but the fluid particles do not have the rigid locations in space that they have in a solid; they continuously move relative to one another. This ability of the fluid to move, but retain its properties, requires that we associate an energy per unit volume to the fluid, that is, an energy density. The energy density has the same dimensions as that of a pressure, so that the sum of the kinetic energy density, the potential energy density, and the pressure is conserved as the inviscid fluid moves. This is the new conservation law.

Flow in the Body

The three principles of Pascal, Archimedes, and Bernoulli play a remarkable role in the locomotion of the human body. We can see how they determine the flow of blood from the heart to the arteries to the capillary beds. There is a smooth laminar flow of blood when everything is working as it should, and tell-tale heart sounds when obstructions in the arteries and/or veins produce more rapid local flow and turbulence in the bloodstream. These principles are not confined to the internal workings of the body; they also determine the interaction of the body with the outside world. The long-distance cyclist is no less aware of the influence of the fluid in which he/she is moving than is the swimmer or diver.

Laplace's Law

The above principles apply to the physical world regardless of application. For example, Bernoulli's principle is always true for an inviscid, irrotational, uncompressible fluid. However, what can change is the way in which the various quantities are calculated. In a rigid pipe, we know that the pressure is defined by the force per unit area. In a membrane, which is squishy rather than rigid, this definition is no longer adequate. In a membrane, the forces acting to balance the pressure of the fluid being forced through it is the surface tension. The balance between tension and pressure is determined by Laplace's law, in which the pressure at the surface of a membrane is given by the ratio of the surface tension to the local radius of curvature of the membrane in two orthogonal directions. Since the radius of curvature is the same in all directions for a symmetric soap bubble, the pressure in a soap bubble is given by twice the surface tension divided by the radius of curvature. Thus, the smaller the bubble, the greater is the interval pressure. The simple proportionality between pressure and surface tension in Laplace's law does not hold generally in biological systems, where a more general nonlinear relation is found empirically.

CHAPTER 13

Fluids in Motion

Objectives of Chapter Thirteen

- Learn the difference between the properties of fluids in motion and fluids at rest.

- Understand that friction internal to a fluid (viscosity) generates patterns in fluid flow and is the single most important physical property of real fluids.

- Much of what we know about fluids in the design of ships and planes is due to the three scientists—Reynolds, Froude, and Euler—and the dimensionless scaling parameters that now bear their names.

- Learn that the reason fluid dynamics is so difficult is because fluid flow is nonlinear, and why this nonlinearity is so important.

- Understand the tyranny of many dimensionless constants.

Introduction

The Importance of Dimensionless Constants

The purpose of this chapter is to gain an appreciation of perhaps the most difficult phenomenon in classical physics: how fluids flow in physical and biological systems. We address this problem in order to understand why mathematics has not been particularly successful in modeling certain biological phenomena. Our approach is to extend dimensional analysis in order to replace extremely complicated fluid equations with dimensionless parameters. Engineers have used this technique to design full-sized aircraft and full-sized ships using scale models in the laboratory. The dimensionless parameters characterize complex fluid flows generated by different mechanisms. One can make decisions about the relative importance of the

Biodynamics: Why the Wirewalker Doesn't Fall. By Bruce J. West and Lori A. Griffin
ISBN 0-471-34619-5 © 2004 John Wiley & Sons, Inc.

different physical processes by comparing the sizes of these different dimensionless parameters.

Different Flows

Pressure and speed, lift and drag, buoyancy and volume are all part of the duality of fluids, both inside and outside the body. When considering water flowing through a pipe or blood surging through an artery, the kind of flow depends on the smoothness of the walls, the variation in the diameter of the structure, the viscosity of the fluid, and the pressure. The same applies to the swimmer in a pool, where the fluid drag on her body is modified as she changes her distance from the wall and how far she is from other swimmers.

13.1 HYDRODYNAMICS

Dynamics Versus Static Fluids

The difference between what we called fluid mechanics in the previous section and hydrodynamics, is that the former addresses the static properties of fluids based on mechanical forces, whereas the latter addresses the properties of moving fluids. Of course, the distinction between the two is not always clear, as in the case of Bernoulli's principle, somewhat like the distinction we have been attempting to make between biomechanics and biodynamics. However, we recall the idealization that had to be made regarding the properties of fluids in order to apply Bernoulli's principle.

Inside and Outside the Body

Here again, we make some attempt to distinguish the dynamics of fluids outside the body and those important dynamical properties within the body. But we take note that some properties of real fluids, like viscosity, are important in both situations. In fact, viscosity is one of the most important properties of real fluids. It is this internal resistance that determines such things as whether the flow of the fluid is smooth and laminar or discontinuous and turbulent.

13.1.1 Viscosity

Fluids Have Internal Friction

Viscosity is a generalization to fluids of the concept of friction that we previously applied to solid bodies. We identified friction as the force between two macroscopic objects in contact with one another that acts to inhibit their relative motion. In fact, as one body slides over the other, the friction force impedes their relative motion and eventually brings them to a stop. We discussed this mechanism in terms of the roughness of the surfaces of the two objects, but cautioned the reader that the effect was actually a molecular one. Recall that rubbing your hands together makes them warm. Friction is a mechanism that transforms kinetic energy into heat.

Friction and Terminal Velocity

In our discussion of diffusion, we learned that friction causes the heavy, erratically moving, Brownian particle to give up energy to the ambient fluid. In this case, there is no surface roughness to produce the slowing down of the heavy particle. In fact, any object moving in a fluid will encounter a similar resistance. Regardless of size, a body experiences a friction in a fluid that is directly proportional to its velocity. A marble falling in oil, for example, accelerates under the influence of gravity. As the velocity of the marble increases, the effect of friction also increases, working against gravity. Since the acceleration of gravity is constant, eventually, the marble will reach a velocity at which the frictional force will be equal and opposite to the gravitational force, and the velocity of the marble will no longer change. The marble then falls at a constant rate called the *terminal velocity*. We discussed terminal velocity earlier, regarding a person jumping from a plane.

Turbulence and Swimming

A swimmer soon learns that the resistance of the water has to be overcome with every stroke if she is to keep from stopping dead in the water. No matter how much hair the swimmer shaves off, or how smooth she makes her body, there will remain some friction between the body and the water. This drag is a consequence of the fact that the swimmer must expend energy to break the cohesive bonds of the water in order to pass through it. The energy required to tear apart and disrupt the water elements depends on the shape of the body, how the water is parted, and how the water comes back together with the passage of the swimmer. The more smoothly the swimmer can move the water around, that is, the less turbulence the swimmer generates, the more energy that goes into the swimmer's locomotion, rather than into heating the water through friction.

Moving Fluid has Structure

The internal friction of fluids can be experienced by walking in waist-deep water. The fact that the water does not move out of the way with the pushing of the legs, but wraps itself around you as you move, is due to this internal friction, that is, the cohesive forces in the fluid. The water close to your leg moves with the leg. Recall that the velocity of a fluid at a solid surface is zero. The water a little bit away from the leg is shed and does not move as fast as the leg does but is dragged along by the leg as it moves. In fact, if you examine the fluid motion in a cup of coffee as you stir it with your spoon, you see the formation of small swirls of fluid behind the moving spoon. This swirling structure of the fluid, or rather the fluid made up by this structure, is all being pulled along by the spoon. The same thing is happening as you walk quickly through the water. It is not only that you are pushing water out of the way, which you are, but you are also pulling a great deal of water along with you in the form of vortical structures. The vorticity of the fluid is a consequence of the cohesive forces producing the viscosity of the fluid, as well as the fluid motion induced by external forces.

Vortex Shedding

Experiments have been done to determine the force exerted on a body passing through a fluid. This force is called *drag* and is produced both by the interaction of the surface of the body with the fluid and by the interaction of the shape of the body with the fluid. In Figure 13.1 is depicted the consequences of a cylinder

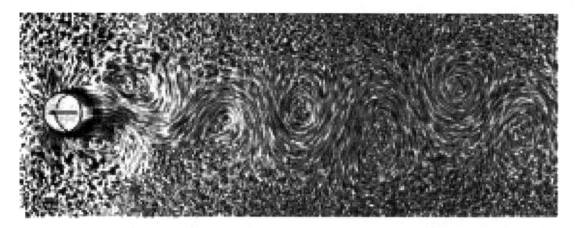

Figure 13.1. Vortex trail in the wake of a moving cylinder (from Rouse [85] with permission).

moving at a constant speed through a tank of still water. Think of your leg in a swimming pool. We can see that in the wake of the moving cylinder there is an orderly trail of vortices, which alternate in position above and below the centerline of the cylinder. This is called the von Kármán vortex trail, after the noted hydro-dynamicist who first provided its physical explanation. The vortices are formed on the surface of the cylinder and then detach, first on one side and then on the oth-er, thereby producing an alternating sequence of clockwise and counterclockwise rotation of the fluid. From Bernoulli's principle, we recognize that this pattern in fluid velocity produces zones of low pressure behind the cylinder that oscillate from side to side. This oscillatory pattern of low pressures produces variable side thrusts, as well as an overall drag on the cylinder, as your body moves through the water.

Resistance to Motion Depends on Shape

It is also useful to examine the dependence of the body drag on the shape of that body. In our technological society, we know that cars have a certain shape so as to reduce air resistance and thereby increase gas mileage; aircraft wings and fuse-lages have a certain shape so as to maximize lift and minimize drag; and swim-mers shave every part of their body, all in an effort to decrease the loss of energy due to drag. We saw above what a cylinder does as it moves through water at a constant speed. Now let us look at a few more bodies. Consider a rectangular plate whose height and width is very much greater than its thickness. If the plate is placed parallel to a laminar fluid flow, there is drag associated with its surface but none associated with its form. However, the reverse is true when the plate is placed perpendicular to the fluid flow, as shown in the uppermost picture in Fig-ure 13.2. From the three pictures in this figure, it is evident that the more abrupt the curvature of the sides of the body, the more pronounced the form effect. The process of streamlining, that is, the easement of boundary curvature at the sides and rear of the body, is entirely for the purpose of reducing the form effect upon the total drag.

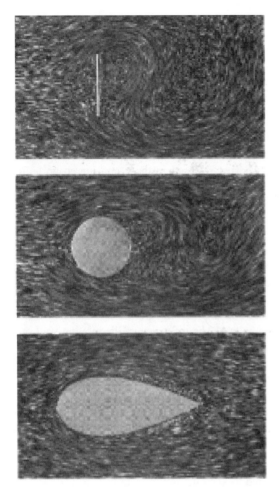

Figure 13.2. Note the different wakes from the different shapes. Here we compare the wakes produced by a plate placed perpendicular to the flow (top), a cylinder (middle), and a streamlined strut. (bottom). (From Rouse [85] with permission.)

Structure Is Due to Viscosity

It cannot be overemphasized that all this beautiful structure seen in the flow of fluids around bodies and even the loss of structure that arises when the flow collapses into white water, as it does in mountain streams, is the result of viscosity. The fact that real fluids possess an internal resistance generates all these phenomena. Of course, the shape of the body may trigger one effect or the other, as we see in Figure 13.2, but it is the internal resistance that enables the fluid to retain the information about the shape downstream from the body. It is also evident from the figure that the more bluff the body, the more structure that is generated, and the greater the trail left by the body in the fluid.

Slow as Molasses

Different fluids have different viscosities that determine the rate at which the various fluids flow. In fluids such as glycerol or sugar solutions, where the viscosity is high, the flow rate is very low. In fact, the viscosity is a function of temperature: the viscosity goes up as temperature goes down. This has lead to expressions such as "as slow as molasses in January," not unlike the first shot of ketchup out of the bottle that you try to put on your French fries.

The Poise Is a Physical Unit of Friction

It should come as no surprise that the law governing viscous friction was originally given by Newton in 1726; it was later generalized by Stokes in 1845. The difficulty was to define a quantifiable measure of the viscosity. Consider a real fluid, which is to say, one having viscosity, rather than those considered up until now, where we have neglected internal resistance. If you push on a real fluid, it will respond and create a shearing stress such as we discussed earlier. Viscosity describes these shearing forces that exist in a real moving fluid. Suppose we have two solid plane surfaces, such as shown in Figure 13.3, with a real fluid between them. The upper surface is moved to the right, parallel with the lower one at a slow speed U, and the lower surface is held fixed. The fluid between the plates does not slip and, therefore, has zero velocity at the lower surface and a velocity equal to that of the upper plate adjacent to the upper surface. The fluid at the upper plate has the velocity of the plate and this moving fluid drags adjacent fluid elements along with it, but not quite as fast. The influence of the moving wall on the motion of the fluid decreases with distance from that wall, until the stationary wall at the bottom is reached where the fluid is also stationary. This, too, is an experimentally proven fact: the motion of the fluid immediately adjacent to the wall is zero relative to the wall. The velocity of the fluid is indicated by the magnitude and direction of the arrows in Figure 13.3.

Fluid Shear Stress

How do we measure the force required to keep the lower plate in Figure 13.3 in a fixed position as the fluid flows by? The force required, so that the lower plate is not dragged along with the fluid, is found to be proportional to the area of the plate and the ratio of the speed of the upper plate to the distance between plates, U/s. The shear stress, the force per unit area, is therefore given by

$$\frac{\text{force}}{\text{area}} = \frac{\text{viscosity} \times \text{velocity}}{\text{distance}}$$

or in symbols,

$$\frac{F}{A} = \eta \frac{U}{s} \tag{13.1}$$

where η is a constant at a given temperature and is called the coefficient of viscosity. Note that Equation (13.1) has the form of the linear stress–strain relation discussed earlier, but rather than a shear on the right-hand side we have, instead, a shear rate, which is to say, a shear per unit of time. This replacement of a shear with a shear rate means that the rate of distortion of the fluid depends on the

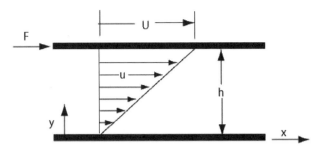

Figure 13.3. The viscous drag between two parallel plates. The lower one is stationary and the upper one is moving slowly to the right. The arrows indicate the fluid velocity at a given distance from either of the two plates. The length and direction of the arrows indicate the velocity of the fluid.

force. This equation allow us to determine the viscosity by reexpressing Equation (13.1) in the form

$$\text{viscosity} = \frac{\text{force} \times \text{distance}}{\text{area} \times \text{speed}}$$

or in symbols,

$$\eta = \frac{Fs}{AU} \tag{13.2}$$

where the product of the force and distance is work and the denominator is the product of an area and a velocity, which is a flow rate. From this equation, it may look like the viscosity changes with changing conditions, but this is, in fact, not true, or at least not true for changes in area, mechanical forces, speed of the plate, or the distance between plates. For example, with a particular area and distance between plates, one finds that changes in speed yield a corresponding change in force, such that their ratio is constant. The dimensional equation for the coefficient of viscosity is

$$[\eta] = \frac{[LMT^{-2}][L]}{[L^2][LT^{-1}]} = [ML^{-1}T^{-1}] \tag{13.3}$$

so that in the cgs system the dimensions of the coefficient of viscosity are gm/cm·sec. One gram per centimeter per second is defined as one poise in honor of the French physician Jean Louis Marie Poiseuille. He was the first scientist to quantitatively study viscosity (in 1843) and since he was a physician and not a physicist, the context of his study was the flow of blood in narrow blood vessels.

A Smaller Unit for Friction

In general, the poise is too large a unit to be of much value in dealing with most liquids, so that one typically refers to the *centipoise* or the *millipoise,* those being one hundreth and one thousandth of a poise, respectively. The viscosity of water at room temperature is approximately one centipoise, whereas the viscosity of glycerol is about 1500 centipoise, or 15 poise. The viscosity of blood is three to four times greater that that of water, so we take it to be 0.04 poise. Note that it is necessary to know the temperature of a fluid in order to specify its viscosity. This is reasonable, since the higher the temperature, the more kinetic energy the particles of fluid have, and the easier it is for them to slip by one another. Therefore, we would expect that the viscosity of most fluids would decrease with increasing temperature.

13.1.2 Dimensionless Constants*

No General Solutions

It should be pointed out that there is no closed-form solution to the hydrodynamic equations in anything approaching the general situation. This means that although we believe we know what the equations are that describe the wind and rain, they are too complicated to solve except on a very large computer using sophisticated numerical techniques, and then for only the simplest of situations. Simple analytic functions are not adequate to describe the behavior of such complex phenomena. So what is the basis of our understanding of the equations of hydrodynamics and the corresponding physical and biological phenomena? That basis is the use of dimensionless constants for design of experiments involving small-scale physical models of the full-sized objects under investigation. The design of airplanes, dams, harbors, canals, and so on, would be impossible without scaling experiments. We choose to describe this strategy using the most general description of viscous fluid flow, that is, the equation of Navier and Stokes.

Mathematical Description of General Fluid Flow

Let $u(r, t)$ be the velocity of a fluid element at the point in three-dimensional space r, at the time t, of an incompressible fluid of density ρ with a viscosity η. The evolution of the velocity of such a fluid in space and time is given by the Navier–Stokes equation in three spatial dimensions:

$$\rho\{\text{mass density}\} \times \left[\frac{\partial u(r, t)}{\partial t} + u(r, t) \cdot \nabla u(r, t) \right]\{\text{acceleration}\} =$$

$$\nabla P(r, t)\left\{ \frac{\text{change in pressure}}{\text{change in distance}} \right\} - \eta\{\text{viscosity}\} \tag{13.4}$$

$$\times \nabla^2[u(r, t)]\left\{ \frac{\text{change in shear}}{\text{change in distance}} \right\} + F(r, t)\{\text{body force acting on liquid}\}$$

Here we write the symbol together with its interpretation; the latter is enclosed in curly brackets. For example, the mass density of the fluid is given by ρ. The acceleration of the fluid is seen to be much more complicated than the acceleration of a point particle because the velocity changes from point to point in space as well as in time. This equation captures the fact that a fluid can be moving rapidly nearby and slowly far away, indicating that the fluid force changes in space. This change in force with distance is familiar to anyone who has stepped into a mountain stream and felt the water tugging at his/her ankles.

*In this section, we draw heavily from Montroll's remarkable 1987 paper "On the dynamics and evolution of some sociotechnological systems" [87].

Flow Depends on the Boundaries

The Navier–Stokes equation is quite general in that it purports to describe fluid flow in typical situations, subject, of course, to the appropriate boundary conditions. A boundary condition is the fluid flow at the boundary of the container holding the fluid. For example, we would call the bed of a stream a boundary for the water, as are the arterial walls for blood. Even though we do not try to solve Equation (13.4), it might be useful to understand from where the various terms in it come. The left-hand side of the equality sign denotes changes in the fluid velocity in both space and time, multiplied by a mass density. We could therefore interpret the left-hand side to be a mass times acceleration, or a force per unit volume within the fluid. On the right-hand side of the equality, we have three terms. The first term is the force produced by the change in pressure between points in space, but at the same time. The second term is the force due to the viscosity of the fluid, and results from the changes in the rate of shear at different locations in the fluid. The third term is called a body force and is applied by some external agent, such as gravity, to the fluid as a whole. Thus, the Navier–Stokes equation is a generalization of Newton's law of motion to viscous fluids.

Blood Flows as a Fluid

Blood flow in arteries and veins can be described using the Navier–Stokes equation, the density and viscosity of blood, and the appropriate boundary conditions, as can air flow in the bronchial tubes by using the appropriate parameter values and boundary conditions. You might think that we have gone back on our word by introducing this equation, but since we have no intention of solving it, we might be forgiven. Rather, what we plan to do is demonstrate how clever scientists and engineers have avoided solving this equation, and others like it, but were still able to gain understanding of fluid phenomena in which they were interested.

The Flow Problem Is Nonlinear

The Navier–Stokes equation is applicable to many processes involving fluids; for example, the flow of fluid through pipes. To completely understand blood flow in the body requires an understanding of Equation (13.4). Or does it? This equation also describes the motion of ships and subsonic flight of airplanes, when the appropriate boundary conditions are specified. We mention that it is the nonlinear term, $u \cdot \nabla u$, that makes the mathematics of Equation (13.4) so difficult. It is also this term that makes the phenomena being described so rich in dynamical structure. Here we distinguish between difficult in the sense of hard to solve but solvable nonetheless, and difficult in the sense that no one has been successful in solving the equation in general.

The Dimensionless Parameters

What we want to do is to transform the Navier–Stokes equation into an equivalent dimensionless equation, whose properties can be deduced from physical or biological, rather than mathematical, arguments. To do this, we measure the local velocity, pressure, and so on, as a multiple of some important basic dimension of the flow pattern being investigated. Suppose we have measured the following quantities:

V = average velocity of the body being studied

L = an important unit of body length

P = average pressure in the absence of the body

Suppose we are interested in modeling a swimmer in a pool. The characteristic quantities would be the average speed of the swimmer, the length (height) of the swimmer, and air pressure above the surface of the water. If we were interested in the motion of a SCUBA diver we might choose the pressure of the water at some depth. All these quantities are measurable. In terms of dimensionless variables, we can rewrite the Navier–Stokes equation as

$$\text{fluid acceleration} = \left(\frac{1}{E}\right)\left(\frac{\text{change in pressure}}{\text{change in distance}}\right) + \left(\frac{1}{R}\right)\left(\frac{\text{change in shear}}{\text{change in distance}}\right) + \frac{1}{F}$$

(13.5)

where we have introduced gravity as the body force. In expressing all the physical quantities in terms of dimensionless variables, all the scaling parameters are pulled out as coefficients in Equation (13.5). These coefficients are actually dimensionless numbers given, in terms of the characteristic quantities measured above, by

$$R = \text{Reynolds number} = \frac{VL\rho}{\eta}$$

(13.6)

$$E = \text{Euler number} = \frac{\rho V^2}{P}$$

(13.7)

$$F = \text{Froude number} = \frac{V^2}{Lg}$$

(13.8)

We shall discuss each of these dimensionless parameters individually soon but, for the moment, we see that their values determine the relative importance of the three force terms on the right-hand side of Equation (13.5). The pressure term in the hydrodynamic equation is measured by the Euler number, the gravity term in this equation is measured by the Froude number, and the viscous term is measured by the Reynolds number. So how do we use this information?

Froude Number Flow

Rather than continuing to present an abstract discussion of the different terms in the hydrodynamic equation, let us construct an example and use some numbers. Consider a 1000 ft (L) long ship operating in the speed range of 40 ft/sec (V), with pressures being measured in units of atmospheric pressure ($P = 14.7$ lbs/in^2). Noting that the ratio of viscosity to the density of water at 15 °C is $\eta/\rho = 1.23 \times 10^{-5}$ ft^2/sec allows us to compute the three dimensionless constants that we introduced above:

$$R = \frac{(40 \text{ ft/sec})(1000 \text{ ft})}{1.23 \times 10^{-5} \text{ ft}^2/\text{sec}} = 10^9$$

$$E = \frac{(40 \text{ ft/sec})^2(0.0361 \text{ lbs/in}^3)}{14.7 \text{ lbs/in}^2} = 3.94$$

$$F = \frac{(40 \text{ ft/sec})^2}{(1000 \text{ ft})(32 \text{ ft/sec}^2)} = 0.05$$

where we have replaced the density of water in cgs units (1 gm/cm^3) with those in engineering units (0.0361 lbs/in^3) and the reference pressure is that at the surface of the water. In this way, we obtain the coefficients in the dimensionless Navier–Stokes equation:

$$\frac{1}{F} = 20; \qquad \frac{1}{E} = 0.25; \qquad \frac{1}{R} = 10^{-9} \tag{13.9}$$

Hence, the $1/F$ term is the most important one on the right-hand side of Equation (13.5) for the conditions given, since it has the largest numerical value. Thus, the flow velocity field for the ship is a function of only the dimensionless quantity F, the Froude number. In this way the small-scale ship model experiments, such as are done at the David Taylor Model Basin for the U.S. Navy, can be made to obtain design data for full-scale engineering of the ship that are based on Froude modeling. This is modeling with a dimensionless constant that depends only on the acceleration of gravity g. Thus, phenomena in which the influence of gravity is dominant are determined by the Froude number. We shall see that this is the case in running animals as well.

Euler Number Flow

Consider an airplane with a wing of width 10 ft (L), designed to operate at a speed of 800 ft/sec (V) or approximately 545 miles/hour. Measuring the pressure in atmospheres and using the ratio of viscosity to the density of water at 15 °C, is $\eta/\rho = 1.59 \times 10^{-4}$ ft^2/sec, allows us to compute the three dimensionless constants:

$$R = \frac{(800 \text{ ft/sec})(10 \text{ ft})}{1.59 \times 10^{-4} \text{ ft}^2/\text{sec}} \approx 5 \times 10^7$$

$$E = \frac{(800 \text{ ft/sec})^2(0.036 \times 10^{-3} \text{ lbs/in}^3)}{14.7 \text{ lbs/in}^2} \approx 0.0158$$

$$F = \frac{(800 \text{ ft/sec})^2}{(10 \text{ ft})(32 \text{ ft/sec}^2)} \approx 2 \times 10^3$$

where we have replaced the density of air in cgs units (10^{-3} gm/cm^3) with those in engineering units (0.0361 $\times$ 10^{-3} lbs/in^3). In this way, we obtain the coefficients in Equation (13.5):

$$\frac{1}{F} = 5 \times 10^{-4}; \qquad \frac{1}{E} = 63.5; \qquad \frac{1}{R} = 2 \times 10^{-8} \tag{13.10}$$

Hence, the pressure factor, expressed in terms of the Euler number, is the most important one on the right-hand side of Equation (13.5) for the conditions given. In this case, the design of small aircraft is determined by wind-tank models based on the Euler number, that is, it is a pressure-dominated phenomenon.

Question 13.1

A grain of sand has a radius of 10^{-4} cm. The grain falls under the influence of gravity in a glass of water and moves at an average speed of 1 mm/sec. Determine which of the three hydrodynamic mechanisms dominates the motion. Explain your result. (Answer in 300 words or less.)

Bernoulli's Equation

The magnitudes of the dimensionless numbers suggest that under certain conditions we can drop all but the pressure term on the right-hand side of the equation for the acceleration of the fluid [Equation (13.5)]. When we consider only the pressure term contributing to the acceleration of the fluid, it is possible to write an equation of the form

$$\frac{\rho u^2}{2} + P = \text{constant} \tag{13.11}$$

which is Bernoulli's equation for a nonviscous fluid. Using Bernoulli's equation in the example of the pressure difference between the bottom and top of a wing section of an airplane, as developed by air circulation around the wing, determines the "lift" of the wing. It is not surprising that the pressure term is most important in the regime of interest. A wind tunnel is the traditional device for measuring the lift and drag (and their ratio) on a model airplane in laminar flow. Since the length L does not enter into the determination of the Euler number, the lift-to-drag ratio would be the same on a small airplane model as on a full-scale object of the same shape. It is this simple fact that makes models useful. The properties of the large aircraft can be determined from the model, with the attendant savings in time, money, and effort.

Question 13.2

What would it mean hydrodynamically if all three of the dimensionless numbers were of order unity ($E \approx F \approx R \approx 1$)? (Answer in 300 words or less.)

13.1.3 The Constants of Industry

One Dimensionless Constant

In the two examples above, the dominant behavior of the fluid can, to a first approximation, be described in terms of a single dimensionless number. Here, by first approximation, we mean that we can rank order the influences affecting a phenomenon and attach a figure of merit to that order. Money is like that. Suppose you have $1,253.24 in a bank account. If someone were to ask how much money is in the account, you would probably give them the number you thought most nearly satisfied their needs. To a friend, you might say it is $1000 (first-order approximation); to a family member, perhaps you would divulge $1200 (second-order approximation); but to the Internal Revenue Service, the sum given would even include the pennies (exact result). Indeed, in most of the areas of technological achievement—in particular those regarding simple questions about ships, airplanes, and flow of fluid through pipes—can be discussed in terms of a single dimensionless constant. Higher-order approximations would require additional constants. Even without an understanding of hydrodynamics, we were able to fly planes and sail ships around the world long before aviators and sailors had Bernoulli's understanding of wind and water.

Planes Fly Due to Lift

The simplest of the difficult technological problems, that of the heavier-than-air flying machine, was solved rather quickly once the principle of lift was shown to be valid. The proof of principle was the success of the Wright brothers flying machine. Igor Sikorsky built a four-engine, 92-foot wing-span, airplane capable of carrying a payload of 4.5 tons within ten years of the first flight of the Wright brothers at Kitty Hawk. In 1914, its successor, the *Ilya Mourometz,* stayed airborne for 6.5 hours and carried six passengers. This early carrier could have served as the prototype for the modern-day commercial airliner and flew nearly a century ago.

Rockets Defy Even Gravity

More difficult technological feats were the first launching of rockets, putting a human in orbit around the earth, and landing on a person on the moon. Robert Goddard first succeeded in shooting a rocket over a mile vertically on May 31, 1935. Twenty six years later, a Soviet cosmonaut was the first human to encircle the globe in a rocket-launched satellite. This accomplishment on the part of the Soviets stimulated the initiation of the United States space program. Thus, only a few decades after the first primitive rocket experiments, the United States had a highly successful space program, as did the Soviet Union, and we put a man on the moon. The space station that is presently being built has also launched us into the era of biology and physiology in a weightless environment.

Nuclear Reactions, the Janus of Power Generation

Perhaps the most dramatic of the technological achievements of the past century was controlled nuclear fission. On December 2, 1942, Enrico Fermi and his colleagues, under the bleachers of the football field at the University of Chicago, established the first sustained fission chain reaction. The fact that they were able to generate power through the mechanism of cascading evermore neutrons, thereby releasing the energy of the nucleus in a controlled way, demonstrated the feasibility of a nuclear chain reaction. This experiment established that both the atom bomb and a nuclear power plant could be built. After three years, the atom bomb had been constructed, tested, and used against Japan. Within fifteen years of their ground breaking research, the first commercial nuclear power plant was operating.

Few Dimensionless Constants

These great engineering successes involved processes that could, to a first approximation, be characterized by a small number of dimensionless constants, such as the Reynolds number, the Euler number, and the Froude number. Hence, only a small number of model experiments were necessary to determine the feasibility of a project and to estimate the cost and difficulties to be surmounted in those experiments. Even the space program was broken down into a number of subprojects, each of which could be analyzed in terms of a small number of dimensionless constants, so that the results of many independent model tests could be used as a basis of the required full-scale engineering designs. But not all physical phenomena, and certainly not most physiological phenomena, can be described using one or a few dimensionless constants.

Failure of Too Many Dimensionless Constants

The above record of successes has led the public, and even numerous scientists, to believe that with a little money (or a lot of money) and ingenuity, any desired scientific goal can be achieved. Unfortunately, this is not always the case. Consider the magnetically confined fusion program that started as Project Sherwood in 1951. Nuclear fusion is the mechanism by which the sun releases its energy. It is also the mechanism used in the hydrogen bomb. The aim of Project Sherwood was to accelerate a plasma of ion-deuterons to a velocity at which, at an energy equivalent to a temperature of approximately 2×10^8 °K, nuclear fusion would occur with a tremendous release of energy. Since 1/6500 of the hydrogen in ocean water is composed of deuterium, it is considered by optimists that, if controlled fusion can be successfully achieved, the energy problems of the world would be solved. Unfortunately, 50 years and hundreds of millions of dollars later, energy by confined fusion seems even further away than it did in 1951. What has happened? Why has this branch of physics failed to live up to technological expectations?

Tyranny of Many Dimensionless Constants

Montroll [87] maintained that the magnetically confined fusion program has fallen victim to the tyranny of many dimensionless constants. Unlike the successes listed above, each of which is dependent on one or two dimensionless constants, the complication of the magnetically confined fusion program seems to be that all the hydrodynamic dimensionless constants (about eight in all), as well as several electromagnetic and nuclear dimensionless constants, are intimately connected in the process of transforming a low-density, low-temperature plasma to a higher-density, very-high-temperature plasma. Since the cost involved in, or the time required for the understanding of, the nature of a process characterized by N interacting dimensionless constants can be expected to grow exponentially with N, we should not be surprised by the slow progress in the field of magnetically confined fusion.

How to Be Free of the Tyranny

The genius of individual inventors sometimes allows them to cut costs and time by going directly to the correct regime of the dimensionless constant of interest, without conducting model tests over a broad range. The probability of an individual's being identified as a genius by going "directly to the point of operation" in the development of a technology that involves N connected dimensionless constants decreases exponentially with N.

They Are Dreamers

Certain social situations and environmental processes might also depend on a large number of dimensionless constants. The understanding of these processes is not exempt from the tyranny of many dimensionless constants; nor is an attempt to make policies exempt, without a considerable insight into the manner in which a change in a single dimensionless constant influences others. Just as the enthusiast for magnetically confined nuclear fusion knows how he/she *would like* to solve the energy problem, so the enthusiast for social and environmental reform knows how he/she *would like* to make our lives full of harmony and beauty. Unfortunately, both of these classes of enthusiasts will remain dreamers until the tyranny of many dimensionless constants is overcome. We shall examine the implications of these ideas applied to biological and physiological systems in a short while but, for the moment, let us continue to build our foundation of knowledge regarding the application of hydrodynamics to the biodynamics of the human body.

13.1.4 More About the Dimensionless Numbers

Three Dimensionless Numbers

In the separation of the hydrodynamic equation of motion into its dominant parts, we introduced three dimensionless constants—the Euler number, the Reynolds number, and the Froude number, each named after the scientist who recognized the physical importance of a particular combination of physical quantities. The Reynolds number measures the relative importance of viscosity to inertial forces in fluid flow, the Euler number does the same for pressure, and the Froude number measures the relative importance of gravity.

Euler Number Can Change

Leonhard Euler, a Swiss mathematician of the eighteenth century, was a friend and colleague of the brothers Daniel and Nikolaus Bernoulli and student of their father, James Bernoulli. He was the first to recognize the important role played by pressure in fluid flow. Recall that Bernoulli's equation [Equation (13.9)] does not contain either viscosity or a body force, and relates the pressure and velocity at one point in a fluid to the same two quantities at another point in the fluid. Suppose the fluid is at rest in a container at pressure P_1 and exits from an orifice to the atmosphere with a velocity V. In such a flow, the ratio of the pressure difference, ΔP, between the inside and outside of the container, to the kinetic energy density is called the Euler number E:

$$E = \frac{\rho V^2}{|\Delta P|} \tag{13.12}$$

Note that conventions vary and sometimes the Euler number is defined as the square root of the definition given here.

Patterns Due to Pressure Differences

Euler understood that the pattern of flow produced by the pressure difference ΔP in a flow field of velocity V and density ρ is determined by the above ratio, independently of the absolute magnitudes of pressure, velocity, mass density, and the boundary value. We can interpret Equation (13.12) as the ratio of the unit acceleration force produced by a difference in pressure ΔP acting on the fluid over a length L (force density $\approx \Delta P/L$) to the inertial reaction force obtained from Bernoulli's equation acting over the same length ($\rho V^2/L$). Thus, the Euler number decreases with an increase in the pressure gradient, that is, the pressure difference over a characteristic distance for a given fluid velocity and density. But note that this term becomes increasingly important as the size of the pressure difference diminishes ($E \rightarrow \infty$) because the velocity term dominates over the pressure gradient. In the latter case, the pressure term in the Navier–Stokes equation disappears, whereas in the former case, this term can become the dominant one.

Ratio of Kinetic to Potential Energy

So long as no other fluid property influences the flow, the characteristic Euler number for any boundary form must necessarily remain constant. We can see that the Euler number can also be interpreted as the ratio of the kinetic energy density to the potential energy density, and, therefore, it determines the relative importance of motion over position. However, when other fluid properties, such as the changes in the body force and viscosity, do influence the flow, we would expect the Euler number to respond to that change.

Reynolds the Hydrodynamicist

The Reynolds number is a dimensionless quantity formulated in 1833 by Osborne Reynolds and named in his honor about 40 years later. This number characterizes the relative importance of inertial flow to that of viscous action, in steady, nonuniform fluid flow. Since the viscosity, η, and the fluid density, ρ, appear in the Reynolds number [Equation (13.6)] as a ratio, it is convenient to treat this ratio of fluid properties as a property itself. The ratio η/ρ is found to be kinematic in nature, since its dimension is $[L^2 T^{-1}]$; it is, therefore, called the kinematic viscosity, ν [84]:

$$\nu = \frac{\eta}{\rho} \qquad (13.13)$$

In terms of this new parameter, the Reynolds number is given by

$$R = \frac{VL}{\nu} \qquad (13.14)$$

The Reynolds number is used to characterize the flow in the transition region from smooth laminar behavior to erratic turbulent flow.

Viscousity-Dependent Flow

It is apparent from Equation (13.14) that the larger the Reynolds number becomes, the less important is the influence of the viscosity on the patterns of flow developed by the fluid. As $R \to \infty$, the flow is more and more dominated by the inertial resistance to acceleration and the viscous resistance plays less and less of a role in the resistance to the deformation of the flow. The smaller the Reynolds number, on the other hand, the more important the role of viscosity. Therefore, $R \to 0$ corresponds to a flow that is dominated by viscosity, and inertial effects are negligible.

Reynolds Number Marks the Transition to Turbulence

Viscous flows are sometimes smooth (laminar), as in a slowly moving, deep river with sloping banks, and sometimes disordered (turbulent), as in the white water of the Colorado River in the spring. Reynolds showed in 1883 the difference between these two kinds of flow, using water flowing in a glass tube. Inserting dye in the center of the tube, he showed that the flow was laminar for $R \leq 2000$, giving rise to a smooth dye streak down the center of the tube. When $R > 3000$, the flow became turbulent and the dye streak spread to fill the entire tube. A primitive strobe light revealed that the turbulent flow was, in fact, a series of whirls and eddies moving too quickly for the unaided eye to see [88]. Turbulent fluid flow remains one of the great unsolved problems in physics and is the athlete's bane.

Froude the Naval Architect

The Froude number was named in honor of William Froude, a junior naval architect, who gained fame as one of the designers of the *Great Eastern*. This ship, from which the first successful trans-Atlantic cable was laid in 1858, was the wonder of its day. Unfortunately, since its design required a great leap from the state of the art, it was plagued by numerous engineering and management problems. His experience in the design of the *Great Eastern* motivated Froude to consider the possibility of estimating power requirements for ships from model tests rather than trying to correct design flaws on the finished vessel. This was a tremendous insight and is the strategy of ship design used today.

Froude Number

All bodies moving in a fluid generate wave patterns as they proceed. A swimmer produces waves as her arms beat the water surface, ships generate characteristic wave patterns that determine their wake in the water, and all bodies, blunt and sleek, produce waves. Since these waves have energy, they sap the energy of the moving body generating them. This is one form of the phenomenon of drag, the conversion of momentum of the body to the momentum of the fluid produced by the passage of the body. Note that wave production is distinct from the kind of drag produced in the fluid at the surface of the body. The dimensionless number that measures wave drag is the Froude number,

$$F = \frac{V^2}{gl} \tag{13.15}$$

The resistance to fluid motion per unit volume of the body of a full-sized system is the same as the resistance to a scale model (assuming skin friction is not important) if the Froude numbers are the same for the scale model and the full-sized system.

Scales for Ships

Consider the case where the length scale of a model ship is 1% of that of the full-sized ship. The condition under which the two systems have the same Froude number is one in which the velocity of the model is 10% of that of the prototype. Thus, if viscosity is negligible under the operating conditions of interest, the model should be able to determine the characteristics of the full-sized ship. But it is not just ships for which the Froude number is useful. Whenever the phenomenon of interest requires a comparison of the energy of motion with gravitational energy, the Froude number is the appropriate indicator.

Dynamic Similarity

Alexander and Jaynes [89] established an allometric relation between the length of stride and the speed of an animal across species using a dimensionless constant, the Froude number. In the present notation, the ratio of the length of the stride (l) to the length of the leg (L) gives $Y = l/L$; the ratio of the animal's speed (u) to a characteristic speed ($\sqrt{gL}$) gives $X = u^2/gL$, where g is the acceleration of gravity, yielding the allometric relation

$$Y = aX^b \tag{13.16}$$

Using the running gaits of humans, kangaroos, dogs, ferrets, and horses Alexander and Jaynes determined an allometric law with parameter values $a = 2$ and $b = 0.4$ in Equation (13.16) (see Figure 13.4). Note that this relation is also of the form of the Froude number [Equation (13.15)]. Alexander and Jaynes based their justification for this relation on what they called the dynamic similarity hypothesis, which stipulates that animals of different sizes tend to move in dynamically similar fashions

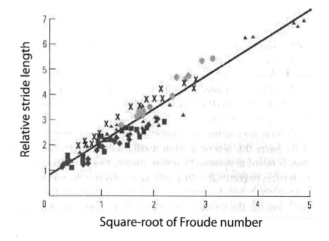

Figure 13.4. A graph of scaled variables—the logarithm of the relative stride length versus the logarithm of the square root of the Froude number—for humans, kangaroos, dogs, ferrets, and horses. (From Ref. [50]) The allometric relation appears as a straight line on this log–log graph paper.

when they walk or run at speeds that give equal Froude numbers. This was discussed earlier in terms of the elastic energy stored in muscle.

13.1.5 Pipe Flow

Laminar Fluid Flow

Let us determine the laminar flow in a straight pipe of uniform cross-sectional area. The pipe walls exert a cohesive force on the fluid adjacent to the pipe walls, causing the fluid to be stationary relative to the pipe wall. Just beyond this layer at the wall, the fluid is moving, but the drag by the stationary layer keeps the fluid from moving too fast. In the same way, drag is handed from one layer of fluid to its neighbors all through the flowing fluid away from the pipe wall to the pipe axis. The flow at the center of the pipe is the greatest, decreasing to zero as the walls are approached. A profile of the typical flow field for both laminar and turbulent flow is depicted Figure 13.5. Note that for laminar flow, the average speed is parabolic and, as we said, is greatest along the axis of the pipe. The turbulent flow, on the other hand, has a velocity profile that is much flatter across the pipe, indicating that the flow speed is fairly uniform in the pipe.

The Velocity Profile in a Pipe

The drag in the fluid is a consequence of viscosity. To change the speed of the steady flow in the pipe to a faster rate all along the pipe requires a change in the overall applied pressure, so that the pressure difference between the ends of the pipe increases. We now wish to determine the mass of a real fluid flowing per second through a pipe of uniform circular cross section. There are a number of ways we could estimate the flow rate but, in keeping with our philosophy, we use a dimensional argument.

Flow Rate Through a Pipe

The equation for the rate of mass flow must involve the change in pressure over the length of the pipe $\Delta P/l$, the density of the fluid ρ, the viscosity of the fluid η, and the radius of the pipe r:

$$\frac{dm}{dt} = C\left(\frac{\Delta P}{l}\right)^{\alpha}\rho^{\beta}\eta^{\delta}r^{\gamma} \tag{13.17}$$

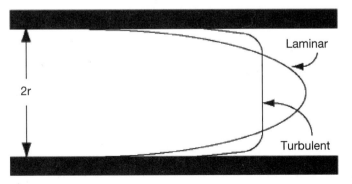

Figure 13.5. A velocity profile of the fluid flowing in a straight pipe of circular cross section.

Huntley [17] has shown that a straightforward solution of the dimensional equations resulting from Equation (13.17) does not yield a useful result. He finds that it is necessary to distinguish between two kinds of mass. We have, until now, assumed that the mass that is a measure of inertia, discussed by Galileo and Newton, is, in fact, the same as the mass that is a measure of the quantity of matter as measured with the force of gravity. It is standard in physics to assume that inertial mass, $[M_i]$, and gravitational mass, $[M_g]$, are, in fact, equal to one another. However, this has not been proven experimentally. The best we have been able to do is to show that the two are proportional to one another. This is one of those rare cases in which this subtle distinction is important.

Dimensional Equation for Drag

The dimensional equation can be constructed from Equation (13.17) using the fact that the mass flow has the dimensions $[M_g T^{-1}]$ using the gravitational mass, the change in pressure over the length of the pipe has the dimensions $[L^{-2}M_i L T^{-2}L^{-1}]$ = $[M_i L^{-2}T^{-2}]$ using the inertial mass, the fluid mass density has dimensions $[M_g L^{-3}]$ using the gravitational mass, the dimensions of the viscosity of the fluid has dimensions $[M_i L^{-1}T^{-1}]$ using the inertial mass, and the radius of the pipe has dimensions $[L]$, so the dimensional equation is

$$[M_g T^{-1}] = [M_i L^{-2}T^{-2}]^\alpha \, [M_g L^{-3}]^\beta \, [M_i L^{-1}T^{-1}]^\delta \, [L]^\gamma \qquad (13.18)$$

Poiseuille's Equation for Fluid Flow in a Pipe

We now assume that Equation (13.18) is dimensionally homogeneous in the two kinds of masses, separately, as well as in length and time. The algebraic equations obtained by equating the exponents of dimensions are:

$$\text{exponent of } L: 0 = -2\alpha - 3\beta - \delta + \gamma$$

$$\text{exponent of } T: -1 = -2\alpha - \delta$$

$$\text{exponent of } M_i: 0 = \alpha + \beta \qquad (13.19)$$

$$\text{exponent of } M_g: 1 = \beta$$

the solution to which yields the parameter values $\beta = 1$, $\alpha = 1$, $\delta = -1$, and $\gamma = 4$. Thus, we obtain for the laminar fluid flow in a straight pipe of uniform circular cross section,

$$\frac{dm}{dt} = C\left(\frac{\Delta P}{l}\right)\frac{\rho}{\eta}r^4 \qquad (13.20)$$

This fourth-power dependence of the flow rate on the radius of the pipe is truly remarkable. If the radius of the pipe is doubled, the flow rate is increased by a factor of sixteen, a very sensitive dependence of the flow on the radius. In physics departments, Equation (13.20), with $C = \pi/8$, is referred to as Poiseuille's equation, after the physician who constructed it. But as Vogel [86] points out, this could also be called the Hagen–Poiseuille equation after its independent discoverers. Hagen had published his results a year prior to Poiseuille (1839 versus 1840), but he expressed them in obscure units, and for many years remained unrecognized. Vogel comments that there is a lesson to be learned here: it is in one's best interest to explain scientific matters in as clear a way as possible. If one's scientific notation is obscure, then very often one's scientific reputation will also be obscure.

Blood Flow

As we have discussed, the property of a fluid that resists shearing deformation is viscosity, or as said so eloquently by Newton: "a lack of slipperiness between adjacent layer of fluid." Many fluids, like water, which has a viscosity of 0.01 poise at 20.2 °C, are called Newtonian fluids. However, blood has a viscosity that is considerably greater than water and is referred to as a non-Newtonian fluid. It was in studying the flow of blood in the venous system that Poiseuille and Hagen derived Equation (13.20).

Dependence on Radius

The dependence of Poiseuille's equation on the fourth power of the radius of a tube has important physiological implications. For example, a decrease of only 16% in radius will reduce the flow of blood by a factor of two. This dynamical sensitivity of blood flow to the radius of the vein implies an extremely effective control of the flow by the arterioles. Special bands of smooth muscle in the walls of the arterioles control the distribution of blood to different parts of the body.

Kinetic Energy Density of Blood

The kinetic energy density of blood flow is produced by a pressure difference, as we learned from Bernoulli's equation. The mean velocity in the aorta for a resting cardiac output (5 liters per minute) is about 20 cm/sec. The kinetic energy density can be expressed in terms of the kinetic energy per milliliter of blood, so that we have $\frac{1}{2} \times (20 \text{ cm/sec})^2 \times 1 \text{ gm/cm}^3$, which works out to 200 ergs/mil or 200 dynes/cm^2. The latter value is clearly a force per unit area, or a pressure. This value of the pressure can be expressed in terms of milliliters of mercury using the result we obtained earlier: 1330 dynes/cm^2, equivalent to 1 mmHg. Thus, we have an average kinetic energy density of 0.3 mmHg, which is the mean difference between the side pressure and the end pressure of the aorta.

Question 13.3

Why is the viscosity of blood greater than that of water? What are some of the implications of blood having a greater viscosity than that of water? (Answer in 300 words or less.)

Question 13.4

Determine how much the radius of a blood vessel must be reduced in order to decrease the rate of flow by a factor of four. How much does this reduce the speed of the blood in the vessel? Explain your answer. (Answer in 300 words or less.)

Question 13.5

Assume that in the ejection period of systole the velocity is three times the mean. What is the pressure difference between the left ventrical and the aorta? During heavy exercise, the cardiac output may be five times the resting level. What is the increase in pressure above the resting level?

**13.2
LOCOMOTION
IN FLUIDS**

What Have We Discussed?

In this chapter we have focused on the physics of fluids, both when the fluid is stationary and when it is in motion. We have briefly discussed how drag influences a person's walking in water, how the viscosity of blood, gravity, and the pumping of the heart affect blood pressure, and have given a brief description of the phenomenon of walking. There are many other activities in which fluids interact with a relatively slowly moving body, such as in water skiing, running, rowing, and swimming. Activities in which fluids interact with a relatively fast-moving body include sky diving, bicycle racing, and ski jumping.

13.2.1   Swimming As a Form of Flying

Different Kinds of Swimming

We shall go over a number of these phenomena in some detail and leave the remainder for others to consider. Our intent is not to be exhaustive, but rather to be representative. In particular, we want to identify the more important physical and biological mechanisms in the locomotion phenomenon being discussed. For example, swimming can be discussed from a number of perspectives. It can be studied as a recreational sport or as a competition, but in either case the science of swimming, if one could give it such a dignified name, should be the same. The physical mechanisms determining how we move in water are not subject to our intent; we are as fast or as slow, as determined by physical law. Our purpose now is to explain, insofar as we can, what those physical laws are.

Competition

A complete discussion of the times involved in competitive swimming is given by Hay [2]. He discusses the time between the sound of the starting gun and the swimmer leaving the blocks, the time the competitor is in the air, and the subsequent glide time in the pool before the swimmer begins his/her stroke. Once initiated, there is the time spent in stroking and the time spent in turning. Each of these times is discussed in great detail by Hay with an eye to assisting in the reduction of the overall time to swim a given number of lengths of a pool. A number of strokes are considered, such as the front crawl, the back crawl, the butterfly, and that old favorite, the breaststroke.

Different Strokes

We do not intend to discuss the speed of a swimmer. We suppose that, if needed, the swimmer's speed could be calculated by taking the distance traveled and dividing it by the total time necessary to cover that distance. There are other more interesting quantities indicative of the efficiency of the swimmer, such as the total distance traveled divided by the number of complete strokes taken. This ratio gives the average distance covered per stroke, which Hay refers to as the *average stroke length*. If you think about it, the same quantity might be used in rowing, with the swimmer's stroke is replaced with the stroke of the oar. Another quantity of the same sort is the *average stroke frequency,* which is the number of complete stokes divided by the total time taken to complete those strokes. These are useful point measures of the efficiency of swimmers and provide a way of directly comparing swimmers in different events using different strokes.

The Physics of Strokes

We, on the other hand, are interested in the forces exerted and overcome during each of the swimmer's strokes. These are the propulsive forces generated by the motion of the swimmer's arms and legs, and the resistive forces of the water opposing the passage of the swimmer. However, this appears to be an active area of research, in that the relative contribution of the arms and the legs to the propulsive and resistive forces in different strokes have not yet been completely determined. In fact, it has not been determined if the legs contribute at all to the propulsive forces driving the swimmer forward. We will be satisfied with identifying the physical mechanisms that determine the locomotion of a swimmer without giving a detailed description of how this is done.

The Magnus Effect

The dominant force contributing to the motion of swimmers is the complement of drag, called lift. It is lift that keeps an airplane from crashing to the ground and none of the innovators of flight were aware of its existence at the time of the first flight, even though the effect had been observed experimentally. The phenomenon of lift was indirectly explained by Lord Rayleigh [90], who was attempting to understand the swerving flight of a "cut" tennis ball. Such a ball does not follow the parabolic path of a bullet or arrow that is predicted by Newton's laws. The reason the ball does not follow such a path is because the racket induces rotational motion of the ball in addition to the translational motion propelling the ball over the net. The peculiar motion of the ball is a manifestation of what is called the Magnus effect. Rayleigh's argument to explain this effect is based on Bernoulli's principle.

Flow Past a Cylinder

Von Karman [91] explains that if a right cylinder is placed in a uniformly moving fluid, and if we neglect viscosity, there will be no forces acting perpendicular to the surface of the cylinder. In Figure 13.6, the fluid flow around the cylinder is depicted by the symmetrical streamlines. Do not be overly concerned about our neglecting the effect of viscosity; we will see from later experiments that this is a reasonable thing to do.

Rotation Induces a Pressure Difference

Now we superpose at the surface of the cylinder a constant circulatory flow that moves in a clockwise direction. At point *A*, we add the circulatory velocity to the ve-

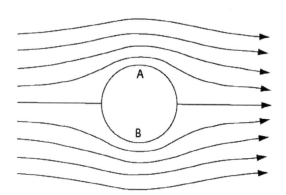

Figure 13.6. Ideal flow of a fluid past a circular cylinder.

locity of the flow, thereby increasing the fluid's velocity, as depicted by the increased density of the flow lines in Figure 13.7; whereas at point B, the direction of circulatory motion is in opposition to the velocity of the flow, so the net velocity of the fluid is decreased. According to Bernoulli's principle, without the rotation, the pressures of the fluid at points A and B, would be the same, as depicted by the symmetry of the streamlines in Figure 13.6. But with the circulatory motion, the pressure is higher at B than at A, because of the changes in velocity, as depicted by the asymmetry in the streamlines in Figure 13.7. Thus, there is an upward force produced by the difference in pressure between points A and B. If the circulatory motion had been counterclockwise, it is evident that the force would act in the downward direction.

Rayleigh's Tennis Ball

This idealized situation is now used to explain the behavior of a spinning tennis ball. Because of the rotational motion of the ball, its rough surface generates a circulatory motion of the air at the ball's surface. The circulatory motion is in the same direction as the ball's rotation and, as in the case of the cylinder, this motion superimposes on the flow of the air relative to the ball. The effect of the superposition of the two flows produces a force perpendicular to the velocity of the ball, that is to

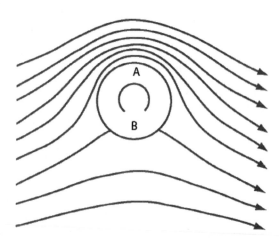

Figure 13.7. Ideal flow past a circular cylinder with clockwise circulatory motion superimposed.

say, perpendicular to the trajectory of the ball, due to the pressure differential. If the spin is opposite to the direction in which the ball is traveling, the lift is positive, and the effect is an apparent reduction in gravity. If the spin of the ball is in the same direction in which the ball is traveling, the lift is negative, and the effect is an apparent increase in gravity. In the first case, the ball will travel farther than without the spin and, in the second case, the distance the balls travel is reduced from what it would have been without the spin.

Flow Past an Airfoil

In the earlier discussion of form drag, we showed the result of fluid flowing around variously shaped bodies. The shape that is usually associated with lift is that of an airfoil, but from the earlier discussion we also know that such a body will also have drag. In general, the equations for the lift and drag forces acting on the airfoil depicted in Figure 13.8 are expressed in terms of the kinetic energy density of the fluid flowing at a speed of u_0 ($\frac{1}{2}\rho u_0^2$), the span, l, and length, c, of the chord of the airfoil:

$$F_L = \frac{C_L l c \rho u_0^2}{2}$$

(13.21)

$$F_D = \frac{C_D l c \rho u_0^2}{2}$$

(13.22)

Here, C_L is an empirical coefficient for the lift and C_D is an empirical coefficient for the drag. These coefficients are determined by the body shape and the angle the body makes with respect to the direction of fluid flow (the angle of attack), respectively. Looking at Figure 13.8, we can see that as the angle of attack increases, a turbulent boundary layer is generated and the drag increases.

Theoretical Velocity Squared Law

The total force due to drag depends on a number of factors, including the size of the moving object, as measured by the cross-sectional area transverse to the fluid flow; the fluid mass density; and the square of the velocity of the object relative to the local fluid [see Equations (13.21) and (13.22)]. Thus, keeping other factors constant, if a cyclist doubles his/her speed, the drag experienced by the cyclist increases by a factor of four. The effect of drag is, therefore, seen to be more important at high speeds than at low speeds in such sports as cycling, speed skating, downhill skiing, and so on [92].

The Net Lift

The direction of the action of drag on a body is in the direction of the fluid flow, since drag acts to inhibit the forward motion of the body. From Bernoulli's principle, we learned that the influence of lift is perpendicular to the direction of the flow field, which at the surface of the body, say the airfoil, is perpendicular to the surface. In Figure 13.9, we schematically indicate the resultant force acting on an air-

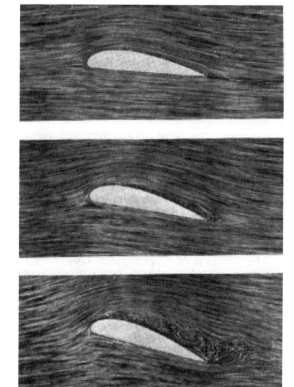

Figure 13.8. Patterns of flow past an airfoil at angles of attack of 5°, 10°, and 15°, with the camera stationary relative to the foil. (From Rouse [85] with permission.)

foil, obtained by adding the lift force and the drag force together as vectors. This is the same mechanism that operates on our swimmer in the pool.

The Resultant Force Is Always in the Same Direction

Hay [2] reviews the swimming literature and provides a plausible discussion demonstrating that during a swimmer's stroke, the motion of the hand is such that the resultant force of the lift and drag produced by the hand's motion is in the direction in which the body is moving. The remarkable thing is that the direction of this resultant force is the same in both the pulling phase and the recovery phase of the swimmer's stroke. The conclusion is that in all four of the competitive strokes men-

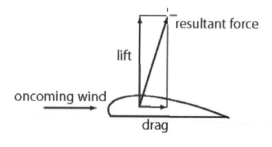

Figure 13.9. A schematic of the forces acting on an airfoil in the wind. Note that it does not matter if the air is moving and the airfoil is stationary, or if the air is stationary and the airfoil is moving; it amounts to the same thing.

tioned above, lift is the dominant mechanism propelling the swimmer through the water. As one would expect, lift was not just accepted by the biomechanics community to be the dominant mechanism in swimming, but was tested by a number of researchers; see Hay [2] for a list of references. The result of those experiments is that lift forces have been shown to contribute significantly in all four of the competitive strokes, which has led to the modifications in the pitch of the hand and arm at various parts of the stoke cycle advocated by swimmers and coaches.

Three Kinds of Drag

We mentioned that there are three kinds of drag that produce the forces resisting the forward movement of a swimmer. These are surface drag, wave drag, and form drag, in inverse order of importance. The surface drag is dismissed by Hay [2] as being of no practical significance in swimming. This, however, has not prevented a generation of Olympians from shaving their bodies in order to gain a fraction of a second edge. The wave drag becomes significant when the body is generating a bow wave as it passes through the water, or the swimmer is pushing up and down rhythmically during the stroke. The latter can be avoided, but probably not the former, although, if the swimmer keeps his/her head down, the generation of the bow wave is reduced. Finally, there is the form drag that depends on two factors—the speed of the body relative to the fluid and the cross-sectional area presented to the fluid flow, that is, the area of the body perpendicular to the direction of the fluid flow. Therefore, for a given cross-sectional area, the faster the swimmer moves through the water, the more form drag there is to overcome. Thus, what the swimmer has available is reduction of his/her cross-sectional area. The form drag can be reduced by changing the body position in the water and the way the strokes and kicking are carried out—the range of movement of the arms and legs.

Swimmers in Drag

Here again, researchers did not merely accept these conclusions; they towed swimmers around pools and measured the appropriate parameters to test these assertions. Some swimmers were towed in a prone position to see how water flows around them and how the ambient forces exerted themselves, such as the influence of the bow wave on drag as a function of body position [93]. Others kicked or performed various strokes while being towed at varying speeds, and the changes in drag forces were recorded [94]. Finally, some swimmers even performed various forms of the back crawl while being towed [95].

Question 13.6

Can you think of a way to discuss lift in terms of dimensionless numbers? A simple "no" is not an acceptable answer. (Answer in 300 words or less.)

Fluids at Rest

**13.3
SUMMARY**

In the last chapter, we presented the oldest and most fundamental of the laws governing the behavior of stationary fluids: Pascal's principle, which asserts that pressure is only a function of depth and does not depend on horizontal position in the fluid; Archimedes' principle, which asserts the existence of a buoyancy force equal to the weight of the volume of the fluid being displaced by a body immersed in the fluid; and Bernoulli's principle, which is a reformulation of the conservation of energy in terms of the pressure, potential energy density, and kinetic energy density. These laws govern the static behavior of fluids and are a consequence of Newton's laws. They also influence the dynamics of the fluid.

Fluids in Motion

We learned that fluids in motion are difficult to understand in their full complexity and so investigators focus on specific physical properties characterized by dimensionless parameters. These parameters are, at bottom, the ratio of forces, so that extremely large or extremely small values of the parameters indicate which of the forces dominate in particular situations. The Reynolds number, $R = VL/\nu$, is the ratio of the inertial to viscous forces, so for large values of the Reynolds number, the inertial forces dominate, and for small values, the viscous forces dominate. The Froude number, $F = V^2/gL$, is the ratio of kinetic energy to gravitational energy and has been found to be useful not only in the design of ships, but also in the characterization of animal locomotion. Finally, we discussed the Euler number, $E = \rho V^2/|\Delta P|$, the ratio of kinetic energy to pressure differences.

Ratio of Flow Mechanisms

We now have two results that we can apply to the flow of blood in elastic tubes. Bernoulli's principle states that the pressure should decrease as the velocity of the fluid increases. The Hagen–Poiseuille equation states that the pressure should increase as the flow rate increases. Note that these mechanisms lead to exactly opposite effects. With an increasing rate of flow, Bernoulli says the pressure should go down, whereas Hagen and Poiseuille say the pressure should go up. Assuming both these effects are present in blood flow, which of them is dominant in the human body? In a recent *Physics Today* article, Vogel [96] suggests that we take the ratio of the flow rates for the two mechanisms. From Equation (13.20), we can determine the average velocity of the blood by dividing by the cross-sectional area of the tube (πr^2) and the mass density to obtain

$$\overline{U} = \frac{\Delta P r^2}{8l\eta} \tag{13.23}$$

The pressure difference in Equation (13.23) can be replaced using Bernoulli's equation to obtain the ratio of the velocities from the two mechanisms:

$$\frac{\overline{U}}{u} = \frac{\rho u r^2}{16l\eta} \tag{13.24}$$

In Equation (13.24), η is the dynamic viscosity of blood, which is typically 0.04 poise, as we mentioned earlier. The density of blood is taken to be approximately 1 gm/cm³, the same as water, so that the kinetic viscosity for blood is $\nu = \eta/\rho = 0.04$

cm^2/sec. Using the example given by Vogel [96], consider a pipe 10 cm long and 0.1 cm in diameter carrying blood at 10 cm/sec. Inserting these values into the above ratio, we obtain approximately 0.01, indicating that the Hagen–Poiseuille average speed is 100 times smaller than that due to the Bernoulli mechanism, that is, $[U] \approx 0.01u$.

Locomotion of Fluids

Finally, we applied some of the above ideas to the understanding of the forces that propel and inhibit the motion of swimmers. The conclusion of experimenters is that the dominant mechanism in swimming is lift, just as it is in the aerodynamics of aircraft. A great deal of research has been done on this subject and we were only able to scratch the surface in our discussion.

PART IV

Data Analysis—
What We Can Learn
from Measurements

In this part of the book, we review the propensity of humanity to predict the future and in so doing we put forth Gauss' supposition that the mean is the most reasonable characterization of a large data set. Further, in addition to the mean, we examine the bell-shaped distribution, which captures the essential features of the variability in measurements (errors). Although true for a large number of data sets, these phenomena, for which the central limit theorem hold, do not span the full range of dynamical systems. In particular, biodynamical time series do not fit into this neat historical paradigm. There are too many outliers in such data sets for them to be described by a bell-shaped curve. Therefore, we introduce the notion of fractal statistical processes.

The idea for the possibility of fractal statistical processes was the outgrowth of the recognition that a simple random process could be constructed in which the variance and even the average value could diverge with increasing data. This lack of a fundamental scale in the stochastic process is one of the defining features of a fractal. We show this to be the case in the St. Petersburg game, where the banker and the gambler cannot decide on a fair ante due to the lack of scale in the game.

In the past, in order to highlight hidden patterns in a given data set, one often increased the resolution of the measuring instrument. What we demonstrate here is that the same kind of amplification of the hidden pattern may be obtained, not by increasing the resolution, but, in fact, by decreasing the resolution, through a process of data aggregation. If a pattern in the data is a consequence of scaling then it makes no difference if one increases or decreases the resolution of the measure-

Biodynamics: Why the Wirewalker Doesn't Fall. By Bruce J. West and Lori A. Griffin
ISBN 0-471-34619-5 © 2004 John Wiley & Sons, Inc.

ments; the pattern will emerge through the process of comparison of quantitative measures at different scales.

The aggregation process gives rise to systematic changes in the mean and variance. Because of the change in these two quantities, we can define a number of parameters that reveal the degree of clustering in the data set. We introduce the Fano factor and Taylor's law. The former was originally developed in the context of cosmic ray physics and later applied in physiology [5], whereas the latter was developed in ecology [97]. There are a number of other measures that exploit the scaling in time series, but since this is neither a book on statistics, nor one on data processing, we limit ourselves to these two.

We show that many, if not all, biomedical time series are random. These time series, with their random fluctuations, are correlated over long time intervals and these correlations show up in scaling. The nature of the scaling strongly suggests that the variations are fractal as well as random. No mechanisms have been identified as producing this variability, except in a very general sense, so we will review a number of possible explanations for this variability. One of the first things that comes to mind for explaining this erratic behavior is the direct nonlinear coupling of dynamical subsystems that results in chaotic behavior of the system. If nonlinear dynamics were, in fact, the cause of the variability, what properties would one expect the system to have? There are at least five functional roles that chaotic variations, the generic solutions to a nonlinear dynamical system of equations, might play in complex biological phenomena, such as heartbeat variability (HRV), stride rate variability (SRV), and breath rate variability (BRV). These are, according to Conrad [98]: (1) search, (2) defense, (3) maintenance, (4) cross-level effects, and (5) dissipation of disturbance .

Search is the first function of chaotic time series. Search would enhance the exploratory activity of the cardiovascular system, in the case of HRV. The heart rate would dance among widely separated values to anticipate possible changes in cardiovascular stress. This picture is also applicable to the variability in the gait interval, as well as in breathing.

Defense is the second function of chaotic time series. The diversity of the heart rate may prevent being ensnared in a situation in which the heartbeat becomes pathologically regular, and therefore cannot adjust to a changing environment. Consider the boxer who dances about in the ring, working to anticipate the next blow of his opponent. Variability in gait similarly prepares the walker for whatever irregularity may lay in waiting. A variety of disease states may alter the function of various subsystems, such as the autonomic control in HRV, and may lead to a loss of physiologic complexity and, therefore, to greater, not less, regularity.

Maintenance arises from the fact that a complex system, whose individual elements act more independently, are more adaptable than one in which the separate elements move in lockstep. Not using the same part of a subsystem each time a task is performed allows for rest and rebuilding. Consequently, the heart can beat strongly, breathing can be maintained, and leg muscles can avoid cramping during period of mild or even strenuous exercise.

Cross-level effects have to do with effects that scale from the shortest to the longest times. Short-term changes in the cardiac control process can have significant long-term effects.

Dissipation of disturbances is the fifth possible function of chaos. If the erratic behavior of time series is produced by a strange attractor, on which all trajectories are functionally equivalent, the sensitivity to the initial conditions is the most effective mechanism for dissipating disturbances. Dissipation occurs because the disturbance is soon mixed with motions that could have been generated by other initial conditions and are therefore consistent with chaotic dynamics.

The interdependence, organization, and concinnity of physiological processes have traditionally been expressed in biology through the principle of allometry. An allometric control system achieves its purpose through scaling, enabling a complex system such as the regulation of heart rate (HR) to be adaptive and accomplish concinnity of the many interacting subsystems. The basic notion is to take part of the system's output and feed it back into the input, thus making the system self-regulating by minimizing the difference between the desired and the sampled output. Complex systems such as those that regulate the HR involve the elaborate interaction of multiple sensor systems, and have more intricate feedback arrangements. In particular, since each sensor responds to its own characteristic set of frequencies, the feedback control must carry signals appropriate to each of the interaction subsystems. The coordination of the individual responses of the separate subsystems is manifest in the scaling of the time series in the output, and the separate subsystems select that aspect of the feedback to which they are the most sensitive. In this way, allometric properties of the cardiovascular control system not only regulate HR, but also adapt to changing environmental conditions.

CHAPTER 14

Biostatistics 1— When Things Converge

Objectives of Chapter Fourteen

- Learn that most phenomena fluctuate in time and that even statistical fluctuations, those that are completely random and uncorrelated, are law-abiding.
- Know why the most frequently used measures of statistical fluctuations of time series, as well as other experimental data sets, are the mean and variance.
- Use the history of statistics to gain an appreciation for why the average is so often used to describe complex phenomena.
- Learn how the uncertainty in the average value is measured by the variance and why this measure is arbitrary but useful.
- Understand linear regression analysis, both what it means and what its application implies about the fundamental properties of the phenomenon being studied.
- Learn that the central limit theorem is central to the understanding of random fluctuations, such as the errors made in measurement.
- Understand how to measure the influence of what is occurring now on the future behavior of random events by using the autocorrelation function.

Introduction

Even Errors Obey the Law

Let us step beyond physical and biological processes to discuss ways of extracting information from data associated with complex phenomena in general. The basis of

Biodynamics: Why the Wirewalker Doesn't Fall. By Bruce J. West and Lori A. Griffin
ISBN 0-471-34619-5 © 2004 John Wiley & Sons, Inc.

uncertainty in outcomes of physical experiments is examined and shown to obey a law. Yes, even statistical errors are law-abiding, most of the time. Statistical distributions (laws) and correlations are the two ways to characterize experimental results. The mean and variance are two measures of the statistics that are most often used. The mean characterizes a data set, and the variance indicates how good that characterization actually is.

The Average Is Not Enough

Consider two athletes, each of whom can run 100 meters in 5 seconds, on average. The first athlete is very consistent and runs within 1/100 of a second of her average on each trial. The second athlete lacks consistency, and his time can vary by nearly a second between trials. How we interpret the two situations is very different, even though the average running time of the two is the same. It is the irregularity in data and how we interpret that irregularity that we explore in this chapter. This is the first time that we quantify the notion of randomness, an idea that has appeared repeatedly in earlier chapters. Quantifying our intuition about what we can and cannot know about complexity is the goal here. Another goal is to learn how the knowledge we gain about complexity is dependent upon the amount of data we collect.

More Data Is Not Always Enough, Either

The underlying assumption in most scientific investigations is that the more data collected, the more we can know about the phenomenon being investigated, and the more certain we are in that knowledge. The notion of quantification is manifest here in the idea that more is better, especially when the measurement varies from experiment to experiment. If a given number of experiments have a certain mean and level of fluctuation, the idea is that by further increasing the number of experiments we can reduce the importance of those results that differ from the mean. We find that this is not always the case.

**14.1
UNCERTAINTY
AND
PREDICTION**

Predicting the Future

Many people, including many scientists, believe that the importance of science lies in the fact that if we have a natural law, and from a knowledge of the present state of a system we can predict the future state of that system, using that law. In fact, to a certain extent, we believe that we only understand a phenomenon, or can gain control over a phenomenon, when we can make predictions concerning its behavior. In science, we strive to understand the underlying processes and to correct deficiencies in both our understanding of these processes and in the limitations of forecasting from existing data. The study of the question of our fundamental limitations in forecasting has been termed "predictability."

We Predict the Future All the Time

It seems that one of the dominant characteristics of being human is curiosity and, in the context of science, this curiosity is about the future. As we become more civilized, our window into the future becomes greater and we attempt to anticipate what will happen for longer and longer periods of time. Parents plan for their children's education, from preschool through university. The purpose of the planning is to make the future more certain, to suppress the unexpected, and ensure that their children's lives are long and harmonious. Perhaps in the caves, mothers fought to get for their child the largest part of the day's kill. In any event, the ability to plan, or to control the future, has certainly contributed to the survival of the species and has perhaps led to a social being with a predisposition for anticipating, if not predicting the future.

Making Forecasts With and Without Theories

In Newton's day, scientists rarely made a living by doing science. They were either gentlemen, the implication being that they had an outside source of income, generally from their families, so they did not have to work, or they did things such as cast horoscopes for the gentry, as did Kepler, and construct games of chance, as did de Moivre. Scientists could make a living at these nonscientific activities because people at all ages and at all times want to know their future. Therefore, we consult a wide range of practitioners of prediction, from fortunetellers to physicists. In some cases, the knowledge we seek is easily obtained and dependable, as is the time the sun will rise and set on a certain day. Almanacs, calendars, and weather reports often contain this kind of information, because the laws governing sunrise and sunset are reliable, Newton's laws of motion and the predictions resulting from them are also reliable. On the other hand, some events in the future are more difficult to predict, like which numbers will win the lottery next week. We can easily obtain lottery number predictions from magazines, horoscopes, and other popular purveyors of good fortune, including fortune cookies. However, these predictions are always worthless in a practical sense.

Completely Random Fluctuations

Unlike the time of sunrise and sunset, the combination of numbers that will win the lottery lack continuity with events, either in the past or in the present. If the set of numbers 5, 15, 89, 54, and 36 win the lottery this week, there is no reliable way to use this information to determine the winning numbers next week. If we know how many times any individual number appeared in past sets of winning numbers, we still cannot use that information to make a better prediction of future sets of winning numbers. The winning set of numbers seems to be not at all determined by identifiable factors. Even when a past winner uses the same "system" to choose the next set of numbers played as he used to predict the winning numbers, there is no possibility that the "system" will produce wining results again. It is the lack of continuity with the past that compels us to label an event of this kind as completely *random*.

Stochastic Phenomena

Not all events are so completely decoupled from the past as the set of numbers that win the lottery. There are some phenomena that have a weak dependence on what has already occurred, but given that knowledge, it is still difficult to accurately predict how they will evolve over time. The maximum temperature that occurs tomorrow in a specific location is an example of such a poorly predicted event. The pattern of weather conditions existing today will certainly influence the temperature tomorrow, but the relationship between the pattern and the resulting temperature can be unstable and additional factors may intervene. The same is true of the maximum speed with which a runner completes the hundred meters or a swimmer completes a certain number of lengths of the pool. The highest speed today is only a partial indicator of what they might do tomorrow, but here again other factors might intervene. In fact, much of training has to do with reducing the variability in this time. We say that such partially decoupled events are a combination of deterministic and random processes. Usually, such processes are also called stochastic, in the same way as are the completely decoupled processes. Other examples of stochastic processes are horse races, total net earnings from investments, and the outcome of serious illness.

14.2 ERROR ANALYSIS

Errors Are Random

When faced with a need to make a prediction of a stochastic event, as defined above, we often "hedge our bet" by attempting to identify the degree to which the event is deterministic and then allow for a compensatory amount of error. This strategy of discriminating between what is predictable and what is unknowable is at the heart of traditional statistical analysis. When developing statistical models of future or unknown events, we propose equations that represent the relationship of the unknown event (dependent variable) to the predictive factor (independent variable) and include a random term to represent the influence of unknown external factors, that is, the influence of the environment. Put another way, we make the best guess concerning the outcome of an experiment and compare the guess with the actual outcome. What we find is that we are never exactly right in our prediction; there is always an error, that is, a deviation between the actual outcome and the guess. Even after accounting for everything we can know concerning the experiment, we still find error and this error is random.

Linear Regression Equation

In Chapter 4, we discussed linear regression, using Huxley's data to fit the allometric growth equation. In order to do this, we introduced the logarithm of the variable. Perhaps that seemed a little strange at the time. What we were attempting to do was to obtain the best linear representation of a given data set. Typical equations that are used to explain fluctuating data sets are assumed to have the form

$$U = a + bW + \xi \tag{14.1}$$

where U is the event we are attempting to predict, b is a parameter that determines the amount of influence the predictive factor W has on the event or process of interest, and a is a parameter that determines the size of the event when the independent variable W is zero. The error term, ξ, represents the magnitude and direction of how much the prediction is wrong. Statistical models of the form of Equation (14.1) can be tested and from the results of these tests explanatory theories are supposed to be either validated or refuted. For example, if the error has as many positive values as negative values, then its average value is zero, so that taking the average of Equation (14.1) we obtain*

$$\overline{U} = a + b\overline{W} \qquad (14.2)$$

This is precisely the equation that Huxley used to fit his data if we associate U with his log Y, W with his log X, and a with his log a in Equation (4.28). Unfortunately, such theories, created to explain stochastic phenomena, often fall short of their goal and the size of the error term can be quite large. One likely reason for this failing is the static nature of the model as opposed to the dynamic nature of the actual phenomenon. The relationship between the independent and the dependent variables could vary over time and additional explanatory variables may impinge on the process only at certain times and at certain levels. This appeared not to be the case in the phenomena investigated by Huxley, so linear regression analysis worked rather well.

Going on a Diet

Suppose you are interested in losing weight, so you put yourself on a diet. Unfortunately, when you weigh yourself, sometimes you weigh more than the last time you weighed yourself and sometimes you weigh less. In spite of this variation in weight, you do not stray from the diet. To determine if the diet is really working you decide

---

*The method of least squares is used to obtain the best linear representation of the data, which is to say, the optimal straight line through the data points obatined by making the defined error between the data points and the regression line a minimum. The method is implemented by defining the overall mean-square deviation $\overline{\xi}^2$ for N measurements,

$$\overline{\xi}^2 = \frac{1}{N} \sum_{j=1}^{N} \xi_j^2$$

where the sum over the elements divided by the total number of data points defines the numerical average. The average defined by this equation gives a global measure of the degree to which the data differ from the linear relation. We can write the mean-square deviation from the data points as an explicit function of the two parameters a and b:

$$\overline{\xi}^2 = \frac{1}{N} \sum_{j=1}^{N} (U_j - [a + bW_j])^2$$

but we do not know the values of these parameters. We find the values of a and b that makes $\overline{\xi}^2$ (the mean-square error) as small as possible. We minimize the mean-square error with respect to both a and b by requiring the variation in the error to vanish. Implementing these conditions, after some algebra, yields the values for the parameters:

$$a = \frac{\overline{W^2}\,\overline{U} - \overline{UW}\,U}{\mathrm{Var}\,W}$$

$$b = \frac{\overline{UW} - \overline{U}\,\overline{W}}{\mathrm{Var}\,W}$$

where the variance of W is defined by

$$\mathrm{Var}\,W = \frac{1}{N} \sum_{j=1}^{N} [W_j - \overline{W}]^2$$

to put all the weekly weight measurements on graph paper, these are the ten ficti-tious points in Figure 14.1.

Fitting the Data

Looking at the data points alone, it is clear that they tend to slope upward, on aver-age, about some theoretical line. Therefore, you might decide that the separate measurements are not important. Instead, it is the overall effect of the diet with which you should be concerned. At this point, you implement linear regression analysis and fit Equation (14.2) to the data points; that is, you find the "best" straight line that goes through the data points. This line, the linear regression curve, gives your average weight loss from week to week. Here, we might interpret the lin-ear regression curve as an indicator of a deterministic weight loss process and the deviations from this average behavior as error.

Success of the Diet

The linear regression curve suggests that you are losing weight at an average rate of 0.84 pounds per week. This is not a dramatic weight loss, but it is the slow kind of weight loss that is sustainable over the long term. If this weight loss is accompanied by a change in lifestyle, then regaining pounds, after the target weight is attained, might be avoidable. The fluctuations away from this average weight loss must be dismissed as a random process that is not strictly controlled by diet.

Unique Is Random

Thus, there are two kinds of processes: those that are predictable and, therefore, are determined by known prior or present events; and those that are unpredictable and are, therefore, independent of known prior or present events. If something is truly unique, which is to say, it does not depend on anything that has already transpired, then it must be stochastic or random. At least a unique event must *appear* random to

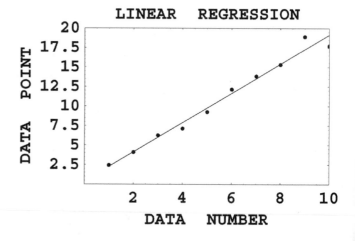

Figure 14.1. The straight line is the best linear regression curve fitting the data points (dots) using Equation (14.2). The slope of the curve gives the parameter $b = 1.88$ and the intercept of the curve with the ordinate gives the constant $a = 0.39$.

us, because we cannot see how it is related to past of present events. It is this lack of continuity with the past that compels us to label this or that event as random. In the standard example of the coin toss, it is not that we think that a coin toss is not deterministic and could not be predicted given sufficient information about how we apply the force to the coin, how long we let it rotate in the air, and how we catch it. It is, rather, that we believe our ignorance of those things does not bias the outcome of the toss and, given our lack of knowledge, the head or tail is equally likely.

Reducing the Uncontrollable

The same unbaised view may be applied to an uncorrelated stochastic process. This is a process for which knowledge of what the process has been doing, prior to the present observation, or what it is doing now, does not help in any way in determining what the process shall do in the future. Of course, not all stochastic processes decouple the past from the future as completely as one determined by a coin toss. In some cases, there is a correlation between the past and future, which is to say, that what occurs now may influence what occurs in the future. This does not mean that what occurs now uniquely determines what occurs in the future but, rather, it biases what may happen in the future. The purpose of the mind-numbing training that a professional athlete undergoes is to reduce the chance of losing, even though it does not guarantee winning. The plans parents make for their children are also intended to increase the chance of future success.

14.2.1   Gauss and His Distribution

Measurement Error

One of the first experiments in a freshman physics lab has to do with experimental error. The "experiment" has the student measure the length of a table a number of times, supposedly obtaining a different number each time the length of the table is measured. The point of the experiment would probably make more of an impression if each student in the class measured the length of the table and secretly recorded that value. Then all the measurements could be compared and it would be found that instead of a single number for the table length, there are as many numbers as there are students in the class. The point is, no two measurements are exactly the same. Each time one measures a given quantity, a certain amount of estimation is required. For example, in reading the markings on a ruler or a pointer on some gauge, there is always some variability from one measurement to the next. So what is the actual length of the table?

Ensemble of Measurements

If one measures a quantity x a given number of times, N, then instead of having a single experimental value X, one has a collection of measurements $X_1, X_2, \ldots, X_N$ In this particular example the collection is an ensemble of measurements of the single quantity of interest. Simpson, the inventor of Simpson's rule in the calculus, was the first scientist to recommend that all the measurements taken in an experiment be utilized in the determination of the quantity and not just those considered to be the most reliable, as was the custom of his time, circa 1730. Simpson was among the first to recognize that the observed differences between results follow a pattern that is characteristic of the ensemble (collection) of measurements. His observations were the forerunner of the *Law of Errors,* which asserts that there exists a relationship between the magnitude of an error and its frequency of occurrence in an ensemble. The observation relating the size of an error and its frequency of occurrence in an experiment was made by Carl Gauss, when he was seventeen years old. Gauss discovered that repeatedly measuring certain characteristics of a population, such as height or weight, yields different results, so he hypothesized that the arithmetic mean would be the "best" way to characterize the data.

Only Distribution Functions Have Meaning

Data is either discrete or continuous. In the discrete case, we might have the number of events that occur on a given day, so in N days, we obtain N data points. In the continuous case, we might have the weight of an individual, since we can always imagine that between any two weights, no matter how close together, we can always squeeze in a third weight. In either case, if the number of data points is sufficiently large, we can characterize the data by means of a distribution function. The general concept is that any particular measurement has little or no scientific meaning in itself; it is only the collection of a large number of measurements, the ensemble, that has a physical interpretation and this meaning is manifest through the distribution function. Of course, if we are talking about people, then it is the individual life that has meaning, but this meaning is very different from that of science. The ensemble distribution function, also called the probability density, associates a probability with the occurrence of an event in the neighborhood of a given magnitude. For example, in Figure 14.2 we indicate the frequency of occurrence of adult males of a given height in the general population of the British Isles. Quantitatively, the probability of meeting someone with a height X in the interval $(x, x + \Delta x)$ is $P(x)\Delta x$, where $P(x)$ is the distribution function depicted in Figure 14.2. The solid curve in Figure 14.2 is given by a mathematical expression for the functional form of $P(x)$. Such bell-shaped curves, whether they are from measurements of heights or

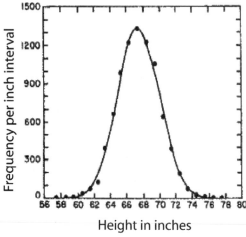

Figure 14.2. Distribution of heights. The points indicate the relative number of adult males in the British Isles having the indicated height in inches. The solid curve is the normal distribution fit to the data for a particular average height and variance. (Taken from Yule and Kendel [99].)

of errors, are described by the well-known distribution of Gauss, also called the normal distribution.

Graphing the Data

Note that in Figure 14.2 the heights are treated as discrete points and we have a mathematical expression for the theoretical normal curve. The normal curve is drawn in the figure and is clearly continuous. In principle, the height is a continuous variable, since growth is a continuous process. In practice, however, we can only comfortably measure the height of a person to within a fraction of an inch. Therefore we "coarse-grain" the data and count the number of people in an interval of height, Δx say. Such an interval is called a bin. Let us suppose that the interval (bin) is one inch ($\Delta x = 1$ in), so that we record the number of people with heights between, say, five feet and five feet, one inch, then between five feet, one inch, and five feet, two inches, and so on. The number of people in an interval is then placed at the midpoint of each bin. That seems to be the way in which the above data was partitioned. Note that the size of the bin, the interval used to coarse-grain the data, is selected to make a smooth gradation of the data. The sharper the gradient, the more important it is for the bin size to be small, since it is only with a small bin size that the full influence of the gradient can be taken into account. The gradient is determined by the ratio of the change in the number of data points between adjacent bins—the distance between data points divided by the bin size. By the same token, if the number of data points changes slowly from bin to bin, then the bins can be made quite large.

The Normal Curve

The second thing to notice about Figure 14.2 is that the normal curve appears to pass through most, if not all, the discrete data points. Because the curve does such a good job of passing through the data, we often say that the distribution of heights is normal. In this way, the underlying process (the one being measured) takes on the characteristics of a distribution because the measurement of the process has those characteristics. However, this need not be the case. Part of our task will be to understand just how much we can learn about a process from measurements, especially from measurements that have large fluctuations.

Central Limit Theorem

The proof that a sum of measurements converges to a random variable, whose statistics are given by a normal distribution or a normal probability curve, is the content of the central limit theorem. The central limit theorem itself rests on a number of important assumptions:

1. Each fluctuation (error) in a measurement process, the error being defined as the deviation of the measurement from the mean value is statistically independent of any of the other fluctuations (errors).
2. The fluctuations (errors) in the measurement process are additive, which is to say that each measurement can be represented as the sum of the quantity of

interest (mean value) and the fluctuation (error). This implies that the error process is a linear additive one.

3. The ensemble distribution function describing the statistics of each of the random fluctuations (errors) is the same.

4. The variance in the distribution of the measurements is finite, which is to say that there is a finite number that the variance of the data will not exceed, no matter how many measurements are made.

Laplace Was the First

The conditions for the central limit theorem may be summarized in the observation that the random influences (errors and fluctuations) are weak, there are many of them, and they are independent of one another. Laplace was the first scientist to calculate correctly the distribution of results obtained by the repetition of a large number of identical observations like measuring the length of the desk. In his case, he was measuring the position of celestial bodies. Under the conditions for the central limit theorem, he proved that the final distribution is always represented by a normal curve centered on the mean value.

14.2.2 Summary

One Equation Is Worth a Thousand Words

Let us briefly summarize the properties of the normal curve given by the mathematical expression

$$P(x) = \frac{1}{\sqrt{2\pi\sigma^2}} \exp\left[-\frac{(x-\bar{x})^2}{2\sigma^2}\right] \qquad (14.3)$$

1. A normal curve is a graph of the function defined by Equation (14.3), with the ordinate $y = P(x)$ and abscissa x. This curve, or function, is a model for events whose outcome has been left to chance and are, therefore, characterized by a probability distribution. This unpredictability is the reason why the variations in measurement are often called errors. The environment has unknown and unpredictable influences on measurement. These influences cause measurements to deviate from what they would be in isolation.

2. The total area under a normal curve is finite. Therefore, one can always divide the function by this finite value and thereby normalize the area under the new function to one. In this way, the fractional area under portions of the curve may be treated as relative frequencies or probabilities.

3. A normal curve is completely specified by two parameters, the mean, $\bar{x}$, and the variance, σ^2. Thus, there are an infinite number of normal curves, all of similar shape, but with different mean and variance. An affine transformation, which is a shift in the data by the mean, followed by a scaling using the standard deviation (the square root of the variance), yields a standard normal curve with mean zero and unit variance. Any other data set, with normal statistics, can be generated from the standard by changing the mean and variance appropriately. How to achieve this variation in the distribution is indicated in Figure 14.3.

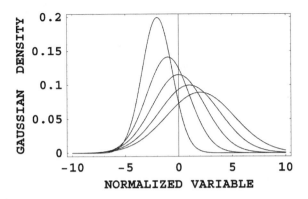

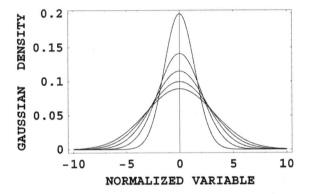

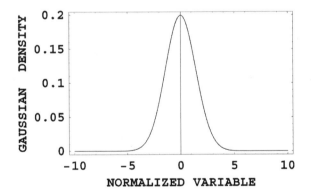

Figure 14.3. All Gaussian distributions are the same. In the top panel, five Gaussian distributions with different means and variances are depicted. In the middle panel, each of the variables has had its mean subtracted, so in the transformed variable, the distributions are all zero-centered. In the bottom panel, the variables for each of the distributions have been scaled by the standard deviation (square root of the variance) and, therefore, all collapse onto a single generic form for a zero-centered Gaussian distribution with unit variance.

4. A normal curve is symmetrical and bell-shaped, with its maximum located at the mean of the curve, so there is as much area under the curve to the left of the mean as there is to the right of the mean: exactly one-half. The area under the curve in the neighborhood of a point is proportional to the probability of the event represented by that point, so the event with the maximum probability has the value of the mean, as seen in Figure 14.3.

5. A normal curve is continuous, so there is a value of the probability density $P(x)$ for every value of x, including rational and irrational values. Therefore,

the use of $P(x)$ as a model frequency distribution assumes the x variable to be continuous.

6. The values of $P(x)$ are positive for all values of the variate x. Although scores may be positive or negative, there is no such thing as a negative relative frequency. Thus, just like the probability, the probability density must be positive definite, that is, the distribution function is never negative.

7. The normal curve approaches, but never touches, the x axis, which is to say that the normal curve is asymptotic to the x axis. The probability of large values of x occurring, either large positive or large negative, becomes smaller and smaller with the curve, but never entirely goes to zero. Therefore, you might meet someone eight feet tall on your trip to Britain, but the probability is very, very small; unless, of course, you go to an American basketball game.

8. The point on the normal curve where the slope changes, the inflection point, occurs one standard deviation on either side of the mean.

Question 14.1

We have two different populations from which we measure heights. In the first population, the average height is 6 ft and $\sigma = 2$ in; whereas in the second population, the average height is 5 ft 5 in and $\sigma = 1$ in, but they both have a Gaussian distribution. Graph both distributions and comment on the differences between the two.

Order Out of Chaos

Even though the individual members of a population may vary in an erratic way, the general characteristics of the population are themselves very stable. The fact that statistical stability emerges out of individual variability (diversity) has the appearance of order emerging from chaos and has inspired a number of metaphysical speculations that we shall not pursue here. We merely note that the normal curve captures an essential feature of a great many phenomena, that being the relative frequency with which a fluctuation (error) of a given size may occur in a sequence of measurements.

14.3 CENTRAL MOMENTS

Generators of Uncertainty

The ambiguities associated with the sources of the irregularities in a time series were touched on above. We do not know in advance whether the irregularities in a time series are a consequence of the process being open to an unpredictable and uncontrollable environment. Alternatively, the erratic behavior could be the preordained outcome of a deterministic nonlinear dynamical process. The phenomenon represented by the time series could be a combination of the two—a superposition of noise and chaos. The nature of these erratic fluctuations is determined by the probability that a fluctuation of a given size occurs within a time series of a given length. The broader the probability distribution, the larger are the random excursions of the dynamical variable, the narrower the distribution, the more like the average is any particular measurement. Thus, we must first learn about probabilities and how to use them before we can understand time series data. So let us go back to fundamentals.

14.3.1 Averages, Means, and Expected Values

The Law of Averages is Illegal

The first fundamental consideration involves a common misconception concerning how we understand the "law of averages." Anyone who has sat through an evening of poker and not obtained a single hand worth betting on has uttered the immortal words, "I'm due for a hand." Anyone who has watched his/her favorite basketball team go down in defeat, because their star player could not find the hole in the basket, has fallen victim to this misconception by uttering the phrase, "He's due for a basket." The meaning of these simple declarative statements is that, over a period of time, the desirable outcomes and the undesirable outcomes balance one another out. But, of course, these interpretations of events are *wrong*. The reason they are wrong is neither obvious nor simply explained, but let us try.

A Fair Wager

Let us consider the simple example of a fair wager, that of flipping a coin. Suppose we bet on a head appearing on the toss of a coin. If the coin is fair, then in a large number of tosses we have a sequence of heads and tails, in no particular order. In fact, if the coin is well balanced and we flip it the same way each time, the sequence should be completely random, which is to say, if N_H is the number of times heads appears in the sequence, out of a total number of N tosses, then

$$\lim_{N \to \infty} \frac{N_H}{N} = \frac{1}{2} \tag{14.4}$$

The relative frequency of the number of heads (tails) in the sequence, as the length of the sequences becomes infinitely long, is exactly one-half. This is an application of the law of large numbers invented (discovered?) by James Bernoulli in 1713 and refined by numerous mathematicians over the centuries. This is the definition of probability, based on two equally probable outcomes of an experiment.

Probability As a Relative Frequency

Let us consider an experiment that has N equally likely, outcomes. By equally likely, we mean that there is nothing in the experiment that favors one outcome over any other outcome. An example of this would be the coin toss above. Let us say that we desire an outcome E, the so-called favorable event, and we obtain n of these favorable outcomes in a sequence of N experiments. The probability $P(E)$ of an event E is defined by the equation

$$P(E) = \frac{n}{N} \tag{14.5}$$

where N is the total number of equally likely outcomes when $N \to \infty$, and n is the number of outcomes associated with the event E, such as the head in the coin toss case above. There are two extremes of Equation (14.5) that are particularly important. The first extreme is if the event E is part of each and every outcome, in which case $n = N$ and $P(E) = 1$, so that the event E occurs with certainty. The second extreme is if the event E is not part of any outcome, in which case $n = 0$, and $P(E) = 0$, so that the event E is impossible. All other cases have a probability that falls between certainty (probability unity) and impossibility (probability zero).

Average or Mean Value

To determine what is meant by an average value we assume the existence of a data set $\{X_j\}, j = 1, \ldots, N$ having N values. The arithmetic average of the data points is given by

$$\bar{X} = \frac{1}{N} \sum_{j=1}^{N} X_j \tag{14.6}$$

What is actually meant by this definition of the average value? In general, an average value is defined as the product of the value of the variable and the probability that the variable achieved that value, so that in terms of our measured data set, we obtain

$$\langle X \rangle = \sum_{j=1}^{N} p_j X_j \tag{14.7}$$

The only way in which these two definitions of an average are the same is if the probability of measuring any value of the variable is the same, which is to say, $p_j = 1/N$. Thus, each and every value measured has the same probability. But, of course, this is not always the situation.

Geometric Interpretation of an Average

Let us construct a physical picture of the averaging process. Suppose we have thirty-six blocks all of equal weight and we distribute them along a very strong, but weightless beam. The point at which one places the fulcrum, such that the beam is in balance, is analogous to the average. In Figure 14.4a, we denote one way the blocks can be symmetrically placed about the origin. The greatest number of blocks, six, are placed at the origin, with five on each side and four at the next farther point on either side, and so on. From our experience with seesaws, we know that if the product of the distance and the weight on either side of the fulcrum is the same, then a balance is obtained. If there is more than one weight on each side of the fulcrum, then it is the sum of the products and the weights on each side that must be equal for a balance to occur. This is not true in Figure 14.4b, where three of the blocks have been moved, so the point of balance, the average, has been shifted from zero to the right.

Sample Statistics

The values of an observable obtained in an experiment are referred to as the sample, and all quantities derived from these values are prefixed with the label "sample." It is assumed (if you sample randomly) that your sample statistics represent those of the population from which the sample is drawn. In a way, you are aiming at a target

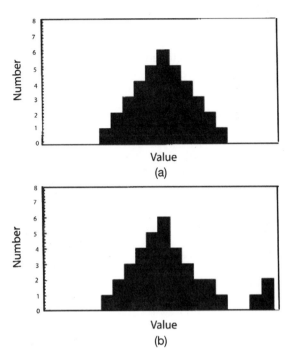

Figure 14.4. Two examples of distributing thirty-six blocks of equal weight along a weightless beam. In (a), a symmetric placement is made so that the point of balance is at zero; this is the same as the average value. In (b), three of the blocks have been shifted to the right, so that the point of balance, and therefore the average, has also been shifted to the right.

(the population) and your sample mean represents one shot. With enough shots, your errors work out and the majority of your results will hit near the center of the target, which represents the mean of the population. Because we usually cannot stand around shooting until we even out our misses, or take hundreds of measurements of our variables, we make the assumption that the best indicator of the population mean, say the average heights of humans, is the sample mean height in a randomly chosen group of people.

Noun or Verb

Note that the word average has been used as both a noun and a verb. As a noun, average denotes the value of an observable that represents a given data set. As a verb, average means to perform the process of taking the weighted sum over the observed values of a data set. In the first case, average is a thing in itself; in the second case, average is a process, something done to the data set. Scientists use these two meanings of the term almost interchangeably.

Mean and Median

Remember that if you are guessing the value of a measurement, your best chance of being correct is to guess the mean value. Often, the only mean you have is your own sample mean. In addition to the mean value, another way to characterize a data set is through the median value. The median value and the mean value can sometimes be numerically equal but they are quite different in concept and are often quite different in value. The best way to define the median is through an example. A data sequence is given by $\{X_j\} = \{1, 2, 3, 4, 5, 3, 4, 1, 2, 3, 5\}$, where these numbers are contrived for the example. We might be interested in the best way to characterize these data without writing them out each time we referred to them. One measure of the set is the arithmetic mean or average value given by

$$\overline{X} = \frac{1 + 2 + 3 + 4 + 5 + 3 + 4 + 1 + 2 + 3 + 5}{11} = \frac{33}{11} = 3 \tag{14.8}$$

The mean value of the data set is therefore 3. If we now reexpress the sequence in order of increasing values of the variate (the variate is a fancy name for the random variable), we obtain $\{1, 1, 2, 2, 3, 3, 3, 4, 4, 5, 5\}$ so that the median is the number in the center of the sequence for an odd number of terms,

$$\text{median} = 3 \tag{14.9}$$

If the number of terms is even, the median falls halfway between the two center values of the sequence. In this example, the mean and median are numerically equal to one another, even though they are arrived at in entirely different ways.

Question 14.2

We obtain a new sequence by changing one value from that in the text example: $\{X_j\} = \{1, 2, 3, 4, 5, 3, 4, 1, 2, 3, 35\}$. So what is the new value of the mean, the median, and the variance. Compare this to the sample calculation.

Effect of Outliers

In the example above, the mean and median values are the same and seem to represent the "center" of the sequence equally well. However, in Question 14.2 we show that this is not the case by changing a single data point. The mean value in Question 14.2 is 5.73, and is larger than any value in the data set except the extreme value 35. The median value remains unaffected by the changed value and represents the midpoint in the sequence of values as well, or as poorly, as it did before the value of the data point was changed. Such a data point, the one changed, is called an outlier, because it lies far from the mean of the data set. Such terms are often dropped from data sets, arguing that they are so different that they must not belong to the underlying process. In the distribution of income, for example, one might argue that all billionaires should be dropped from the data set. However, such people are certainly part of the process and to drop them would bias our conclusion regarding the underlying mechanisms in the phenomenon of making money. Here, we note that there is no valid statistical reason for throwing away such terms and although they may skew the mean to one side or the other, the only correct remedy for balancing their weight is to obtain more data.

Why Use the Mean Value?

One reason we rely on the mean is that it is one of the few measures of central tendency or midpoint of the data that can be calculated algebraically. We have been taught from our earliest years that what constitutes fairness is to appropriate an equivalent amount to each person of whatever is being divided. In numerical terms, this means that we add everything together and then divide by the total number of individuals sharing in the spoils. Historically, the idea of fairness came first, followed by the notion of the mean value. It is quite understandable that we have little or no trouble accepting the idea of the arithmetic mean as the appropriate way to characterize the data. We shall have reason to challenge this belief somewhat later.

Center of Gravity

Unlike the mean, both the mode (peak of the probability density) and median (point at which 50% of the data is above and 50% of the data is below), are determined by individual counts and are not amenable for use in formulae for various tests and descriptions of data. Another reason for the preferred use of the mean value is that it is the unique point to which distances from each value in the data set squared will sum to the smallest value. Said another way, when the mean is subtracted from each data point and the result is squared term by term, the sum of those squared values will be less than if the median value or any other quantity was subtracted from the individual values and the results squared and summed. In this way, we determine that the average or mean value is the center of gravity of the data, which is to say that the mean value is the point from which all the data extend with equal weight in every direction. This is depicted in Figure 14.4a, where the point of balance is the center of gravity.

14.3.2 Sample Time Series

Texas Teen Births

Let us apply our discussion of average values, mean values, and expectation values to a real data set to determine how these ideas may be later applied to human locomotion data sets. In order to focus on the processing alone, we use the number of babies born to teenagers in the state of Texas, obtained from birth certificates at the Texas Department of Health. Data included dates and number of births to women 10 to 19 years old on each day from January 1, 1964 to December 31, 1990. These data, the number of births on a given day versus the day number, are plotted in Figure 14.5.

Many Time Series Have a Regular Part

The time series depicted in Figure 14.5 has two major characteristics. One is an apparently deterministic variation over time, a background of seemingly regular oscillations; a yearly variation in the number of births. This kind of deterministic behavior is often seen in time series and is usually removed using a simple deterministic model, such as a harmonic function with a period of one year. In the context of births to teenagers, this yearly variation might correspond to the school year, or to the change in seasons. This would suggest that the mating habits of teenagers is season dependent.

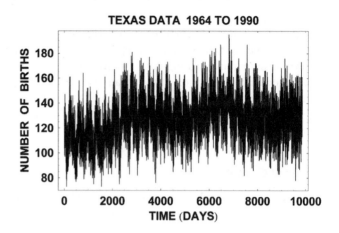

Figure 14.5. The daily teen births in Texas from 1964 to 1990. There were a total of 1,243,276 births in the 9862 days of this 27-year period. Note the background of apparently regular oscillations on which the daily fluctuations in the number of teen births occur.

Many Time Series Have an Erratic Part

The second property of the time series depicted in Figure 14.5 is the erratic fluctuation in the number of teen births. Such random fluctuations are what make time series interesting to analyze. For many decades, it was believed that these random variations constituted noise, which is the influence of the environment on the system, and was, therefore, of no interest, except that it masked the signal, the deterministic part of the time series. From this perspective, the task of the investigator is to smooth these fluctuations by means of some kind of filtering process. Today, there is an alternative interpretation of the fluctuations: they can be the result of nonlinear dynamical interactions among the teenagers resulting in chaos. From this latter perspective, the fluctuations contain vital information about the system of interest. In fact, this is where the information is often hidden, the deterministic part of the time series being an artifact. We shall introduce techniques that allow us to distinguish chaos from noise in our subsequent analysis.

More Complicated Averages

If we calculate the average of the data in Figure 14.5, we find that the number of babies born daily was 126. However, this is not very useful because of the seasonal variation in the number of births. A somewhat more useful way to present the data might be to average the data on a weekly, monthly, or yearly basis. For example, Figure 14.6 depicts the same data as in Figure 14.5, with each seven successive data points averaged together, resulting in a weekly average value of the data points. It is clear by comparing these two figures, that the degree of irregularity is less in the weekly averaged data than in the daily data. The process of averaging, therefore, suppresses fluctuations in the data.

Aggregating Is Not Averaging

We should point out that aggregating a data set and averaging a data set are not the same thing. Averaging a data set means to add a certain number of data points

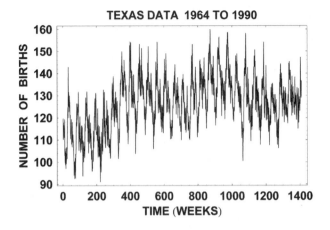

Figure 14.6. The weekly averaged number of births to teens in Texas for 1408 weeks starting in 1964. The line is intended to aid the eye in following the fluctuations in the time series.

together and then divide by that number; for weekly averages, that divisor is seven. The averaged data give a single-number representation of the seven values that occur during the week. Recall that scientists argued that the average is the best characterization of a number of data points. Thus, even though weekly birth data is more coarse-grained than daily birth data, many would argue that it contains the same information. We do not concur. It is clear by comparing Figures 14.5 and 14.6 that the averaging smoothes the data, and in that smoothing process we may lose information about the underlying dynamics. Therefore, rather than averaging, we aggregate. Like averaging, we add a certain number of data points together when we aggregate, but we do not divide by the number of points being added. Thus, aggregated data does not smooth in the way that averaged data does. However, the data may scale as we aggregate, but we will discuss that later.

What Does Smoothing Mean?

The loss of irregularity observed between the daily and weekly data sets has two possible interpretations. If the fluctuations are produced by a coupling of the teen-birth process to the environment, which is to say the fluctuations are noise, then smoothing the time series reveals hidden information regarding teen births. If, on the other hand, the fluctuations are generated by an intrinsic dynamical property in the teen-birth phenomenon, then smoothing the time series throws away information that might be crucial to the understanding of the teen-birth phenomenon. At this point in our discussion, we do not know which of these two interpretations is correct. We could skip ahead and use the results given by West [9], but that would interrupt the logic of the discussion. The point is that the problem of identifying the source of the fluctuations—noise or chaos—is not specific to teen-birth data but relates to all time series derived from complex phenomena.

14.3.3 Variance and Standard Deviation

Limitations of the Average

The average or mean value of a data set is often thought to be the best characterization of the measured quantity, which was certainly the opinion of Gauss. Of course, the average value itself can vary from one set of measurement to another, but if the statistics of the process are well behaved, then, as more and more data are included in the averaging procedure, the less the value of the mean varies from set to set. Suppose we consider the average number of beats per minute of the human heart. We may find that the average number of beats per minute of a healthy adult is 60. However, does this number remain constant over time? We know that the heart rate increases if we exercise or make love, but what about when we are sitting comfortably reading a book? Yes, even then our heart rate varies over time, independently of what we are reading. Is this heart rate variability (HRV) large or is it small? We find that this variability can be quite large and, in fact, it can be used as a diagnostic for a number of pathologies [100], which is to say that it can be used as an indicator of disease.

Importance of Variation

From the teen birth example, we see that the more important question in many practical situations turns out to be the deviation from the mean in the values that a variable actually achieves, that is, how extensively these values are strewn, scattered, or dispersed, than is the mean value itself. Will these observations be, for the most part, closely grouped around the mean value, or, on the contrary, will the majority of them differ markedly from the mean value? Here we wish to construct a reasonable measure of the dispersion of the random variable, which would indicate how large the deviation from the mean one might expect.

The Average of the Deviation Squared

Here, the measured values of the process are denoted by the set of values, $\{X_j\}$, and the mean value by the sum of the measured values divided by the total number of measurements. One of the most important averages is the square of the deviation of the variable from the mean, that is, the set of values $\{X_j\}$ is shifted using the mean to the set of values $\{X_j - \overline{X}\}$. The variance is the mean square value of the zero-centered data set:

$$\mathrm{Var}\, X = \frac{1}{N} \sum_{j=1}^{N} (X_j - \overline{X})^2 \qquad (14.10)$$

In statistics texts, the variance is often defined, for technical reasons that need not concern us here, by dividing by $(N-1)$ rather by N, but this difference is inconsequential for large N. There is a certain amount of arbitrariness using the variance as a measure of the degree of variability in a data set, but Equation (14.10) has withstood the test of time. In large part, the variance has been successful because it both satisfies our intuition about the properties of variability and is amenable to algebraic manipulations.

Standard Deviation of Teen Birth Data

Here again, we consider a real data set—the teen-birth data depicted in Figure 14.3. Using a standard program for the calculation of the mean value and variance (*Mathematica*) we obtain 126 as the mean number of births and the variance is given by

$$\text{Var}\, X \approx 296.89 \tag{14.11}$$

One test of how well the mean characterizes the data is determined by taking the square root of the variance to obtain the standard deviation:

$$SDX = \sqrt{\text{Var}\, X} \tag{14.12}$$

The standard deviation has the same dimension as the variable of interest and is, therefore, directly comparable with it. For the teen-birth data, the standard deviation is given by the square root of Equation (14.11):

$$SDX \approx 17.2 \tag{14.13}$$

So for this particular time series, the variation in the number of births, as measured by the ratio of the standard deviation to the mean, is approximately $17.2/126 \approx 14\%$. When the standard deviation is much smaller than the mean value, $SDX \ll \overline{X}$, the mean value is taken to be a good measure of the data set. A 14% standard deviation would indicate that the mean number of births alone is not sufficient to characterize the phenomenon because the variation around the mean value is much too large.

Variables Without Dimension

It is not possible to directly compare the mean value and standard deviation of one process with another different process because the units are different in general. Therefore, it is useful to transform the data to a representation in which units have been eliminated and the variability of the data has been *standardized*. This procedure is similar in spirit to the dimensionless variables we considered in fluid dynamics. First of all, we note that each data set can take the mean as the point of reference, or zero point, so that, we define a zero-centered variable:

$$\Delta X_j = X_j - \overline{X} \tag{14.14}$$

and the deviation from the mean value becomes the important variable. Second of all, recall that the standard deviation has the same units as the variable itself. A natural choice of standardized units is therefore the dimensionless quantity

$$z_j = \frac{\Delta X_j}{SDX} = \frac{X_j - \overline{X}}{\sqrt{\text{Var}\, X}} \tag{14.15}$$

Thus, the normalized variable provides the number of standard deviations in the interval of the variable being measured. Here ΔX_j is the deviation of the jth observable from the mean value, SDX is the standard deviation of the x data, and z_j is the standardized deviation. Let us apply this standardization technique to the teen-birth data in Figure 14.5 to obtain the data shown in Figure 14.7.

The Number of Standard Deviations

In Figure 14.7, we can see that the variability of the teen births is no longer given in terms of the number of births but in terms of the number of standard deviations, which is what this normalized variable z measures. From the figure, we can see that the data is slightly skewed above the mean value, with more excursions to large values of the time intervals than to small values of the time intervals. The overall variability is 14%, a significant number of births, being nearly four standard deviations from the mean. The fact that the data has values so far from the mean is indicative of some, yet to be determined, underlying pattern in the data [9]. There would be extremely few data points so far from the mean in a Gaussian distribution. To be precise, if a Gaussian distribution had a standard deviation of one and a mean value of zero (the bottom curve in Figure 14.3), the probability of having a value greater than four standard deviations would be 3×10^{-5}.

The Influence of One Variable on Another

In terms of the standardized deviation, two different observables can now be compared. In particular, we can determine if two biodynamical quantities are correlated with one another, say the metabolic rate and the length of the stride interval. Let us evaluate the strength of the relationship between variables, where the mean value and variance play a major role. The underlying premise on which the idea of a correlation rests is that variables that are related vary in a coordinated fashion around their mean values. For example, in a classroom one gives two tests. The scores on the two tests are said to be positively correlated if scores are high on one and high on the other, or if scores are low on one and low on the other. The tests are said to be negatively correlated if high scores on one are related to low scores on the other.

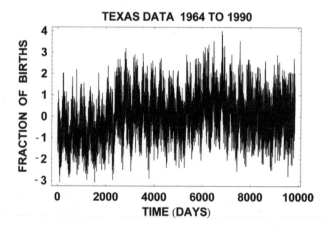

Figure 14.7. These are the same data shown in Figure 14.5, transformed using Equation (14.15).

If all but a few of the scores on each test are at the mean value, the tests are also said to be correlated, positively or negatively, depending on whether they are mostly on the same side of the mean value or on opposite sides of the mean value.

Cross-Correlation Coefficient

Let us consider two variables X and Y for which we have the same number of data points, say N. For simplicity, we say the two data sets are time series. To determine if the two variables influence one another, we first express each in dimensionless form, to suppress any variability that is due to the units of the measurements. In this way we write

$$\Delta \hat{X}_j = \frac{X_j - \overline{X}}{SDX}$$
$$\Delta \hat{Y}_j = \frac{Y_j - \overline{Y}}{SDY} \tag{14.16}$$

and the correlation between the two time series is measured by the quantity

$$\text{Corr } XY = \frac{1}{N} \sum_{j=1}^{N} \Delta \hat{X}_j \Delta \hat{Y}_j \tag{14.17}$$

Here we see that the correlation function is dimensionless, since it is expressed in terms of the standardized variables. The maximum value of the correlation coefficient is given when the two time series are in fact the same time series, in which case Equation (14.17) reduces to

$$\text{Corr } XX = \frac{1}{N} \sum_{j=1}^{N} \Delta \hat{X}_j \Delta \hat{X}_j = \frac{\text{Var } X}{SDX^2} = 1 \tag{14.18}$$

Using the definition of the standard deviation, we see that Corr $XX = 1$ or, in other words, a variable is perfectly correlated with itself. An alternative notation for the correlation coefficient is

$$r_{XY} \equiv \text{Corr } XY \tag{14.19}$$

or r alone and $r = 1$ mean a completely correlated process.

Another way the above correlation coefficient is used is to take the same time series and lag the observations by a given amount to define a second time series. The original time series is $\{X_j\}, j = 1, 2, \ldots, N + n$, from which we construct a second time series $\{Y_j = X_{j+k}\}$ that is lagged by k steps, so that the correlation function defined in Equation (14.17) can be used to construct

$$r_k = \frac{1}{N-k} \sum_{j=1}^{N-k} \Delta \hat{X}_j \Delta \hat{X}_{j+k} \tag{14.20}$$

In this case, we have only a single time series and not an ensemble of data sets, so Equation (14.20) is an estimate of the autocorrelation function, that, in the statistics community, is called a *correlogram*. A plot of the sample autocorrelation coefficient r_k versus the lag time k is called the correlogram of the data. When there is no separation between the time series in Equation (14.20), $k = 0$ and $r_0 = 1$. In general, the subscript k denotes the time lag between the two variables. Note that the maximum magnitude of the correlation coefficient is one, and if $r_k = 1$ this would be a horizontal line across the graph. In Figure 14.8 depicts the correlogram for the teen-birth data for time lags up to $k = 30$ days.

A Measure of Memory

One use of correlograms is to determine if there is evidence of any serial dependence in an observed time series, that is to say, if the fluctuations in the time series at one point in time "remember" what they did at earlier points in time. If so, then how much earlier can the time series remember? Note that it is difficult to interpret the correlogram for large values of the lag time but, then again, the notion of "large" depends on the size of the data set and the phenomena that is being measured.

Apparent Correlation in the Teen Birth Data

In Figure 14.8, we see a precipitous drop in the correlation coefficient from a zero time lag, where the correlation is one, to one or two days, where the correlation plummets to 0.3. However, the correlation coefficient increases again at a lag of seven days, followed by a decrease and then an increase at a lag of fourteen days, and so on. We find that this pattern of oscillations persists for very long lags. This measure of the influence of the time series between successive days indicates memory. What we mean by memory is that what happens in the fluctuations at one day carries forward to other fluctuations at successive days, so that the variations in the number of births are apparently dependent across time. The short-term memory is the weekly increase in the correlation, a manifestation of the weekly schedule in maternity wards; doctors try to avoid deliveries on weekends, if at all possible. On the other hand, the long-term form of the memory is contained in the fact that the amplitudes of the weekly oscillations are not all equal, but decrease over time. In

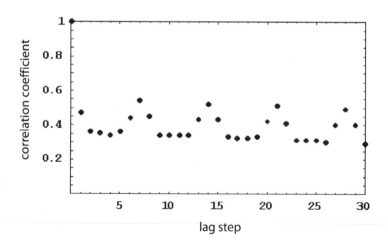

Figure 14.8. The correlation coefficient evaluated using Equation (14.20) is plotted versus the lag time, using the normalized teen-birth data from Figure 14.7.

fact, we could draw a curve connecting the peaks of these oscillations and we would find that such a curve would be an inverse power law in time [9]. However, the correlogram is too blunt an instrument to capture the full implications of the memory contained in the teen-birth time series, much less the human locomotion time series in which we are interested in these lectures. Therefore, we shall develop additional statistical measures.

Question 14.2

Consider the time series generated by the oscillatory function $\sin[0.2\pi t]$ obtained by sampling every unit of time, that is, $t = 0, 1, \ldots, 10$. The time series is, with only two-place accuracy, $\{0.00, 0.59, 0.95, 0.95, 0.58, -0.01, -0.06, -0.96, -0.96, -0.57, 0.02\}$. Calculate the mean value and variance of this time series. Also calculate and graph the correlogram.

Question 14.3

Consider the time series generated by random numbers using a computer program, $\{-2.26, -1.28, 0.52, -0.43, 0.68, -1.22, 0.22, -0.20, 0.06, -0.78\}$. What statistical properties can you determine from this limited sample?

Question 14.4

Calculate the cross-correlation coefficient of the time series in Question 14.2 with that in 14.1. Explain why this result is reasonable. (Answer in 300 words or less.)

Mean, Variance, and the Central Limit Theorem

We have discussed well-behaved data sets in which the mean is used to characterize the data set and the variance is used to indicate what the expected variability of the data around the mean value might be. In such data sets, it is expected that the central limit theorem is applicable and, therefore, the statistics of the data are determined by a normal distribution, often called the Gaussian distribution in the physical sciences after Gauss. The implication for biodynamics is that the time series one measures, such as the heartbeat interval, the breathing interval, and the interstride interval, can be best characterized by their mean value. Thus, the physician talks about the heart rate, meaning the average time between beats, and the breathing rate, the average time between breaths. We did not discuss these time series in this chapter, however.

Diverging Variance and Self-Similarity

We did examine the births to teenagers in Texas from 1964 to 1990 and these teen-birth data seem to be well behaved. So what happened to those strange data sets we talked about at the end of the last section, the random fractals? If a data set does not have a mean value, then it most certainly does not have a variance. The divergence of the variance with increasing amounts of data is a consequence of a property of fractals known as self-similarity. Self-similarity is defined in this context as the small irregularities of small scales being reproduced as larger irregularities at larger scales. These increasingly larger fluctuations become apparent as more data are collected. Hence, as additional data are analyzed from a fractal object or process, these ever-larger irregularities increase the measured value of the variance. In the limit, as the amount of data analyzed continues to increase, the variance continues to increase; that is, the variance becomes infinite asymptotically. What does this have to do with the teen-birth data? Or any other data set for that matter?

Where Are Diverging Variances Seen?

This surprising behavior, the unlimited increase in the variance with increasing amounts of data, has been observed in many different contexts. The phenomena in which this behavior have been observed include the measured density of microspheres deposited in tissue to determine the volume of blood flow per gram of heart tissue [101], the growth in the variance of the mean firing rate of primary auditory nerve fibers with the increase in window length [102], and many other examples that may be found in the physiological context in West and Deering [38] and in Bassingthwaighte et al. [5].

CHAPTER 15

Biostatistics 2— When Things Diverge

Objectives of Chapter Fifteen

- Use what was learned in the previous chapter to understand that certain random phenomena cannot be described by a mean and variance.
- Learn how we can use the failure of the traditional measures of fluctuations to gain insight into complex phenomena.
- The fractal dimension of a time series can help us understand the properties the generator of a time series must have.
- The distribution of the frequency of error is not always normal and the best fit to the data is not always given by linear regression, contrary to what was said in the previous chapter.
- Central moments, such as the mean and variance, can diverge with increasing amounts of data, so how these moments diverge is important.

Introduction

Things Are Not as Simple as We Thought

We start this chapter with a direct contradiction of the assumptions underlying the previous chapter. The goal here is to understand that in some situations, more data does not necessarily improve one's knowledge of a phenomenon. The average, as a measure of knowledge, and variance, as a measure of certainty, discussed in the previous chapter, are not necessarily well defined in some cases. In complex phe-

Biodynamics: Why the Wirewalker Doesn't Fall. By Bruce J. West and Lori A. Griffin
ISBN 0-471-34619-5 © 2004 John Wiley & Sons, Inc.

nomena these measures may not converge with increasing amounts of data, but may, in fact, diverge to infinity for an infinite amount of data. We propose a way to harness these divergences and make them work for us. Thus, we introduce new ways of examining the properties of complex phenomena through time series measurements.

Statistical Fractals

One measure that has proven to be of value in characterizing physical and biological systems is the fractal dimension introduced in earlier chapters through the discussions of scaling. The closer the fractal dimension is to integer values, the more regular is the phenomenon under study. For example, as the fractal dimension increases above 1.0, the process becomes increasingly irregular and less like a line, which has a topological dimension of one. This increasing irregularity continues until the fractal dimension achieves the value 1.5. As the dimension increases beyond 1.5 and approaches 2.0, the process more closely resembles a surface. Techniques successfully used to extract information regarding the scaling properties of biomedical random time series are used to determine the fractal dimension of the time series. Of particular importance is the interpretation of the statistical measures that enables us to dispel some of the mystery surrounding the physical basis of locomotion.

Not So Fast with that Distribution

The great mathematician and astronomer Poincaré in 1913 observed that, regarding the law of errors, "All the world believes it firmly, because the mathematicians imagine that it is a fact of observation and the observers that it is a theorem of mathematics" [14]. He was, of course, alluding to the fact that phenomena do not necessarily satisfy conditions 1 to 4 of the central limit theorem given in the last chapter for a given sequence of measurements. Therefore, one must examine separately each process that is measured. However, this observation is by no means unique; it was also made by the biostatistician Karl Pearson in 1900: "The normal curve of error possesses no special fitness for describing errors or deviations such as arise either in observing practice or in nature" [103].

**15.1
OTHER THAN
GAUSSIAN**

The Mean Value Is the Best Representation of the Data

Prior to the ideas of Gauss about the mean value and measurement error, the "best" measurement was the one thought to be the most reliable. It was reasonable to throw out measurements that were ambiguous or unreliable and keep only those in which the experimenter had confidence. The problem with such an approach to experimental data is that it presupposes that we understand the nature of error in measurements. However, ideas about error in the 18th century were not the primary considerations in drawing conclusions from observations in research. The recognition that error appears in all measurements and that laws govern such error set the stage for drawing conclusions about one's own observations and, more abstractly, about observations not yet made but presumed possible in a world where error follows a definite law. The mean value is the most probable value, with large deviations from the mean value occurring with lower and lower frequency in any sample as one deviates more and more from this value. The probability of making errors of varying sizes (deviations from the mean) became the basis for hypothesis testing as it is carried out in most biomedical and social research today and serves as the foundation for all parametric statistics.

A Deeper View of Normality

Let us again consider the distribution of height mentioned in the last chapter. We may conclude that a bell-shaped distribution for a phenotypic characteristic, such as height, can be ascribed to a genetic factor only if this factor operates *randomly* and *independently* on each of a large number of genes that work together to produce the phenotypic characteristic of interest. In this case, the physical size of an individual is determined by the sum of the size of over two hundred bones. In a large population of adult males, such as in the British Isles example used earlier, the small accidental differences in the average size of a particular bone fluctuates randomly from person to person and independently from bone to bone over the population. These variations are produced by a large number of both environmental and genetic factors. Some bones are larger than the population average value and some are smaller, so those above the average value are more or less balanced by those below this value. Even bones with large deviations from the mean value make a statistically negligible contribution to the variance of the sum over the population.

Dynamics in the Life Sciences

The types of phenomena of interest to most life scientists differ from the height example in that they are dynamical processes embedded within a changing environment. Consider, for example, an individual with the flu. The illness has a certain pathogenesis from contraction to eventual resolution. The ultimate resolution depends on the period in history when the disease was contracted. If the flu was contracted in the time period 1914 to 1918, then it was quite possibly a death sentence and the resolution was the grave. Today, with the appropriate medication, there are aches and pains and a few days missed from work, and the resolution is a return to health for all but the most vulnerable. Thus, the dynamical course of the disease is dependent upon how the social/medical environment of the individual has changed over time. Such considerations must be taken into account in the development of models of epidemiology and other complex dynamical social/medical phenomena.

Linear Regression Analysis

The idea underlying time series analysis is that there exists one or more mechanisms that determine the evolution of an observable over time and that, by appropriately processing the time series, we can understand these mechanisms. The theory of choice has historically been linear regression analysis, in which the data are fitted by three terms. The first term is the overall level that is fit to the data by the parameter a in Equation (14.1). The second term is linear in the quantity of interest with a coefficient that is determined by the data, the parameter b in Equation (14.1). The size of this second parameter determines the degree of response of the dependent variable to changes in the independent variable. The third term is the error that is made by means of the linear fit and is assumed to be a random variable, the ξ in Equation (14.1). The procedure is then to characterize the statistics of the error in order to determine how well the linear model represents the data.

Nonlinear Transformation of the Data

The techniques chosen to characterize the fluctuations in the linear regression equation swirl around various measures of the error, such as the average value, the variance, the correlation coefficient, and the confidence interval. But certain questions are often left unanswered. How well does the linear regression technique work when the data are not linear? In such a case, one often assumes that a simple transformation of the data might exist that makes the process linear. For example, a logarithmic transformation of the data is used, assuming that the true dynamical variable is not the one measured but the logarithm of the one measured. But what of the situation when the data cannot be made linear by means of such a simple nonlinear transformation of the independent variable? What can be done when the character of the data changes during the time interval of observation, that is, when the data is nonstationary?

Errors Now and Then

The reason that the deviation of the regression from a linear fit is called an error is because it appears to vary in an unpredictable or random manner and the underlying mechanisms are believed to be deterministic, that is, regular. This erratic behavior is, therefore, thought to be a consequence of the complexity of the process in which the deterministic observable is embedded. We argue that even if the process is completely deterministic, if it has a sufficiently large number of variables and we examine only one or a few of these, those variables may appear random in the above sense, due to the influence of the unmeasured variables. This type of irregularity is known as *noise* and for a long time this was the only kind of randomness considered in physical and life science phenomena. That this should be the case is a reasonable assumption. Biology is very complex, with many variable quantities, most of which are unseen and not amenable to measurement and control. The influence of these hidden variables on the observables of interest cannot be determined directly, but must be inferred. It is upon such inferences that most biological theories are based. Thus, it is crucial to understand that noise is not the only source of randomness in time series data, because if noise is not the explanation of the erratic behavior, the suggested theory can be entirely wrong.

Nonlinear Dynamics and Chaos

Our conceptualization of relations is often limited by the constraints of the statistical methods we use in analysis. For example, as part of exploratory data analysis, researchers examine the form and direction of a relationship. These forms have definite implications for subsequent analysis of the strength of the relations. When distributions are nonnormal or bivariate relations are not linear, the researcher transforms individual variables in order to meet the special assumptions of specific statistical analyses. This process of transformation is necessary in order to gain confidence in the statistical outcomes, but the transformation may mask important features of the true nature of the relations. Assumptions about error in the process under investigation are also addressed in deciding what statistical tests are appropriate and what conclusions may be drawn from them. Errors are almost always assumed to be random, additive, linear, and uncorrelated among the different variables. Error is also assumed to be separate from the relations of interest and its removal does not affect the conclusions drawn. These assumptions about error have in the past been necessary for interpretation of statistical findings; however, they may greatly oversimplify the complexity of the phenomenon being studied and in some situations are flat wrong.

Confusion and Chaos

This leads us to one of the main themes of our discussion of time series—the detection, analysis, and prediction of *nonlinearities* in complex physical and biological phenomena. In the context of randomness, one of the major manifestations of nonlinearity is *chaos*. There is no three-sentence definition of chaos that will be useful for the typical investigation. Like noise, chaos appears as erratic variations in time series, but unlike noise, it does not require an environment consisting of many degrees of freedom. Chaos can emerge in a continuous system described by as few as three dynamical variables, but these variables must have a nonlinear interaction among themselves. It can also be found in a discrete dynamical system as we found in Chapter 3 for a one-dimensional nonlinear equation. Chaos and nonlinearity are intimately related, chaos being a general property of nonlinear dynamical systems. Here we find the first obstacle for linear regression to overcome. If a time series is chaotic, how can linear regression capture its essentially nonlinear nature? Do the techniques of linear regression fail utterly, do they fail partially, or can they be salvaged?

Statistics and Chaos

We mentioned that chaos is a consequence of nonlinear deterministic, dynamical equations. The dynamical equation is a prescription by which a variable changes over time. The dynamical rule can be either continuous or discrete. In the physical sciences, such rules as Newton's equations of motion are continuous differential equations that predict the evolution of a system away from a given initial state to some knowable final state. In physiology, respiration and the beating of the heart are examples of continuous processes. In biology, the equations are sometimes discrete, for example, in describing the growth of a population in which the unit of time may be expressed as the generation of a species. The change in population of *Drosophila,* the fruit fly, is thereby accurately followed from generation to generation. These dynamical equations are deterministic in the sense that they determine with absolute certainty a unique final state from a precisely given initial state. There is no random component to the evolution of the systems, at least not in the sense of noise. However, there may be randomness in the sense of sensitive dependence on initial conditions leading to chaotic behavior in deterministic systems; see, for example Ott [104].

Statistics of Chaos

The chaotic solution to a nonlinear map or a system of nonlinear equations of motion is a time series that for all intents and purposes appears to be random. If the time series were, in fact, random, we would further expect that we could characterize the fluctuations by means of a probability density function, like the Gaussian distribution. Various ways of determining the statistics of such chaotic time series have shown that the probability density of the fluctuations are exponential or even inverse power law, but they are rarely Gaussian. Intermittent fluctuations can be generated by nonlinear maps that have inverse power-law statistics and, therefore, have diverging central moments.

Question 15.1

Consider the nonlinear map $X_{n+1} = 4X_n(1 - X_n)$ where the initial condition is $X_0 = 0.6$. Use the computer to generate the 200 data points. Plot these 200 data points X_n versus n and interpret the result. (Answer in 300 words or less.)

Question 15.2

Without writing out the mathematics, explain what the probability density of the data in Question 15.1 will be for $n \to \infty$. Apply the central limit theorem in your discussion. (Answer in 300 words or less.)

15.2 WHEN CENTRAL MOMENTS DIVERGE

Complex Phenomena

There is no universally accepted definition of complexity or of complex phenomena, but we can set forth a working definition that will be sufficient for our purposes. We define a complex phenomenon as being one that contains both regular and random behavior. In the limit where the phenomenon becomes either totally regular or totally random, we then classify it as being simple. A dynamical system with closed-form solutions that can be used to make predictions is simple. For example, the solutions to Newton's equations of motion, predicting the orbit of the earth's moon, can be considered to be simple. A dynamical process whose properties can only be predicted using a probability density can also be simple, for example, the motion of Brown's pollen mote in water, the paradigm of diffusion. A phenomenon stops being simple when nonlinearities become important, for example, when Newton's equations yield chaotic motion, as they do for the moons of Jupiter.

Some Averages are Infinite

Most texts on statistics end the discussion on averages where we left it in Chapter 14, leaving the impression that all data sets of interest have a well-defined arithmetic mean value. This, however, is not the case. There are a surprising number of processes for which the mean value does not exist. Take, for example, a typical college class in which a take-home assignment is given by the professor. Now suppose the professor is new to the course and wants to calculate the average length of time it takes to complete the assignment in order to determine if she has allowed sufficient time for the students to do the work. After a few semesters, the professor learns that there are always some students who do not turn in the assignment. Therefore, the average length of time to complete the assignment, if all the data points are included in the arithmetic average, is infinite. On the other hand, the professor could argue that only the time taken to complete the task, for those that were turned in, should be included in the average, and, therefore, the average does exist. But the latter result would not answer the initial question. It would be the right answer to a different question. Using this simple example, a case might be made for ignoring some of the data, based on our understanding of the overall process of education. However, that is not the situation with most complex phenomena. Our understanding is generally not sufficient to warrant throwing out any subset of the data, no matter how counterintuitive the resulting average value.

Sample Mean Values and Population Mean Values

Of particular interest to us is the so-called fractal random process for which the moments, such as the arithmetic mean, may not exist. How can a moment not exist? It does seem that it ought to be possible to repeat a given experiment a large number of times, take the measurements of a given quantity, and punch these measured values into a hand calculator, sum them, divide by the number of experiments, and thereby determine their average value. This procedure gives us the mean value of the process itself. In statistical theory, this one realization of the measured values is one *sample* taken from the entire *population* of all the possible experimental values. The mean determined from that sample of data is called the *sample mean value* and the mean of the entire population is called the *population mean value.* Whether the two are the same is moot.

A Diverging Mean Value

The sample mean value can always be determined using the available data to carry out the averaging. The population mean value, on the other hand, is a theoretical concept, and we use the sample mean value to estimate it. If the population mean value is finite, then as we collect more and more data, the sample mean converges to that fixed value, the value we identify as the best estimate of the population mean value. This need not be the case, however. If, as we collect and analyze more and more data, the value of the sample mean value keeps changing, we must conclude that the population mean value of the process may not exist. This is exactly what happens in some fractal processes. In Figure 15.1, a schematic of the results of experiments sampling the values from a population of possible values are sketched, where the sample mean value is the arithmetic average of the experimentally measured values. If, as we collect more data, the sample mean value converges to a fixed value, we identify that value with the population mean value, as shown by the dashed line in the figure. However, for many fractal processes, as more data are analyzed, the sample mean value continues to increase or to decrease as shown, and never converges to a nonzero constant.

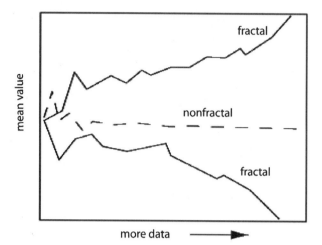

Figure 15.1. Here we show the behavior of a normal mean value with increasing data. It settles down to a finite value as the size of the data set increases. The mean values of two hypothetical fractal statistical processes are shown to diverge.

St. Petersburg Game

In 1713, Nicolas Bernoulli (a nephew of James and John), introduced an unusual game of chance in which the mean winnings were infinite. In 1724, this game was again discussed by Nicolas' bother Daniel Bernoulli in the *Commentarii* of the St. Petersburg Academy. As Shlesinger et al. [105] points out, this game has since become known as the St. Petersburg Game or the St. Petersburg Paradox. For example, suppose a player wins $1 if a coin toss comes up heads and wins nothing if the coin falls tails. The average winnings per game are the sum over all possible outcomes, that is, the product of the probability of the outcomes times the winnings of that particular outcome. Thus, the average winnings per game are ($1)(0.5) + ($0)(0.5) = $0.5. As the number of games played becomes very large, the expected winnings converge to $0.5, as shown in Figure 15.2, lower curve. So, now what is the new game?

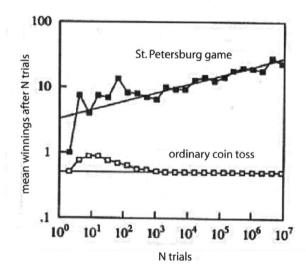

Figure 15.2. Computer simulation of the average winnings per game after N games of an ordinary coin toss and the St. Petersburg game are compared. The mean winnings per game of the ordinary coin toss converges to a finite value. The mean winnings per game of the St. Petersburg game continues to increase as it is averaged over an increasing number of games; that is, it approaches infinity rather than a finite well-defined value. (Taken from Bassingthwaighte et al. [5] with permission.)

Bernoulli's Game of Chance

Let us consider the following game of chance, originally due to Bernoulli. A player tosses a coin and a second player agrees to pay the first player $2 if a head occurs on the first toss, $4 if a head occurs on the second toss but not the first toss, $8 if a head appears for the first time on the third toss, and so on, doubling the prize every time a winning head is delayed an additional toss. What is the player's expected winnings in this game? The chance of a head on the first toss is ½. The chance of a tail followed by a head is ½ × ½ = ¼. The chance of a tail occurring on the first $n - 1$ tosses and then a head on the nth toss is $½^{n-1} × ½ = ½^n$. The prize, if there is no head until the nth toss, is therefore given by

$$\text{average} = \frac{1}{2}\$2 + \frac{1}{2^2}\$2^2 + \cdots + \frac{1}{2^n}\$2^2 + \cdots \tag{15.1}$$

which increases with the number of tosses. If we multiply out each term in the series we obtain

$$\text{average} = \$1 + \$1 + \$1 + \cdots \tag{15.2}$$

an indefinite sum of single dollars. Thus, the expected winnings of the player is n for n tosses of the coin. But n is arbitrary and can always increase; in particular, it can be made infinite. This is the type of process that is depicted in Figure 15.2, where the greater the number of tosses, the greater is the average value. In fact, there is no well-defined expectation value for this game of chance, since there is no limiting value for the mean. A very readable and entertaining account of such games can be found in Weaver [106].

St. Pertersburg Paradox Occurs Because of a Lack of Scale

The St. Petersburg game leads to a paradox. The paradox arises because the bettor and the gambling house (second player) cannot agree on a fair ante to play the game. The player favors a low ante because the median winnings (those that occur with probability up to ½) are only $2. The banker favors an infinite ante because these are his expected (mean) losses. The paradox arises because neither the player nor the banker can convince the other of his idea of a fair ante. The problem occurs at a fundamental level because one is trying to determine a characteristic size from a distribution that does not possess a characteristic size. If a billion people play the game, then, on the average, half will win $2, a quarter will win $4, an eighth will win $8, and so on. Winnings occur on all scales, which is the paradigm of fractals.

The Use of Other Methods

There are many more data sets that manifest the type of behavior depicted in Figure 15.2 than most scientists were previously willing to believe. In fact, much of time series analysis was dedicated to the development of techniques that would suppress just the kind of behavior we have described. It was a widely held belief that processes with diverging average values were and are anomalous, and the "real" process is one with a well-defined mean value. This certainly was the belief put forward by most of the physical and biological scientists of the 19th and 20th Centuries. This is, in fact, the basis of traditional statistical analysis. We must now learn to use the traditional methods when they are appropriate and to develop and implement new approaches when the traditional ones are no longer applicable.

What is Nonlinear?

The adjective nonlinear is not particularly enlightening, since it defines what we intend to examine by specifying what we shall not examine. It means that we are interested in the analysis of those random processes for which the system response at a specified time is not directly proportional to the system excitation at that time. For example, the system response may decrease with an increasing applied force or may even turn off. Consider, for example, the child's toy handcuffs that cover the index fingers of your hands with a cylinder of woven straw. If you pull your hands apart rapidly, the straw clings and your fingers are locked together. No matter how hard you pull, the handcuffs will not release their hold. By contrast, if you move your fingers together, the increase in the force at the ends of the cuffs causes the weave to open, gently releasing their hold. Thus, there is a threshold above which the action of the handcuffs becomes independent of the applied force; clearly a nonlinear effect.

What Is Stochastic Analysis?

By stochastic analysis we mean the study of the properties of random time series or, more properly, complex erratic phenomena. This, of course, includes the study of the mean, variance, and correlation function, but these are the traditional linear measures of the properties of random data. When time series are such that the variance diverges, or the mean diverges, then we can no longer effectively use these traditional quantities, but must, instead, consider other measures that are not as well known. For our purposes here, we need to go beyond the traditional, but in doing so we try not to impose an avoidable mathematical burden, if we are to keep our word.

15.2.1 Aggregating the Data

Decay of Influence

One characteristic shared by nearly all biological time series is that they fluctuate erratically in time. Some fluctuate a great deal, whereas others change only slightly. Furthermore, we have found that correlations in such time series decrease with increasing time delays. What this means, of course, is that the farther apart in time two events occur, the less influence they have on one another. We often need to characterize correlations, that is, the degree and extent of similarity of a measured property with itself as a process varies in time. It is often assumed that the random shocks that drive a system are independent of one another, but that is not the case in general. Real time series, such as the interbeat intervals of the human heart, the interbreath intervals of the human lung, and the interstride intervals for human gait, all have some level of internal correlation [9]. But we delay discussion of these time series until subsequent chapters. For the moment, we shall restrict our attention to the sample time series of teen-birth data introduced in Chapter 14. Let consider the peculiar form of the correlogram we uncovered in Figure 14.8. We want to determine if these data scale and, consequently, if the teen-birth process has a long-term memory.

Scaling in Time Series

The correlations in biomedical time series decay in time, that is, the correlation starts at one, for zero time lag, and approaches zero in some way as the time lag (time interval between data points being compared) gets longer and longer. In the physical sciences, the functional form for such decay is usually taken to be exponential. The percentage change in a variable over time for an exponential function is linear in the lag time. This equal percentage would mean that if the correlation coefficient r_k decreased by an amount $\Delta r_k = r_k - r_{k-1}$, over the lag time between k and $k - 1$, we would have

$$\frac{\Delta r_k}{r_k} = -\lambda \tag{15.3}$$

where λ is the as yet unknown constant rate of decay of the correlation. However, such correlations are now considered to be fairly short range in time, meaning that events with this kind of correlation would lose their mutual influence very rapidly with increasing time separation. This kind of decay, although often discussed, is actually, rarely seen in physiological time series. What is usually observed is a kind of self-similarity upon scaling in the correlation. This means that correlations between neighboring data points separated by a given unit of time, say seconds, are the same as correlations between neighbors in time intervals with much greater units, say minutes, even though the unit of time has changed by a factor of 60 between the two representations of the data. In terms of equations such as Equation (15.3), such scaling means that we do not have one parameter such as the decay rate λ to characterized the data. This kind of scaling is amply discussed in the following chapters for heart rate variability, stride rate variability, and breath rate variability.

Multiple Scales

The assumption of an average correlation time or average correlation distance, although quite common, is often unwarranted because the phenomenon of interest need not posses a characteristic scale. The zigzag pattern of the field mouse eluding the barn owl, the foraging of the albatross [107], the discharge of electricity from a cloud to the ground [26], and the beating of the human heart [108], are all examples of phenomena that cannot be characterized by a single scale. What they have in common is a multiplicity of scales, either in space, in time, or both, that is required to capture the full range of dynamics. These phenomena may be called complex, for want of a better word, and most, if not all, sociophysical [9, 109] and biomedical [8] phenomena are complex. We distinguish between a complex and a simple process in statistics, as we have elsewhere in this book, in terms of the number of scales necessary to characterize the phenomenon. A simple process (phenomenon or object) is one that has one or a few characteristic scale sizes. For example, all objects that can be described by Euclidean geometry are simple in this sense, so all machines resulting from the industrial revolution are simple according to our definition.

Normal Statistics are Simple

This definition of simplicity may appear to be overly restrictive, so let us consider an uncorrelated statistical process. Is such a process simple or complex? It is true that an uncorrelated random process has no memory of the past and is completely unpredictable. However, for a sufficiently large number of independent samples of this process, we can determine the average value and variance and the statistics will become normal. This is the content of the Central Limit Theorem, which states that even uncorrelated random behavior is subject to law and order. The underlying process can therefore be characterized by a mean value, variance and normal distribution, provided that the variance of the process is finite. Thus, like the complicated, but man-made, objects of Euclid, such statistical phenomena are simple, according to our definition.

The Influence of Scales

Bassingthwaighte et al. [5] point out that most modern elementary statistics textbooks have almost no discussion of how to analyze data when the variance depends on the choice of units. They go on to say that the standard texts give useful methods for assessing estimates of means, for testing differences between regions, and so on, but, in general, these approximations do not recognize that the estimates of variance may be dependent upon assuming a particular unit of time. We should emphasize here that scaling is quite different from the correlations that are observed when there is seasonality or some other periodicity in the data. These latter correlations do not scale the time series, they merely move it up and down and can be subtracted out, just as one does with a trend. What we mean by scaling is something different; scaling is a process that so interleaves the data that no amount of differencing can remove its effect. The fractal concept captures this kind of interrelatedness of the data across multiple time scales and is manifest in the autocorrelation function.

15.3 THE FANO FACTOR

The Size of Units

We analyze the heterogeneity properties of a measurement by determining how the variance depends on the size of the units used to measure the time series. We focus our attention on basic scale-independent methods of assessing temporal heterogeneity. The Fano factor (FF) is defined by the ratio of the variance to the mean

$$FF = \frac{\text{variance}}{\text{mean}} \tag{15.4}$$

How can we interpret this ratio? For a review of the application of this index to the kinds of time series in which we are interested, we refer the reader to the excellent review article of Thurner et al. [110].

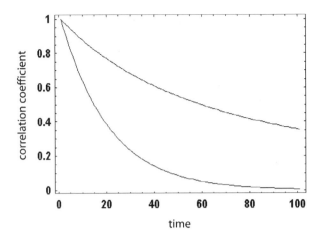

Figure 15.3. Here we compare an exponential distribution (lower curve) with an inverse power law distribution (upper curve). It is clear that the exponential distribution decays more rapidly than does the inverse power-law distribution.

A Fractal Time Series and the Fano Factor

We mentioned that a time series that scales can be characterized as fractal because it lacks a dominant scale. Therefore, fractal relationships can often be used to describe the apparent heterogeneity in a time series over some fairly large range of units, that is, multiplicity of scales. The general relationship was observed in an empirical context by Bassingthwaighte et al. [5] to exhibit a linear relationship between the logarithm of the variance and the logarithm of the size of the unit of observation. We can express the logarithm of the Fano factor, using the empirical observation, as

$$\log FF(m) = \alpha + \beta \log m \tag{15.5}$$

Thus, the size of the deviation of the Fano factor from a constant value is determined by how much the slope in Equation (15.5) differs from zero. A graph of the logarithm of the Fano factor versus the logarithm of the aggregation interval m would be a straight line with a slope of β.

How to Use Data Aggregation

We use data to estimate the parameters in the phenomenological equation for the Fano factor [Equation (15.5)]. The procedure is to partition the data into blocks of ever-increasing numbers of data points and, at each level of aggregation, calculate the Fano factor and mean value. We then determine how the Fano factor and mean value of each group change as the aggregates are enlarged to contain more and more data points. We follow the aggregation procedure outlined in Bassingthwaighte et al. [5].

Defining the Data Set

First of all, we define the one-dimensional time series by sampling a continuous variable according to some prescription. For example, we could sample a continuous voltage $V(t)$ at equal time intervals, such that the data set consists of $\{V_j = V(t_j)\}$, where the sampling times are $t_j = j\Delta t$, Δt is the time interval between samples, and $j = 1, 2, \ldots, N$, for a total of N data points. Of course, this is not the only way to define the data set and, in fact, it is not the usual way in physiology. In dealing with the systems in the human body, we allow those systems to establish their own time scales. Therefore, the data of interest for the human heart is the time interval between heartbeats, the data of interest for human gait is the time interval between strides, and the data of interest for breathing is the time interval between breaths. So the data set consists of the intervals between successive maxima in whatever process one is examining. Note that in such a data set there is no Δt, since the time interval itself is the quantity of interest. We discuss this more completely in subsequent chapters.

How Do We Use the Data?

The best way to describe our method of analysis is to actually use a data set. For this purpose, we generate a sequence of uncorrelated random data points on a computer. The statistics of the data points are chosen to be Gaussian, that is, when the computer selects the data points it is programmed to choose them from a Gaussian distribution of values. In Figure 15.4 we show the original set of 256

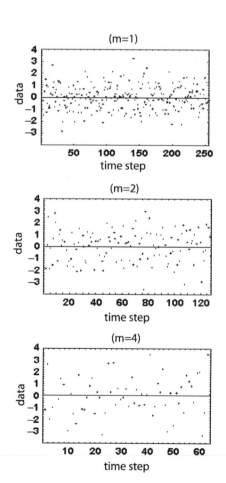

Figure 15.4. An example of data aggregation. Adding successive pairs of data points for a computer-generated set of data with Gaussian statistics. The number of elements added to form a single data point is indicated at the top of the frame.

computer-generated data points. We use these data to calculate a mean value and variance, from which the Fano factor is calculated and recorded in Table 15.1 in group 1. The next step is to add the nearest neighbor data points together and calculate the Fano factor and mean from these data. For this $m = 2$, $N = 128$, and $m \times N = 256$, as it should since we use all the data at each step. The resulting Fano factor is recorded in Table 15.1 in group 2. We now repeat the process and obtain the first through third graphs shown in Figure 15.4, and record the corresponging Fano factors in Table 15.1.

The Fano Factor

We now have a list of Fano factors in the final column of Table 15.1 whose values decrease with increasing aggregation number. We now test our hypothesis and fit the Fano factor data with an equation of the form of Equation (15.5). Note that Equation (15.5) is just an application of linear regression to the logarithm of the variable of interest, that being the Fano factor. The least-squares fit of Equation (15.5) to the Fano factor data in Table 15.1 yields the additive constant $\alpha = 3.27 \pm 0.16$. The slope of the fit in Figure 15.5 is found to be $\beta \approx -0.09$, a value consistent with a statistical fluctuation of a Gaussian random variable from zero. Thus, we see that the fractal dimension of the Gaussian test time series is determined to be $D = 1.45$. This is remarkably close to the theoretical value of 1.5 for Brownian motion, as it should be for an uncorrelated Gaussian process.

Fluctuating Fano Factor

The fluctuations in the Fano factor recorded in Table 15.1 and depicted in Figure 15.5 are a consequence of the fact that we have so few data points remaining when we are aggregating 32 of the original data points. These fluctuations in the Fano factor farther away from a constant value are drastically reduced when we go to a large number of data points. The remarkable thing is not that there are fluctuations in the Fano factor; the remarkable thing is that the fluctuations are so small for only 256 original data points.

Table 15.1. The Fano factor of subgroups of a 256 point, uncorrelated time series with Gaussian statistics

Group	m	N	$FF(m)$
1	1	256	3.18
2	2	128	3.21
3	4	64	3.35
4	8	32	3.19
5	16	16	3.22
6	32	8	2.88
average			3.27

Note. The group number indexes the level of aggregation. m is the number of data points in an aggregate. N is the number of groups. $FF(m)$ is the Fano factor with m of the original data elements in each aggregate. The standard deviation in the value of the Fano factor is 0.16, so that we can write the average as 3.27 ± 0.16.

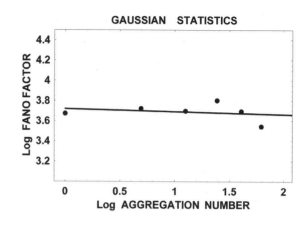

Figure 15.5. The Fano factors for increasing levels of aggregation of the 256 computer-generated data points are indicated by the closed dots. The straight line is the best mean-square fit to these points using Equation (15.5).

15.4 ALLOMETRIC SCALING AND TAYLOR'S LAW

An Aggregation Process

An apparently different kind of scaling from that of Fano was observed by Taylor in 1961 [97, 111]. He examined the statistical properties of the number of biological species that were heterogeneously distributed in space and found that the mean value and the variance were not independent quantities. The relationship he found between these two statistical measures was

$$\text{Var } X = a\overline{X}^b \tag{15.6}$$

We note that this is of the same form as the allometric growth relation discussed in Chapter 2. We draw your attention to the fact that in ordinary diffusive processes, the statistics are Gaussian, with the mean value $(\overline{X})$ and variance (Var X) statistically independent of one another. That is not the situation in Equation (15.6).

The Distribution Can Be Important

In the ecology literature, where the mean value and variance are spatially aggregated quantities, Equation (15.6) is referred to as the power-law curve. The reference curve in the ecological literature is that of a Poisson distribution, in which case the mean value and variance are equal, so that the power-law curve becomes a straight line with unit slope when plotted on log–log graph paper. Because of this equality, the Poisson case is referred to as random, since the spatial distribution in the number of species is homogeneous. In the homogeneous case, it does not matter where in space the number of species are sampled; statistically, things will be the same. Of course, the same can be said in a physiological context. The perfusion of blood throughout a muscle and the dispersal of oxygen in the blood are two situations, among many others, where the nature of the spatial distribution can be important [5].

Clumping Versus Non-Clumping

If the power-law index in Equation (15.6) is greater than one ($b > 1$), the distribution of species using Taylor's law, is spatially clumped, relative to the Poisson distribution. Such a distribution could describe cows in a meadow or flocks of migrating birds. On the other hand, if the power-law index is less than one ($b < 1$), the distribution is regular, that is, the individuals are more evenly dispersed in space. One might describe the trees in an orchard using such a distribution or perhaps a military formation on a parade ground. Thus, the slope of the power curve provides a measure of the degree of spatial heterogeneity of biological species, and is called the "evenness" by the ecological community. The greater is the slope, the greater is the variety of species [97]. The assumption is that the greater the deviation of the variance to mean value ratio is from one on log–log graph paper, the greater the extent of the departure of the process from uncorrelated randomness.

Determining Clustering

Taylor graphed the variance and mean value by increasing the resolution of the spatial areas and recalculating these two quantities as a function of the resolution cell size. We follow the same procedure with time series by calculating the variance and mean value for N data points, that is, Var $X(1)$ and $\overline{X}(1)$. We then add adjacent data points together to from a data set with $N/2$ data points, and from this calculate the variance and mean vlaue, Var $X(2)$ and $\overline{X}(2)$. Going back to the original data set, we now add three adjacent data points together to form a data set with $N/3$ data points, and from this calculate the variance and mean value, Var $X(3)$ and $\overline{X}(3)$. In this way, we continue to aggregate the original data set and after aggregating m neighbors in the original data set, the variance and mean value of the aggregated data are Var $X(m)$ and $\overline{X}(m)$. Thus, if we graph the variance versus the mean value of the aggregated data on log–log graph paper, and the underlying time series is fractal, we obtain a straight line given by

$$\log \text{Var } X(m) = \log a + b \log \overline{X}(m) \qquad (15.7)$$

The slope of the empirical curve, b, would allow us to determine the fractal dimension. The relationship between the power-law exponent in Taylor's law and the fractal dimension is given by

$$D = 2 - \frac{b}{2} \qquad (15.8)$$

You may recall the relation between the fractal dimension and the Hurst exponent H, in which case you would conclude that $b = 2H$, which is true for an infinitely long data set.

A Test Data Set

For orientation, before we examine experimental data, we examine the computer-generated time series with uncorrelated Gaussian statistics used to evaluate the

Fano factor. We generate 10^6 such points and then begin the aggregation process, calculating the variance and mean value at each level of aggregation. In Figure 15.6, we see that allometric scaling persists over more than two orders of magnitude variation in both the mean value and the variance. The slope of the computer-generated curve is $b \approx 1$, so that the fractal dimension is $D \cong 1.5$, as it should be for an uncorrelated random process with Gaussian statistics.

Interpretation of Allometric Scaling

We emphasize that the power-law index cannot be interpreted for the time series data in the same way as Taylor did for the spatial data. In Taylor's interpretation, $b = 1$ was the value for a Poisson spatial process, whereas we have seen above that $b = 1$ is the value for an uncorrelated Gaussian time series. In both cases, however, $b = 1$ corresponds to an uncorrelated random process. A value $b > 1$, interpreted as spatial clustering by Taylor, corresponds in our case to the bursting in time for a time series. However, as $b \rightarrow 2$, the correlation of distant events in the time series becomes stronger and stronger, resulting in a nonrandom regular process in the limit $b = 2$. To avoid confusion, we refer to the relation between the mean value and variance for time series as the aggregated allometric relation (AAR), rather than Taylor's law.

Measures of Scaling

We have argued that Gaussian statistics, discussed in the last chapter regarding the law of frequency of errors, does not always appropriately describe the random fluctuations in a time series. We provided an example of a process in which the average value (expected winnings) diverges—the St. Petersburg Paradox. When the central moments diverge, the statistics of the process cannot be Gaussian. This led us to the more general notion of a fractal random process in which the central moments diverge. This latter type of process requires a new way to get at the scaling behavior of the time series. The Fano factor and Taylor's law were used for that purpose.

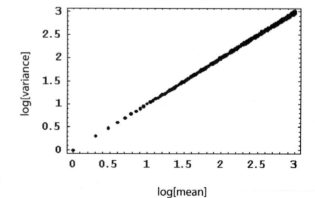

Figure 15.6. Here we graph the variance versus the average value of a computer-generated data set with Gaussian statistics. The aggregation process clearly shows that Taylor's law [Equation (15.7)] is applicable to these data.

Aggregated Allometric Scaling

The aggregation process consists of first calculating the mean and variance using the N data points, then calculating the mean and variance using the $N/2$ data points obtained by adding together the nearest neighbor data points, and continuing this procedure to obtain the mean and variance using the $[N/m]$ data points, where each one is an aggregate of m of the original data points. Both the mean and variance can diverge as $N \to \infty$, but the scaling is determined from the systematic way these two quantities diverge with increasing N. Graphing the variance versus the average on log–log graph paper yields a straight line for a statistical process that scales.

Fractal Random Processes

It is only recently that the influence of nonlinear dynamical systems theory, has been systematically applied to the biological sciences [5]. One of the limitations of these applications has been the failure to provide unambiguous evidence for the existence of chaos in biomedical data sets. This evidence is presented in subsequent chapters for a number of biomedical phenomena whose time series manifest scaling and, therefore, may be described as fractal random processes. Such fractal random processes interconnect multiple time scales, indicating clusters within clusters of activity in time of the underlying process. Any biological phenomenon described by fractal random time series is, therefore, not simple, but consists of biological organizations that respond interactively across a broad spectrum of time scales. Any deterministic generator of such a process must be dynamically unstable, which is to say that the solutions to the low-dimensional, nonlinear, deterministic, dynamical equation, when they exist, are chaotic. One of the properties of a chaotic system is resistance to change and the ability to absorb random shocks without modifying its fundamental nature. Thus, if an undesirable biological phenomenon or system (disease) could be demonstrated to be scaling, it would require subtle changes in protocol to modify its growth. On the other hand, identifying the control parameters that determine the global properties of an attractor on which the dynamics of the system unfold would pinpoint the protocol changes to which the system would be the most responsive.

The Next Three Chapters

In the next three chapters we apply the techniques discussed in this chapter that are appropriate for time series that have long-term memory, time series that may not be stationary, and time series that may not have characteristic time scales. In short, we approach all biomedical time series as if they were generated by fractal random phenomena. The analysis employs graphical methods that are easy to follow, again with the emphasis being on the interpretation of results and not on the mathematical formalism. The conclusions we draw from these analyses, contained in the last chapter of the book, is that health is the result of variability. We remain alive, active, and alert for the same reason the wirewalker does not fall. Movement and the interleaving of the fast and slow act so that we retain the harmony of the dance for as long as we live.

CHAPTER 16

Biodynamic Time Series 1— Stride Rate Variability

Objectives of Chapter Sixteen

- Understand the properties of human gait through the study of the fluctuations in the period of a stride leading to a measure of the stride rate variability (SRV).

- Learn that the body measures the control of walking through the degree of variability in gait from one step to the next.

- Determine that human gait is a process that changes as we mature, becoming less variable as we get older.

- Understand that we can learn more from the variability in gait than we can from the average value of a stride period.

- Learn that the fractal dimension is a robust quantitative measure of health and is easily obtain from the allometric scaling of the aggregated variance and mean of the SRV time series.

- The variability in the SRV time series as measured by the fractal dimension decreases with age.

Introduction

The Different Data Sets We Use

We now turn our attention to a number of physiological phenomena of interest in biodynamics. We use these data sets for a number of reasons, not the least of which is that there are a lot of them. We examine three biophysical time series to gain insight into the control mechanisms that regulate how we walk, breathe, and

Biodynamics: Why the Wirewalker Doesn't Fall. By Bruce J. West and Lori A. Griffin
ISBN 0-471-34619-5 © 2004 John Wiley & Sons, Inc.

pump blood through our bodies. The first time series we consider consists of the length of the time interval in our stride as we walk along in a relaxed manner. This may not seem very interesting, but the data holds the key to our direct control over locomotion. The second time series is that of the intervals between heartbeats as we sit in a relaxed condition and rest, compared with increasing states of exercise. Here again, the time series registers the health of one of the most important physiological systems: the cardiovascular system. The third and final time series consists of the time interval between breaths during normal relaxed breathing, again contrasted with increasing states of exercise. Here again, the respiratory control system is crucial for good health. We refer to the time series of the changes in the time intervals as the stride rate variability (SRV), the heart rate variability (HRV), and the breath rate variability (BRV) because, in each case, we are interested in the variability of the intervals between events.

Variability in the Stride Interval of Children

Anyone who has ever watched a child learn to walk recognizes the extreme variability in their step sizes and the unstable character of their sway as they struggle to remain upright. The lack of control exhibited by the very young, say ages three years old or younger, appears to be absent in preteens as they run around, climb trees, ride bicycles, and in most respects appear to have mastered locomotion. Of course, with the onset of puberty, instability returns, and the cause of this kind of instability is not so different from the cause in the very young. The variability in stride interval steadily decreases from three years old to eleven years old, the latter still not being fully mature compared with the gait of a fourteen year old, whose stride dynamics resembles those of adults. Hausdorff et al. [112] found significant age-dependent changes in the interstride intervals of children, even after adjusting for such things as height. We look for allometric scaling in SRV using the aggregation procedure described in Chapter 15.

16.1 INTERSTRIDE INTERVAL DATA

Walking Is Organized Falling

Walking is one of those indispensable processes that we do every day without giving it much thought. Most people walk rather confidently, with a smooth pattern of strides and without apparent variations in their gait. This pattern is remarkable when we consider the fact that the process of walking is created by the destruction of balance. In his 15th Century treatise on painting, Leonardo da Vinci points out that nothing can move by itself unless it leaves its state of balance and, with regard to running, a "... thing moves most rapidly which is furthest from its balance." So in one respect, walking can be viewed as a sequence of fallings and, as such, has been the subject of scientific study for nearly two centuries; see for example, Weber and Weber [113].

Variations in Stride Intervals

The regular gait cycle, so apparent in our everyday experience, is not as regular as we think. Cavanagh [114], in his recent brief history of running, makes the following observation with regard to the 19th century experimenter Vierordt:

His principal contribution appears to have been the realization that there is considerable variation in the stride parameters of normal locomotion (time of swing, stride length, left and right step lengths, etc.). This finding has certainly not been exploited by subsequent generations of researchers.

Follow-up Experiments

It has taken over 150 years for the follow-up experiments on the variation in stride intervals to be done and for the degree of irregularity in walking to be quantified. This quantification of walking irregularities has been done in a series of experiments by Hausdorff et al. [115, 116] involving healthy individuals, as well as elderly subjects and those with certain neurodegenerative diseases that affect gait [117]. Additional experiments and analysis have been done by West and Griffin [118, 119] that both verify and extend the results of Hausdorff et al. These experiments demonstrate for the first time that the fluctuations in the stride-interval time series exhibit long-time correlations in the form of an inverse power law, indicating that the phenomenon of walking is a self-similar fractal activity.

Standard Deviation of Gait Intervals

In the Griffin and West data, a stride interval is determined by the extension of the knee and the time interval between successive maximal positive extensions of the same knee define the stride interval. The variation of the gait time series, during a typical 15 minute walk, depicted in Figure 16.1, is found to be only 4%, indicating that the stride pattern is very stable and the SRV is quite small in an absolute sense. It is this stability that historically led investigators to assume that to a good first approximation the stride interval is constant and SRV is negligible. However, the fluctuations around the mean gait, although small, are nonnegligible, because they are indicative of some underlying complex structure and cannot be treated as uncorrelated random noise, as we shall demonstrate.

Average Plus a Fluctuation

We denote the interval between strides by the dynamic variable X_j, where j denotes the jth interval in the sequence $\{X_j\}$ of interval magnitudes. In analogy with

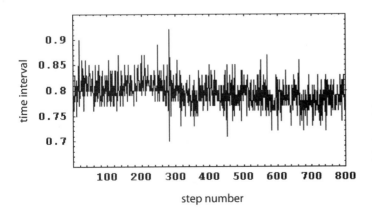

Figure 16.1. The stride-interval time series for a typical member of the experimental group. The number of points shown (800) corresponds to approximately 15 minutes of walking, with the lines drawn between points to aid the eye.

linear regression, we write each of the data points as a mean value plus a fluctuation:

$$X_j = \overline{X} + \xi_j \tag{16.1}$$

where the mean is determined by an average over all the values in the time series. The variation in Figure 16.1 is, therefore, the fluctuation away from the mean $[\overline{X}] \approx 0.79$ sec, depicted in Equation (16.1). The ξ fluctuations can be both positive and negative, that is, a person can step either faster or slower, so that the time interval can occur both above and below the average value. The magnitude of the random fluctuations, the SRV, is determined by the standard deviation.

How Well Does the Averge Characterize the Gait Interval Data?

We use the gait intervals depicted in Figure 16.1 to evaluate the variance in the fluctuations of a typical walker. Here, we assume that the data set consists of a number of discrete and distinct values of the time interval required for a complete stride. Using a standard program for the calculation of the mean and variance, or using Equation (14.6) for the average and Equation (14.10) for the variance, we obtain, as before, a mean stride interval of 0.79 sec, which is used to obtain a variance of

$$\text{Var } X \approx 0.00073 \text{ sec}^2 \tag{16.2}$$

One test of how well the mean characterizes the data is determined by taking the square-root of the variance, called the standard deviation, which has the same units as the variable of interest and, therefore, is directly comparable with it:

$$SDX = \sqrt{\text{Var } X} \approx 0.027 \text{ sec} \tag{16.3}$$

So for this particular walker, the variation in the interstride intervals, as measured by the ratio of the mean to the standard deviation, is approximately 4%. When the standard deviation is much smaller than the mean, $SDX \ll \overline{X}$, the mean is traditionally taken to be a good measure of the data set. The value of the variability, $SDX/\overline{X}$, given by this data set is sufficiently small that one would take the variation in the data to be negligible, as has been done historically in biomechanics. We shall show that a great deal of information regarding gait is lost when this criterion is adopted.

16.2 GETTING THE SRV DATA

The Experiments

Time series for the dynamics of gait are obtained by recording an individual's stride as he/she walks at a relaxed pace. The data consists of the time interval for each stride and the number of strides in a sequence of steps. One such data set is obtained from the maximal extension of the right leg [120]; another is obtained from the sequential heel strikes of a given foot [115]. In Figure 16.1, the stride interval versus the stride number is plotted on a graph. We can see from this figure that SRV has all the characteristics of a time series. In the Griffin and West experiment, relaxed walking was monitored for ten participants, and an electrogoniometer was used to collect kinematic data on the angular extension of the right leg. In these experiments, subjects walked, at a self-determined normal pace, around a square track with rounded corners, 40 feet on a side. The signal from the electrogoniometer was recorded at 100 Hz by a computer contained in a "fanny pack" attached to the walker's belt.

The Stride Interval Data

These data on the continuous angular movement of the right leg were downloaded from the computer on the walker's belt to a PC after twelve to fifteen minutes of collection. The interval between successive maximal extensions of the right leg in the analog signal was digitized and used as the time series data, which was then processed. The SRV time series for the interstride interval, for a typical walker, is depicted in Figure 16.2a. In this figure we can see that the fluctuations closely resemble that of a simple random walk process, shown in Figure 16.2b. These experimental data are from the first 900 steps of a particular individual in the study. Analysis by West and Griffin [118, 119] has shown that the control of walking is apparently a fractal random process that can be characterized by a power-law index, the fractal dimension. Similar results were found somewhat earlier by Hausdorff et al. [115, 116], using a very different processing technique.

Not Quite Independent

Figure 16.2a shows typical data of the time interval between successive maximum extensions of the leg for a healthy individual waking in a relaxed manner. The fluc-

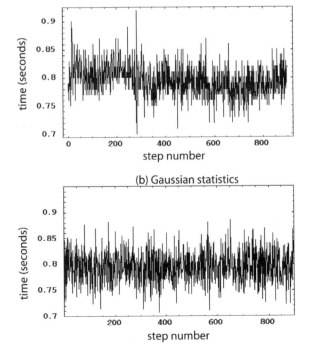

Figure 16.2. Real data compared with computer-generated data. The time interval between strides for the first 900 steps made by the seventh walker in the experiment are depicted in (a). In (b) a computer-generated time series having uncorrelated Gaussian statistics, with the same mean and variance as in (a), is depicted.

tuations about the average value in this time series, approximately 0.8 seconds, looks like a random process. However, we can detect a somewhat richer structure in the SRV time series by comparing these experimental data with a computer-generated data set. In Figure 16.2b we show such a data set for an uncorrelated Gaussian random process with the same mean and variance as the data in Figure 16.2a. This computer time series was generated using *Mathematica*. The question is whether there is a quantifiable difference between the two time series since they have the same mean and variance and whether we can determine this measure in a simple way. Herein, we provide evidence that the human gait time series is a modified random fractal process and that a relatively simple data processing technique is all that is necessary to provide quantitative information about this time series to establish this fact.

16.2.1   Processing Using Allometric Scaling

The Fit to Allometric Scaling

We can see in Figure 16.3 that the successive aggregation of the data, explained in Chapter 15, yields a relation between the mean and variance that is a straight line when the logarithm of the variance is graphed versus the logarithm of the mean. The equation used to provide the best mean-square fit to the data is given by the logarithm of the aggregated allometric relation, AAR, an allometric relation between the aggregated mean and aggregated variance, to obtain

$$\log \text{Var}\, X(m) = \log a + b \log \overline{X}(m) \qquad (16.4)$$

The overall constant in Equation (16.4) is determined by the intercept of the linear fitting curve with the horizontal axis, and the power-law index is determined by the slope of the straight line. For the SRV data in Figure 16.3, we obtain from the least-square fit of Equation (16.4) to the SRV data, using base ten logarithms, $a = 10^{-3}$ and $b = 1.43$. The empirical fractal dimension of these data is determined, using $D = 2 - b/2$, to be $D = 1.28$, and the form of the allometric relation is, taking the antilogarithm of Equation (16.4)

$$\text{Var}\, X(m) = 10^{-3}\, \overline{X}(m)^{1.43} \qquad (16.5)$$

Figure 16.3. The SRV data for walker number 7 in the experiment is used to construct the coarse-grained variance and mean as indicated by the dots [118]. The logarithm$_{10}$ of the variance is plotted versus the logarithm$_{10}$ of the mean, starting with all the data points at the lower left to the aggregation of twenty-five data points at the upper right. The SRV data curve lies between the extremes of uncorrelated random noise (lower dashed curve), and regular deterministic process (upper dashed curve).

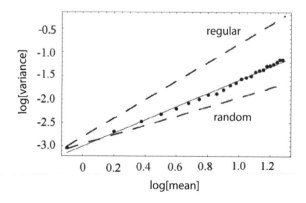

It is evident from Figure 16.3, where the solid curve is determined by Equation (16.4) with the appropriately fitted parameters, that the data is well bracketed by the two limits of regular behavior ($b = 2$) and uncorrelated random behavior ($b = 1$), the two limits being indicated by the dashed curves.

The Interpretation of the Power-Law Index

In the previous chapter, we argued that if the power-law index, b, is greater than unity, then the data are clustered. In the present time series context, this clustering, indicated by the form of the allometric equation in time, means that the intervals between strides change in clusters and not in a uniform manner over time. This result suggests that the walker does not smoothly adjust his stride from step to step. Rather, there are a number of steps over which adjustments are made followed by a number of steps over which the changes in stride are completely random. The number of steps in the adjustment and the number of steps between adjustments are not independent. The question concerns the universality of this result. Is such clustering in the time series typical of a healthy gait, or is it unique to this particular walker? We found that this scaling behavior is typical of the Griffin and West data, in agreement with the earlier, but distinct analysis of Hausdorff et al. [115], who used a modified random walk processing technique to analyze their data. Let us now turn our attention to the SRV data from these earlier experiments.

A Different Data Set

The research group at Beth Israel Hospital in Boston and Harvard Medical School, consisting of a long line of graduate students and postdoctoral researchers, and lead by J. M. Hausdorff, C. K. Peng, and A. L. Goldberger, have been gracious enough to post their experimental gait data on the internet at http:www.physionet.org. This is a National Institutes of Health sponsored *Research Resource for Complex Physiologic Signals*. We downloaded the sixteen stride interval time series for young healthy adults and applied the allometric scaling procedure to analyze their data. Hausdorff et al. [115] define the stride interval as the time between the heel strike of one foot on successive steps. In their experiment, the subjects walked continuously on level ground around a circular path 130 m in circumference. The stride interval was measured using ultrathin force-sensitive switches taped beneath one shoe of the walker. The data were registered on an ambulatory recorder and subsequently digitized at 416 Hz, and the time between successive heel strikes was automatically computed.

Seeing the Parameters

In Figure 16.4, we show the results of the processing using allometric scaling in conjunction with the aggregation argument for three of the sixteen SRV data sets obtained from the web site. We fit each of the sixteen time series using Equation (16.4) and record the fractal dimensions, Hurst exponents, and correlation coefficients in Table 16.1. The Hurst coefficient was the first of the scaling parameters used systematically in this kind of statistical analysis. Hurst was a civil, geographical, and ecological engineer working in the Nile River basin. He needed to deter-

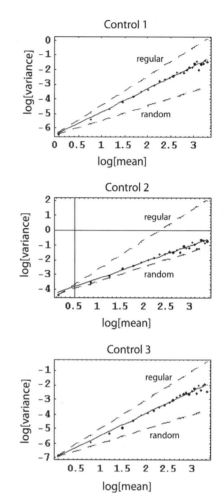

Figure 16.4. The stride interval data processed using the allometric scaling for three of the sixteen SRV time series obtained from *http:www.physionet.org* are depicted. The logarithm$_{10}$ of the variance is plotted versus the logarithm$_{10}$ of the mean, starting with all the data points at the lower left to the aggregation of twenty data points at the upper right.

mine what minimum height of the then proposed Aswan dam would provide sufficient storage capacity to govern the flow of the Nile downstream. The project in which he was involved used data covering over a millennium of annual flows, and required prediction of what reservoir capacity would be needed to smooth out the flows. Hurst observed that records of flows or levels did not vary randomly, but showed series of low-flow years and high-flow years. The correlation between successive years meant that the reservoir would fill or empty more completely than would occur with strictly random variation. This observation of Hurst led Mandelbrot and Wallis [121] to see unusual events such as extended rainfall or prolonged drought as natural phenomena that are likely to occur when there is "memory" in the system. The same is true of the memory we observe in gait dynamics data.

The Nearest-Neighbor Correlation Coefficient

Even though the power-law relation in allometric scaling may not give the complete story, the parameters obtained in this way can be very useful in determining the

Table 16.1. Data from sixteen SRV time series

Control	D	H	r_1
1	1.17	0.85	58%
2	1.27	0.73	38%
3	1.27	0.73	38%
4	1.30	0.71	32%
5	1.31	0.69	30%
6	1.21	0.79	50%
7	1.26	0.74	39%
8	1.21	0.78	49%
9	1.28	0.73	36%
10	1.24	0.76	43%
11	1.26	0.74	39%
12	1.26	0.74	39%
13	1.23	0.78	45%
14	1.25	0.76	41%
15	1.22	0.79	47%
16	1.20	0.81	52%

Note. The two coefficients from the least squares fitting formula for allometric scaling [Equation (16.4)] are not listed in the table. Instead, the derived fractal dimension, $D = 2 - b/2$, the Hurst exponent $H = 2 - D$, and the autocorrelation coefficient, r_1, are listed. The average fractal dimension is $D = 1.25 \pm 0.04$, the average Hurst exponent is $H = 0.75 \pm 0.04$, and the average autocorrelation coefficient is $r_1 = 0.42 \pm 0.07$.

properties of the SRV time series. Consider the autocorrelation function. The significance of the fractal dimension can be determined using the autocorrelation function, which for nearest-neighbor data points can be written as [5, 122]

$$r_1 = 2^{3-2D} - 1 \qquad (16.6)$$

This simple equation relates the two-point correlation coefficient to the fractal dimension of the stochastic process; here that process is the interstride interval.

Extremes of Dimensions

If the fractal dimension is given by $D = 1.5$, the exponent in Equation (16.6) is zero, and, consequently, $r_1 = 0$, indicating that the data points are linearly independent of one another. This would be the case for a slope of $b = 1$ in the log–log plots of allometric scaling. This fractal dimension corresponds to an uncorrelated random process with normal statistics. The variation in the gait interval would, in this case, change in a completely uncorrelated random manner from stride to stride. If, on the other hand, the nearest neighbors are perfectly correlated ($r_1 = 1$), the irregularities in the time series are uniform at all times and the fractal dimension is determined by the exponent in Equation (16.6) to be $D = 1.0$, which would be the case for $b = 2.0$. A fractal dimension of unity implies that the time series is regular, such as it would be for simple periodic motion. This would mean that the gait interval would change in a completely systematic manner from stride to stride. Here again, these two extremes are indicated by dashed curves in Figure 16.4, where we see that in actual walking there is neither complete regularity in variation nor are there uncorrelated random changes.

Range of Correlations

The range of fractal dimensions listed in Table 16.1 is between the extremes of regularity $D = 1.0$ (upper dashed curve) and uncorrelated random noise $D = 1.5$ (lower dashed curve) depicted in Figures 6.3 and 6.4. In fact, the average fractal dimension is exactly midway between the two extremes, whereas the individual walkers may be closer to one extreme than to the other. We can now use Equation (16.6) to deduce the correlations implied by the fractal dimensions. The range of correlations obtained using allometric scaling and the aggregation of SRV data is given by $0.30 \leq r_1 \leq 0.58$, indicating that the time series are all very strongly correlated. This strong correlation implies that the fluctuations in the interstride intervals do not act independently of one another. A change in the time interval that the control system regulating gait induces at one point in time strongly influences any changes made in the stride interval far into the future.

The Variation in the Fractal Dimension

We could include the graphs for each of the sixteen walkers, but nothing new would be learned about SRV from them. Each of the plots look pretty much like those in Figure 16.4, but with slightly differing slopes. The differences in slope would not be easy to detect with the naked eye. The values of the fractal dimension listed in Table 16.1, obtained from the slopes of these graphs, are more instructive than the graphs themselves. In fact, if we graph these sixteen empirical values of the fractal dimension, as we do in Figure 16.5, we obtain a fairly clear impression of the properties of the SRV in the overall behavior of the sixteen adult walkers.

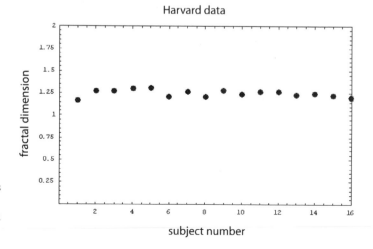

Figure 16.5. The fractal dimensions obtained using the allometric scaling for the sixteen stride interval time series from *http:www.physionet.org* are depicted. The average fractal dimension for SRV is $D = 1.25 \pm 0.04$.

The Level of Variability

Recall that the fractal dimension is a measure of the level of variability in a given time series. In Figure 16.5, it is evident that all sixteen of the walkers have very nearly the same level of variation in their gait dynamics, as measured by the fractal dimension, with a standard deviation of approximately 3%. It should be equally clear that the variation in the fractal dimension from person to person is very slight, indicating that this quantity may be a good quantitative measure of an individual's dynamical variability. We suggest that the fractal dimension is a quantitative measure of how well the motor control system is doing in regulating locomotion. It appears that there may be a fairly narrow region of values of the fractal dimension for gait in healthy individuals and excursions outside this interval may be indicative of pathologies.

The Fractal Dimension of Children's SRV

To test the remarkable speculation regarding the possibility of the fractal dimension being used as an indicator of health, we examine another data set on gait from the same Web site. The new time series are for the gait of young children, a total of fifty of them. We use the allometric analysis of the children's gait dynamics to emphasize the difference in variability between those of children and those of mature adults. In Figure 16.6, representative stride interval time series are given using a four-, a seven-, and an eleven-year-old child. Here again, the SRV data was obtained from the Web site *http:www.physionet.org*. The typical behavior of the variability in gait dynamics decreasing with increasing age, established by Hausdorff et al. [116], is evident. In the four-year-old, we can see that there are large changes in the time intervals between steps, with apparent modulations of the intervals over short time intervals. In the seven-year-old, there might be an overall increase in the stride intervals in the data shown. Neither of these is evident in the time series for the eleven-year-old child. We can, using allometric scaling, determine that the variability in the fractal dimension is larger in young children than in mature adults. Thus, with the available data, we do not see a significant change in average fractal dimension between children and adults; however, we do see a significant change in the variability of the fractal dimension.

The Variation in the Fractal Dimension of Children

In Figure 16.7, the fractal dimensions for the SRV time series of the first eleven children, from the fifty available, are recorded. These eleven are the children under five years of age and they provide a fairly strong contrast with the homogeneous group of mature adults recorded in Figure 16.5. The average fractal dimension of these eleven children is determined to be $D = 1.24 \pm 0.12$, a value not too different from that of the adults. However, what is different in the two data sets is the level of variation in the fractal dimension from child to child compared with that from adult to adult. The comparison between the average fractal dimensions of the children and adults is similar to that between two students having a C average. The one student has this average because she obtained a C in each of her courses. The other student has a C average because he received an equal number of As and Fs in his

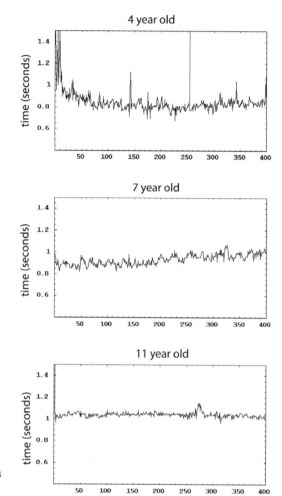

Figure 16.6. Representative walking time series of four-, seven-, and eleven-year-old children. The stride-to-stride fluctuations (SRV) are largest in the four-year old child and smallest in the eleven-year old child. The SRV data is taken from *http:www.physionet.org*.

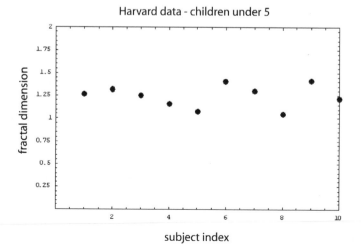

Figure 16.7. The fractal dimensions obtained using allometric scaling for the first eleven stride interval time series out of the fifty children from *http:www.physionet.org* are depicted. The average fractal dimension is $D = 1.24 \pm 0.12$.

courses. In the first case, the average represents the data well but, in the second case, the mean does not characterize the data well. For children, the variability in the fractal dimension is on the order of 10%, whereas for adults the variability in the fractal dimension is a factor of three smaller, only about 3–4%.

Variability Decreases with Age

The stride interval time series has been shown to be scaling, that is, to be a fractal random process. The observation that the stride-interval time series is a random fractal is consistent with the results of Hausdorff et al. [115–117], who used a more elaborate method of analysis in processing their data. In terms of a random walk model, the phenomenon of fluctuations in human gait is persistent. The stride interval fluctuations are not completely random, like Brownian motion, nor are they the result of processes with short-term correlations. Instead, the inverse power-law form of the correlation function reveals that the stride intervals at any given time are influenced by fluctuations that occurred many tens of strides earlier. This behavior is a consequence of the fractal nature of the stride interval time series.

The Interstride Intervals Are Clustered

The value of the power-law index in the AAR is greater than one, which, in keeping with Taylor's interpretation, indicates that the changes in the stride interval cluster in time. What this means in the context of human gait is that the intervals between strides do not change smoothly or continuously in time, but in clusters. This would suggest that the walker's control system does not adjust his stride in a smooth manner from step to step. As we said earlier, there are a number of steps over which adjustments are made followed by a number of steps over which the changes in stride are completely random. The number of steps in the adjustment and the number of steps between adjustments are themselves random, but not independent. It seems that these results may be universal. By universal, we mean these properties may be characteristic of human gait and independent of the individual walker.

There Is Order in the Randomness

Note that the correlation in the interstride intervals implies that the change in the stride interval and not the stride interval itself is correlated for an extraordinary length of time, on the order of 100 steps. The measure of the variability of SRV is the fractal dimension, which appears to be midway between an absolutely predictable process with $D = 1$ and a completely uncorrelated random process with $D = 1.5$. In addition, the variation in the fractal dimension from individual to individual seems to diminish with age, thereby making D an increasingly reliable indicator of health with advancing age.

The Persistence of Memory

Failure to accommodate changes, resulting from an inability to regulate high-frequency motor control in a coordinated way with low-frequency motor control, occurs in certain pathologies and in the elderly [117]. The suppression of high-frequency response is indicated by a decrease in the value of the power-law index, H, below the range of that of healthy individuals. Note that as H decreases, the fractal dimension increases toward the Brownian motion value of 1.5 and the memory in the gait fluctuation process disappears. The implication is that the control vanishes with the memory. This is consistent with a hypothesis made by West and Goldberger [123] concerning the use of inverse power laws in physiology as diagnostics for the health of physiological systems.

CHAPTER 17

Biodynamic Time Series 2— Heart Rate Variability

Objectives of Chapter Seventeen

- Understand the properties of the human heartbeat through the study of the fluctuations in the period between beats of the heart leading to a measure of heart rate variability (HRV).
- Learn how the body measures the control of heart rate through the degree of variability from one beat to the next and the role played by the branchings of physiological systems.
- Determine that the heartbeat is a process that changes as we mature, becoming less variable as we get older.
- Understand that we can learn more from the variability in heartbeat (HRV) than we can from the average heart rate.
- Learn that the fractal dimension is a robust quantitative measure of the health of the cardiovascular system and is easily obtained from allometric scaling of the aggregated variance and mean of the HRV time series.

Introduction

Heartbeat Variation Contains Information

One of the oldest indicators of the state of health of an individual is the heartbeat. Both the degree of regularity and the extent to which the intervals between beats change convey information to the physician. What we focus on here are quantitative measures of the fluctuations in the sequence of intervals between heartbeats and how such measures can be used as diagnostics of cardiovascular disease and well being. This approach to processing the intervals between the peaks in the electro-cardiogram waveforms, that is, the so-called RR-interval sequence, is now called *heart rate variability* (HRV) analysis. There are a great many measures of HRV that have been proposed over the past decade, some scale-dependent and others scale-independent, all in an effort to develop inexpensive and noninvasive measures of cardiovascular function; see for example [101]. We again use aggregated allo-metric scaling in the analysis of HRV.

Cardiac Chaos

In the past two decades there have been a number of research efforts focused on un-covering any possible relationship between cardiac activity and chaos in dynamical systems. It probably bears repeating that chaos is a type of nonlinear dynamical process that appears as an erratic time series in which the fluctuations are con-strained to some finite interval. As Goldberger and colleagues [123–125] point out, the extent to which chaos relates to physiological dynamics is under investigation and is controversial. West and Goldberger [123] hypothesized that the subtle but complex heart rate variability observed during "normal sinus rhythm" in healthy in-dividuals, even at rest, is a consequence, at least in part, of deterministic dynamical processes. Further, various diseases, such as those associated with congestive heart failure syndromes, may involve a paradoxical decrease in this type of nonlinearly generated variability [125]. The role of chaos in physiology or pathology, however, remains unresolved, but is certainly one of the more exciting areas of research.

Fractal Form and Function

Although the application of nonlinear deterministic randomness, or chaos, to physi-ological form and function may still be controversial, the physiological application of the geometric concept of fractals is less so. We have observed that venous and ar-terial trees have the kind of self-similarity that is indicative of fractal structure, as do the tracheobronchial tree and the His–Purkinje conduction network in the heart [8]. The physiological function of these fractal networks is rapid and efficient trans-port over a complex, spatially distributed system. Goldberger and West [124] list various organ systems that contain fractal structures that serve functions related to information distribution (nervous system), nutrient absorption (bowel), as well as collection and transport (biliary duct system, renal calyces). Let us take a closer look at one of these systems, the cardiac conduction system.

**17.1
HEARTBEATS**

The Cardiac Signal

Normally, each heartbeat is initiated by a stimulus from pacemaker cells in the si-nus node in the right atrium of the heart, and the activation wave then spreads through the atria to the AV junction. Following activation of the AV junction, the cardiac impulse spreads to the ventricular myocardium through a ramifying net-work, the His–Purkinje conduction system. The branching structure of the His–Purkinje conduction system strongly resembles the bronchial fractal we discuss in the next section (cf. Figure 17.1). In both structures, one sees a self-similar tree with finely scaled details on a "microscopic" level. The spread of this depolarization wave is represented on the body surface by the QRS complex of the electrocardiogram; see Figure 10.7.

Fractal Branching Networks

The simplest type of branching tree is one in which a single conduit enters a vertex and two conduits emerge. This dichotomous process is clearly seen in the patterns of biological systems, such as botanical trees, neuronal dendrites, lungs, and arteries, as well as in the patters of physical systems, such as lightening, river networks, and fluvial landscapes. The quantification of branching through the construction of the mathematical laws that govern them can be traced back to Leonardo da Vinci (1452–1519). We shall discuss these networks more fully in the next chapter, where we talk about the transport of fluids rather than the conduction of electrical impulses [126].

Fractal Conduction System

To explain the inverse power-law spectrum associated with a single heartbeat, Goldberger et al. [127] have conjectured that the repetitive branching of the His–Purkinje system represents a fractal set in which each generation of the self-similar segmenting tree imposes greater detail onto the system. At each fork in this network (see Figure 17.1), the cardiac impulse activates a new pulse along each conduction branch, thus, yielding two pulses for one. In this manner, a single pulse entering the proximal point of the His–Purkinje network with N distal branches generates N pulses at the interface of the conduction network and myocardium. In a fractal network, the arrival times of these pulses at the myocardium will not be uniform. The effect of the finely branched fractal network is to subtly decorrelate the individual pulses that superpose to form the QRS complex [127]. This decorrelation

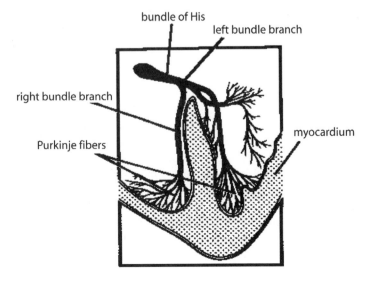

bundle of His

left bundle branch

right bundle branch

myocardium

Purkinje fibers

Figure 17.1. The ventricular conduction system (His–Purkinje) appears to be a fractal-like structure demonstrating repetitive branching on progressively smaller scales.

broadens the initially sharp pulse to the form that one observes in the time trace of an ECG (see Figure 10.7).

So Why a Fractal?

A fractal network is one that cannot be expressed in terms of a single scale, so that one cannot express the overall decorrelation rate of impulses measured at the myocardium by a single time scale. Instead, one finds a distribution of decorrelation rates in the time trace in direct correspondence to the distribution of branch lengths in the conduction network. These rates are based on an infinite series in which each term corresponds to higher and higher mean decorrelation rates. Each term, therefore, represents the effect of superposing finer and finer scales onto the fractal structure of the conduction system. Each new "layer" of structure renormalizes the distribution of mean decorrelation rates. This renormalization procedure eventually leads to a transition in the distribution of decorrelation rates to a power-law form in the region of high decorrelation rates. The spectrum of the time trace of the voltage time pulses resulting from this fractal decorrelation cascade of N pulses will also show inverse power-law behavior [8].

Heart Rate Variability (HRV)

We have argued that the shape of the individual pulse of a single cardiac cycle is determined by the fractal structure of the cardiac conduction system. Now let us examine the statistical properties of the time interval from one heartbeat to the next. In Figure 17.2, we compare the geometrical self-similarity of a fractal tree with the statistical self-similarity of a heartbeat interval time series. On the left of Figure 17.2 is a fractal tree that is geometrically self-similar. The geometric self-similarity is determined by comparing the observed tree structure, at successive levels of magnification, and observing that they differ by only an overall scale factor. The same is true of the heartbeat interval time series on the right of the figure but, rather than geometrical self-similarity, we have statistical self-similarity, that is, the time series has the same statistical properties on different time scales. In the figure, the time series for 300 minutes is compared with that for 30 minutes, which in turn is compared with the time series for 3 minutes. We see that more and more fluctuations are revealed as we go to smaller and smaller scales. This uncovering of structure with magnification implies that the HRV time series scales.

Two Kinds of Fractal Time Series

There are two kinds of fractal time series—those that are random and those that are not—but both may have a fractal dimension. In addition, when the time series is a random fractal, it can have two distinct sources of randomness, one being noise and the other being chaos. In the case of noise, the system of interest is coupled to an infinite-dimensional environment and the influence of the environment on the system is random, as is familiar from the discussion of Brownian motion and diffusion. The random fluctuations in this case contain no information about the system, so one is justified in smoothing (filtering) the time series. In the case of chaos, on the other hand, the erratic behavior of the time series is a consequence of deterministic, non-

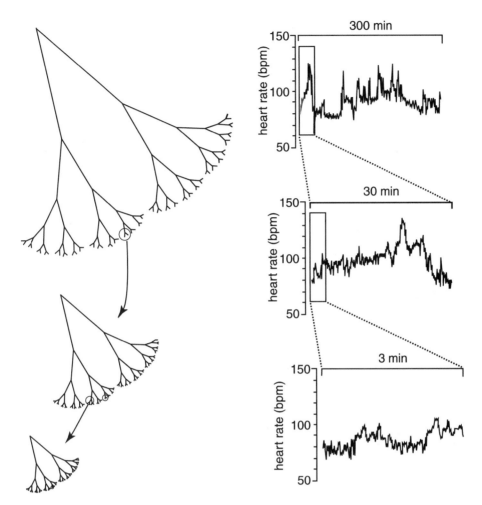

Figure 17.2. Self-similarity in cardiac time series. On the left is depicted a self-similar tree, as an exemplar of a fractal object. In such an object, the small-scale structure is virtually indistinguishable from the large-scale structure. This becomes apparent as more and more of the small-scale features are revealed. A fractal process, on the other hand, like the heart rate regulation on the right, generates statistical fluctuations on all time scales and is statistically self-similar. (From Goldberger [100] with permission.)

linear, dynamical interactions among the components of the system. Chaotic motion includes an infinite number of unstable periodic motions. A chaotic system never remains long in any of these unstable orbits, but continuously switches from one periodic motion to another, thereby having the appearance of randomness. Thus, the random fluctuations, in the case of chaos, contain information about the system and we wish to extract this information from the observed time series, not suppress it. In this section, we attempt to determine the source of the fractal behavior observed in cardiac time series, and thereby determine if the existence of such behavior is useful in the characterization of the health and control of the cardiac control system.

Feedback Control System

We argue here, as we have elsewhere [108], that a random fractal time series in a physiological context is a consequence of a feedback control system that operates over multiple time scales such that there is no fundamental time scale in the control process. Thus, a random fractal time series manifests scaling in the distribution function. The time scaling property of heart rate variability (HRV) is analyzed by using the aggregated Taylor relation as an indicator of the fractal nature of the time series. One reason for choosing this method over and above others, such as the spectral analysis of Kobayashi and Mushi [128], who were the first to find an inverse power law for an HRV time series, is that it requires little preprocessing of the original data and is fairly accurate, even using relatively small data sets.

17.2 GETTING THE HRV DATA

The Experiment

The first experimental data used in the presentation below are discussed in [108]. Six healthy subjects (5 men and 1 woman) with a mean age of 29 ± 8 yr, height of 177 ± 7 cm, and weight of 76 ± 14 kg, voluntarily participated in the study. All were nonsmokers and were free of known cardiovascular, pulmonary, and cerebrovascular disorders. Each subject was informed of the experimental procedures and signed a written consent form approved by the Institutional Review Boards of The University of Texas Southwestern Medical Center and Presbyterian Hospital of Dallas [108].

Procedures and Experiments

All experiments were performed at least two hours postprandial, and more than twenty-four hours after the last caffeinated beverage or alcohol, in a quiet, environmentally controlled laboratory, with an ambient temperature of 25 °C. After at least thirty minutes of supine rest, an analog electrocardiogram (ECG Monitor, Hewlett Packard) was monitored continuously for two hours during spontaneous respiration. Beat-to-beat HR was obtained with peak detection of R waves using a voltage trigger circuit (Cardiotachometer, Quonton Instrument) and then sampled at 100 Hz and converted into RR interval series for off-line RR analysis. During the two hours of data recording, each subject was quiet and awake. Therefore, one unique feature of this study is that we obtained a relatively long data set of HRV recordings without external perturbations, and are therefore able to explore intrinsic nonlinear properties of cardiovascular regulation under stationary conditions, should they exist. A typical HRV data set is depicted in Figure 17.3.

Other Data Sets

We emphasize that most previous studies attempting to demonstrate an inverse power-law spectrum for HRV have used a twenty-four hour ambulatory recording device to obtain data points over sufficiently long time periods or decades of frequency; see for example [129]. However, the physiological state of the subjects is rarely controlled and large amounts of noise (uncontrolled environmental influences) are injected into the time series by talking, eating, physical activity, arousal,

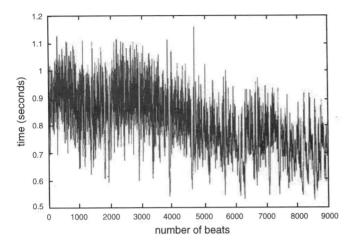

Figure 17.3. The sizes of the interbeat intervals are plotted as a function of the beat number for two hours of HRV data, using a typical member of the group of six subjects. (From West et al. [108] with permission.)

etc., which affect heart rate independently of any fundamental control system. We would, therefore, interpret the fractal character, reported in these studies, as being the result of the internal dynamics and the remainder being a consequence of the coupling of the cardiac system to the environment. In contrast, the data discussed here is obtained over a prolonged time period under carefully controlled conditions and thus may be uniquely able to assist us in discriminating between chaos and noise in the HRV time series.

The Fit to Allometric Scaling

We analyze the HRV data in the same way we did the gait data in the preceding chapter—by calculating the mean and variance as a function of the level of aggregation and plotting the results on log–log graph paper. The AAR is then fit to the data points using a mean-square minimization procedure based on Equation (16.4):

$$\log \text{Var}\, X(m) = \log a + b \log \overline{X}(m) \tag{17.1}$$

In Table 17.1, we record the fractal dimension, $D = 2 - b/2$, Hurst coefficient, $H = b/2$, and correlation coefficient, $r_1 = 2^{b-1} - 1$ resulting from this fitting procedure.

It Is Probably a Bit More Complicated

In Table 17.1, we see that the average fractal dimension for the six subjects is $D = 1.13 \pm 0.07$, a value that is consistent with values obtained for healthy subjects by previous investigators [129], using a very different and more complicated data processing technique. Here again, we see a high level of correlation in the data, spanning the range $0.45 \le r_1 \le 0.84$, suggesting a strongly correlated component in the HRV of the six resting individuals. It is probably worth mentioning that the time series of a regular process, like the motion of a harmonic oscillator, would have a fractal dimension of one, so the correlation we see in the data is a consequence of the low value of the fractal dimension.

Table 17.1. Fractal dimension, Hurst coefficient, and correlation coefficient resulting from fitting procedure

Control	D	H	r_1
1	1.09	0.91	77%
2	1.23	0.77	45%
3	1.06	0.94	84%
4	1.06	0.94	84%
5	1.19	0.81	54%
6	1.14	0.86	65%

Note. The two coefficients obtained from the least squares fitting formula for allometric scaling Equation (17.1) are not listed in the table. Instead the derived fractal dimension, $D = 2 - b/2$, the Hurst exponent $H = 2 - D$, and the auto-correlation coefficient, r_1, are listed. The average fractal dimension is $D = 1.13 \pm 0.07$, the average Hurst exponent is $H = 0.87 \pm 0.07$, and the average autocorrelation coefficient is $r_1 = 0.68 \pm 0.16$. (From West [9] with permission.)

A Different Data Set

We can again go to the Web site *http:www.physionet.org* to obtain a number of data sets on heart rate variability. We use the *Fantasia Database* of human heart rate data, which as stated there, was constructed as a teaching resource for an intensive course ("The Modern Science of Human Aging"), conducted at MIT in October, 1999. The data collection was done on five young (21–34 years old) and five elderly (68–81 years old) healthy subjects, who underwent two hours of supine resting while continuous ECG signals were collected. The name of the data file derives from the fact that the subjects watched the Disney movie *Fantasia* (1940) so as to remain wakeful, but in a resting state in sinus rhythm. Figure 17.4 shows the time series for the first data set in each category. It is apparent from the figure that the time series for the young is very different from that for the elderly, with the younger having a great deal more variability.

The Old Are More Predictable

In Figure 7.4, we observe that the amplitude in variation of the hearbeats is significantly greater in the young than in the old. This relative increase in the HRV of the young is typical of all ten data sets. Seen from the other perspective, the loss in variability with age is the same property that was observed with the stride interval time series in the previous chapter. We expect to quantify this reduction in variability with a reduction in the fractal dimension.

Seeing the Parameters

In Figure 17.5, we show the results of the processing using allometric scaling in conjunction with the aggregation argument for the two data sets depicted in Figure 17.4. It is clear from Figure 17.5 that the data lie along straight lines and, therefore, the variance and mean satisfy an allometric relation. The slope of the top curve is given by $b = 1.66$, corresponding to a fractal dimension of $D = 1.17$, whereas the slope of the lower curve is $b = 1.85$ yielding a fractal dimension of $D = 1.08$. The

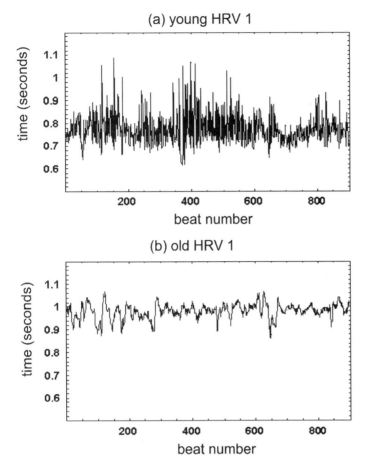

Figure 17.4. Comparing young and old heart rate variability. The time interval between heartbeats for the first 900 beats made by the first young and first elderly subject in the *Fantasia Database* (taken from *http:www.physionet.org*) are depicted. Lines are used to connect the data points to aid the eye in discerning patterns.

parameter values are clearly consistent with the intuition that the elderly have less variability in their HRV time series and, therefore, the fluctuations in heartbeat are more tightly correlated. Recall that a regular curve has a fractal dimension of one (upper dashed curve).

Processing the Other Eight Time Series

We fit each of the ten time series using Equation (17.1) and record in Table 17.2 the fractal dimensions, the Hurst exponents, and the autocorrelation coefficients. The first thing to notice is that the fractal dimensions obtained for the younger subjects are consistent with those obtained by West [9], with an identical average fractal dimension and Hurst exponent, as can be seen by comparing the entries in Tables 17.1 and 17.2. The fractal dimension of the five elderly people is less than that of the younger people, indicating a loss of variability with age. Although the data set is probably too small to draw any definite conclusions, this result is consistent with that of other investigations. For example, Pikkujämsä et al. [130] have done an extensive comparison of the power-law indices for HRV data to study the effects of aging from early childhood to advanced age (114 subjects ranging in age from 1 to

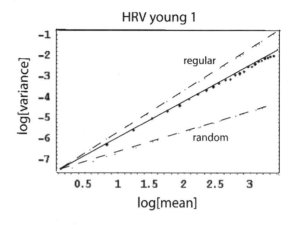

Figure 17.5. The HRV data processed are depicted using the allometric scaling for two of the time series taken from *http:www.physionet.org*, from the *Fantasia Database*. The logarithm$_{10}$ of the variance is plotted versus the logarithm$_{10}$ of the mean, starting with all the data points at the lower left to the aggregation of twenty data points at the upper right.

Table 17.2. Fractal dimensions, Hurst exponents, and autocorrelation coefficients

Subject	Young			Old		
	D	H	r_1	D	H	r_1
1	1.17	0.83	58%	1.08	0.91	80%
2	1.12	0.88	69%	1.06	0.94	84%
3	1.14	0.86	66%	1.10	0.90	74%
4	1.12	0.88	71%	1.13	0.87	68%
5	1.11	0.89	73%	1.06	0.94	84%

Note. The two coefficients obtained from the least squares fitting formula for allometric scaling [Equation (17.1)] are not listed in the table. Instead, the derived fractal dimension, $D = 2 - b/2$, the Hurst exponent $H = 2 - D$, and the autocorrelation coefficient, r_1, are listed. The average fractal dimension for the youths is $D = 1.13 \pm 0.03$, whereas that for the elderly is $D = 1.08 \pm 0.03$; the average Hurst exponent for the youths is $H = 0.87 \pm 0.03$ whereas that for the elderly is $H = 0.92 \pm 0.03$; and the average autocorrelation coefficient for the youths is $r_1 = 0.67 \pm 0.06$, whereas the average autocorrelation coefficient for the elderly is $r_1 = 0.78 \pm 0.07$.

82 years of age) and found a loss in variability with age. These latter data sets are twenty-four hours in length rather than the two hours we use here.

The Interaction of Subsystems

As observed by West et al. [108], the specific mechanisms underlying HRV are not always clear. However, it is known that heart rate is regulated by a complex feedback control system. Data suggest that this control system has the following features. First, it is nonlinear. This property is not only manifest as nonlinear interactions between parasympathetic and sympathetic nerve activity, which directly controls beat-to-beat changes in heart rate, but also appears in the central nervous system (CNS), which determines the output level of autonomic nerve activity [131]. Second, the system has multiple input signals originating from sensors located in different parts of the body for monitoring changes in blood pressure, HR, and tissue oxygenation. These input signals are converted into and interact with each other at the cardiovascular center in the CNS. Third, the cardiovascular system is a closed feedback and feedforward control system.

The Effect of Scaling

We hypothesized [125] that the observed HRV time series generated from nonlinear dynamical properties of the cardiovascular system may be chaotic. Chaotic variations in HRV would be consistent with the fractal characteristics of HRV revealed in the above analyses. Furthermore, the long-term correlation or memory observed in HRV time series indicates that although different regulatory mechanisms may act independently on different time scales, their effects on dynamical changes in HR may be tied together through scaling. Thus, impairment of one individual component of HR regulation may influence other regulatory mechanisms via interdependence.

Loss of Variability Indicates Dysfunction

A corollary to the hypothesis that cardiac dynamics is nonlinear and possibly chaotic, explaining why HRV time series are scaling and fractal, is that a variety of disease states that alter autonomic function may lead to a loss of physiologic complexity and, therefore, to greater, not less, regularity. Decrease in HRV has been described in numerous settings, including multiple sclerosis [132], fetal distress [133], and in certain patients at risk for sudden cardiac death [134]. Presumably, the more severe pathologies will be associated with the greatest loss of spectral power, analogous to the onset of the most serious arrhythmias, which begin to resemble "sine wave" patterns. Such spectral narrowing has been referred to as a loss of spectral reserve [127]. These and other such findings were anticipated by the West–Goldberger hypothesis, which states that a decrease in healthy variability of a physiological system is manifest in a decreasing fractal dimension [31]. Normal HRV depends on the integrity of autonomic regulation, and the stability of cardiovascular control. On the other hand, loss of HRV may indicate impairment of autonomic function and instability of the cardiovascular control system. It has been shown that the absence of HRV after an acute myocardial infarction is a risk factor for the development of significant morbidity, including arrhythmia and death. Furthermore, HRV is known to be diminished in patients with heart failure [135]. Thus, reliable universal measures of HRV are very important but, in addition, a testable theory of the mechanism producing HRV is potentially more important.

Return of the Wirewalker

The engineering paradigm of signal plus noise has its equivalent in medicine in the principle of homeostasis. This principle asserts that physiological systems self-regulate so as to retain a constant output, independently of their input. In this way, the desired constant output is interpreted as signal and the undesired variable response to external perturbations is interpreted as noise. Here, we have seen that healthy physiological systems, even at rest, display highly irregular dynamics through their time series. Furthermore, the coupling of the variability at different time scales, rather than conforming to signal plus noise, brings to mind the various interrelated movements of the wirewalker and his balancing act.

CHAPTER 18

Biodynamic Time Series 3—
Breathing Rate Variability

Objectives of Chapter Eighteen

- Understand certain properties of human breathing through the study of the fluctuations in the interval between breaths, leading to a measure of breath rate variability (BRV).
- Learn that the body controls BRV, in part, through the branching of the bronchial tree and, in part, through the degree of variability in the rate of breathing.
- Determine that BRV is a process that changes as we mature, becoming less variable as we get older.
- Understand that we can learn more from the variability in breath rate than we can from the average interval between breaths.
- Learn that the fractal dimension of the BRV is a robust quantitative measure of the health of the respiratory system.

Introduction

Breathing Is Not Regular Either

A somewhat less obvious process to examine than heart rate and stride interval is that of breathing. It seems fairly clear that when we are healthy and sitting quietly, we breathe at regular intervals, in and out. On the other hand, both during and after exercise, our breathing can occur in short, deep gasps. The connection between

Biodynamics: Why the Wirewalker Doesn't Fall. By Bruce J. West and Lori A. Griffin
ISBN 0-471-34619-5 © 2004 John Wiley & Sons, Inc.

breathing and the body's immediate need for O_2 intake and CO_2 expulsion seems obvious, as obvious as the smoothness of our walking or the beating of the heart. The more oxygen we need or the more CO_2 we need to eliminate, the faster and deeper we breathe. We shall find, however, that, like the variability in the dynamics of gait and the variability of heartbeat, there is significant information in the deviation of breathing from regularity. Therefore, we refer to the variation in the interval between maxima in breaths as the breath rate variability (BRV) and use aggregated allometric scaling in its analysis.

The Complexity of the Lung

Perhaps the most compelling feature of physiological systems, such as the cardiovascular system for heartbeat, the motor control system for gait, and the respiratory system for breathing, is dynamic complexity. An example of another kind of complexity is static or structural complexity, which is manifest in the bronchial system of the mammalian lung. The lung may indeed serve as a useful paradigm for anatomic complexity. This tree-like network of complicated hierarchy of airways begins with the trachea and branches down on an increasingly smaller scale to the level of tiny tubes called bronchioles, ending in sacs called alveoli. The area of the alveolar surface in adult totals 50 to 100 square meters, or about the size of a tennis court.

The Function of the Lung

The smooth muscle cells in the bronchial tree, innervated by parasympathetic and sympathetic fibers, contract in response to stimuli such as increased CO_2, decreased O_2, and deflation of the lung [136]. During inspiration, fresh air is transported through some twenty generations of bifrucating airways of the lung down to the alveoli in the last four generations of the bronchial tree where there is a rich capillary network for the exchange of gas to the blood. At this tiny scale, the movement of air can no longer be described by hydrodynamics, so diffusion takes over to transport O_2 and CO_2 in a fraction of a second across the alveolocapillary membrane. This is the mechanism for dissolved gas exchange between alveolar air and capillary blood.

18.1 THE ARCHITECTURE OF THE HUMAN LUNG

The First Quantitative Observation of Branching

The quantification of branching through the construction of the mathematical laws that govern them can be traced back to Leonardo da Vinci (1452–1519). In his *Notebooks,* da Vinci writes [29] "All the branches of a tree at every stage of its height when put together are equal in thickness to the trunk [below them]." With the aid of da Vinci's sketch (reproduced in Figure 18.1), this sentence has been interpreted as follows: if a tree has a trunk of diameter d_0 that bifurcates into two limbs of diameters d_1 and d_2, the three diameters are related by

$$d_0^\alpha = d_1^\alpha + d_2^\alpha \tag{18.1}$$

Simple geometrical scaling yields the diameter exponent $\alpha = 2$, which corresponds to "rigid pipes" carrying fluid from one level of the tree to the next, while retaining

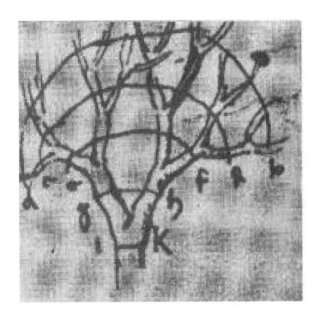

Figure 18.1. da Vinci saw it first. The sketch is taken from da Vinici's *Notebooks,* and indicates, along with Equation (18.1), his remarkable understanding of the relationship between successive levels in a botonical branching system.

a fixed cross-sectional area through successive generations of bifurcation. Although the pipe model has a number of proponents from hydrology, the diameter exponent for botanical trees was determined empirically by Murray, in 1927, to be independent of the kind of botanical tree and to have a value 2.59 rather the value 2. Equation (18.1) is referred to as Murray's law in the physiology literature [30].

The Branching Laws Became Very Important

The significance of Murray's law was not lost on D'Arcy Thompson. In the second edition of his work, *On Growth and Form* [36], first published in 1917, Thompson argues that the geometrical properties of biological systems can often be the limiting factor in the development and final function of an organism. This is stated in his general "principle of similitude," which was a generalization of certain observations made by Galileo regarding the trade-off between the weight and strength of bone, circa 1638. Thompson goes on to argue that the design principle for biological systems is that of energy minimization, a principle that is often used today in the discussion of the thermodynamic properties of both microscopic and macroscopic systems [125]. The energy minimization principle stipulates that the function is performed using the minimum amount of energy, whether the dynamics refer to particles in closed orbits, or the relaxation of stress in a material body.

The Lungs As a Cascade of Pipes

Fritz Roher, a German physiologist, reported his investigations on scaling in the bronchial tree in 1915 [137]. The remarkable feature of his research was that he did not study any biological system; instead, he investigated the properties of the flow of a Newtonian fluid in systems of pipes of varying lengths and cross-sectional areas arranged in cascades of different kinds. His purpose was to determine the properties of the flow in branched systems and, from this "experimental" reasoning, to derive formulae for the flow in terms of the length and diameter of a conduit as a function of the stage (generation z) of a sequence of branches. He explored a number of assumptions regarding the scaling properties of conduits in one branch of a cascade to those in the next generation of the cascade. For example, if there were only a scaling in length, but not diameter, he could build the assumed relation into the cascade of pipes and measure the difference in the flow at the output. Each of his assumed properties led to different scaling relations between the flow at successive generations of the branching system of pipes. Although much of the data could be connected with different assumptions, no single set of assumed properties was recognized at that time as being clearly superior to the others in terms of properly describing the mammalian lung.

Classical Scaling of Bronchial Trees

The mechanical arguments of Roher, overall, indicated a geometric decrease of the airway diameter with generation number z. The next major attempt to apply scaling concepts to the understanding of the respiratory system was made in the early 1960s by Weibel and Gomez [34]. The intent of their investigation was to demonstrate the existence of fundamental relations between the size and number of lung structures. They considered the architecture of the bronchial tree to consist of the division of a parent airway into two offspring, thereby constituting a dichotomous branching process. The volume of the total airway is the same between successive generations, on the average, but with significant variability due to the irregular pattern of dichotomous branching in the real lung. Their formal results are contained in the earlier work of Roher if one interprets the fixed values of lengths and diameters at each generation used by Roher as the average value used by Weibel and Gomez. As the bronchial tree branches out, its tubes decrease in size. The classical theory of scaling predicts that their diameters should decrease by about the same ratio from one generation to the next. If the ratio were ½, for example, the relative diameter of the tubes would by ½, ¼, ⅛ and so on, a geometric decrease in diameter with generation number. Weibel and Gomez measured the tube diameters for 22 branchings of the bronchial tree, and fit the reduction in the size of the bronchial airway with the functional form $e^{-\gamma z}$. On semilog graph paper, the scaling prediction is that the data should lie along a straight line, (Figure 18.2) with z the generation number and γ the reduction in bronchial diameter per generation.

Classical Scaling of the Bronchial Tree is Not Successful

The theoretical prediction using classical scaling, according to Figure 18.2, might explain the first ten generations of the Weibel-Gomez data (the decrease in the average diameter of the bronchial airway for the first ten generations), but above $z = 10$ there is a systematic deviation of the data away from the classical prediction. Weibel and Gomez attribute this deviation from the geometric reduction in size to a change in the flow mechanism in the bronchial tree from that of minimum resistance to that of diffusion. An alternative explanation involves the variability in the linear scales at each generation rather than just the average length and diameter used in classical scaling [8]. The problem is that the seemingly obvious classical

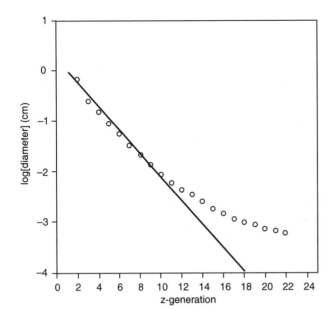

Figure 18.2. The human lung cast data of Weibel and Gomez for 23 generations are indicated by the circles and the prediction using the exponential form for the average diameter is given by the straight line. The fit is quite good for $z \leq 10$, after which there is a systematic deviation of the anatomic data from the theoretical straight-line curve [31].

scaling assumes a characteristic scale size and this assumption clearly fails for complex systems where no characteristic scale exists. We find that the fluctuations in the linear sizes are inconsistent with classical scaling, but are compatible with a more general scaling theory involving fractals.

Self-Similar Bronchial Tree

The classical model of bronchial diameter scaling, as we saw, predicts an exponential reduction in the average diameter of the airway with generation number. However, this prediction is only good for the first ten generations. The arguments of classical scaling assume the existence of a simple characteristic scale governing the decrease in bronchial dimensions across generations. If, however, the lung is a fractal structure, no characteristic smallest scale should be present; instead, there should be a distribution of scales contributing to the variability in diameter at each generation. Introducing a self-similar distribution of diameters, the subsequent dominant variation of the average bronchial diameter with generation is found to be an inverse power law in generation number, not an exponential [8, 31]. The comparison of the fractal model of the size of airways with the experimental data from four species of mammal are depicted in Figure 18.3.

Comparing Different Species

We find that the average diameter is an inverse power law in generation number modulated by a slowly oscillating function and rather than pursuing the matter further here, we refer the interested reader to the literature [31]. We merely note that the fractal model provides an excellent fit to the lung data in four distinct species: dogs, rats, hamsters, and humans. The scaling differences between mammalian species studied appear to be related to postural orientation—quadruped as opposed

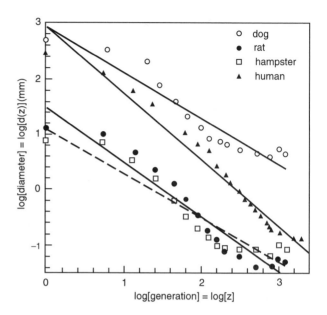

Figure 18.3. The triangles are the Weibel–Gomez data from Figure 18.2 plotted on log-log graph paper. We can see that the best fit to the average bronchial diameter follows an inverse power-law, not classical, scaling. In addition, there appears to be a harmonic (periodic) variation of the data points about the power-law regression line. The inverse power-law scaling and periodic variation are also present in measurements of bronchial tube diameters obtained for dogs (open circles), rats (filled circles), and hamsters (open squares).

to biped—but the selection advantage of the particular scaling patterns remain a mystery.

Braching Trees as Fossil Remnants

Until now, we have restricted our discussion to a static context, one describing the relevance of inverse power-law scaling and fractal dimensionality to anatomy. Such physiologic structures are only static in that they are the "fossil remnant" of a morphogenic process, that is, they are the structures remaining after the growth process is completed. It would seem reasonable, therefore, to suspect that mophogenesis itself could also be described as a fractal process, but one that is time dependent [38]. From the viewpoint of morphogenesis, the newly identified scaling mechanisms have interesting implications regarding the development of complex but stable structures using minimal genetic codes. One of the many challenges for future research will be the unraveling of the molecular and cellular mechanisms whereby such scaling information is encoded and processed.

Morphogenesis of the Mammalian Lung

The morphogenisis of the human lung was modeled by Nelson and Manchester [139] using computer simulation of growth, in which the computer codes are defined by fractal algorithms. Variations in the limits imposed by simple constraints, such as the size of the chest cavity, generate structures that are in good agreement, in two dimensions, with actual structural data. We have examined the fluctuation tolerance of the growth process of the lung and found that its fractal nature does, in fact, have a great deal of survival potential [139]. The growth of error for the fractal model of the lung is significantly slower than that due to uncorrelated random fluctuations. This suppression of error is a consequence of a fractal not having a charac-

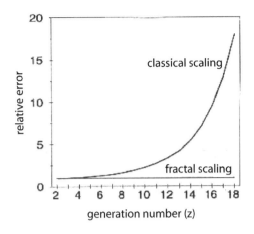

Figure 18.4. The error between the model prediction of the size of a bronchial airway and the prediction with a noisy parameter is shown for the classical scaling model and the fractal scaling model.

teristic scale, so that errors produce scale sizes that are already present in the lung. In particular, it has been shown that fractal structures are much more error-tolerant than those produced by classical scaling (Figure 18.4). Such error tolerance is important in all aspects of biology, including the origins of life itself.

A Little Help from our Friends

The authors had not been directly involved in any experiments that measure the time intervals between breaths, so we walked across campus and met with medical colleagues at the Duke University Medical Center. Richard Moon, M.D., the Director of the Hyperbaric Chamber Laboratory, was more than happy to supply us with data under the condition that we share our results with him and his colleagues, and work together on any conclusions that might be drawn. The Hyperbaric Laboratory consists of a number of large chambers, each of which, as we walked through, resembled a small submarine or diving bell with portholes. The pressure in these chambers can be increased substantially above atmospheric pressure and an individual's vital statistics can be monitored and recorded, both at rest and during various stages of controlled exercise. This joining of complementary talents is how a great deal of research is done. One group has the experimental apparatus and data collection skills, another has the data processing tools and analytic skills, and they work together, and sometimes they even solve problems of mutual interest.

How the BRV Data Was Collected

The data we analyze was collected in the dry hyperbaric chamber, first at 1 atmosphere of pressure absolute (ATA) and subsequently at 2.82 ATA [140]. The size of the cohort group was 18 subjects, both male and female, at a variety of ages. All 18 subjects were required to be free of clinical pulmonary or cardiac disease, as determined by history, physical exam, chest radiograph, and pulmonary function tests, and to have no contraindications to diving as indicated in [140]. In other words, this was a healthy group of people. Data were recorded for an individual under three conditions of activity—rest, light exercise, and heavy exercise—on a Monark 818 cycle ergometer under the two pressure conditions. Picture the scene: a subject has

18.2 GETTING THE BRV DATA

wires running from his/her chest, tubes coming from his/her mouth, is peddling a stationary bicycle with wires connected to a variety machines sitting off to one side. All this is contained within a chamber that looks every bit like a spherical diving bell.

How the BRV Data was Processed

We refer the reader to the literature to find out more details of the experiment and here turn our attention to the processing of BRV data, such as that shown in Figure 18.5, for a single individual. Typically, we obtained on the order of one hundred data points, that is, breath intervals, for each individual under each of the different conditions of exercise and pressure. One hundred breath intervals is a rather small data set, but it appears to be sufficient to indicate the qualitative properties of the respiratory system. We can see from the figure that the level of breath interval variability appears to decrease when the individual goes from sea level atmospheric pressure to 2.8 ATA. This observation is made quantitative by the fitting parameters to the BRV data recorded in Table 18.1 and averaged over the 18 participants in the experiment.

Correlations for BRV Data Seem to be Large

The fitting parameters recorded in Table 18.1 are interpreted in the following sections. However, we cannot help but notice that the correlation coefficient seems to

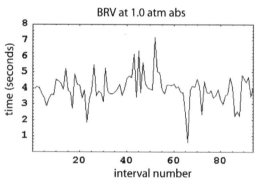

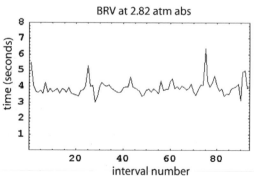

Figure 18.5. Breathing rates at one atmosphere and 2.82 atmospheres pressure. The size of the interbreath intervals are plotted as a function of the breath number for approximately one hundred breath intervals. These data are for the individual at rest under 1 ATA of pressure for the upper and 2.82 ATA of pressure for the lower curve.

Table 18.1. BRV data for a variety of experimental conditions

HRV	D	H	r_1
SR	1.32 ± 0.11	0.68 ± 0.11	28%
SE1	1.33 ± 0.12	0.71 ± 0.12	33%
SE2	1.22 ± 0.16	0.81 ± 0.16	54%
DR	1.54 ± 0.13	0.67 ± 0.13	27%
DE1	1.46 ± 0.13	0.75 ± 0.13	41%
DE2	1.32 ± 0.14	0.82 ± 0.14	56%

Note. The two coefficients obtained from the least squares fitting formula for allometric scaling [Equation (16.4)] for BRV data are not recorded in the table. Instead, the derived fractal dimension, $D = 2 - b/2$, the Hurst coefficient $H = 2 - D$, and the autocorrelation coefficient r_1 are listed. There were 18 individuals that participated in the experiment, so the recorded values are the average quantities $\pm$ standard deviation of the measurements.

be quite large under most conditions. This would suggest that the BRV has long-time memory, in a way similar to that of HRV and SRV. If this is the case, then a question that immediately comes to mind is whether the variation in breathing and the variation in these other quantities, such as heart rate, are correlated with one another. If they are correlated, does this indicate an interdependence of the systems controlling these different physiological functions?

Simultaneous HRV Data Collection

In addition to the breath interval data collected under the conditions outlined above, there was a simultaneous measurement of the individual's heartbeat. From the data collected, we are able to determine the heart rate variability obtained at the same time as the breath interval variability. In Figure 18.6, we depict the heartbeat interval as a function of time, for the individual resting at both 1 ATA and at 2.82 ATA. To the naked eye, two things are evident. First of all, the average heartbeat interval is longer for the higher pressure. Second, the degree of heart rate variability appears to be approximately the same for both pressures. These observations are quantified with the fitting parameters recorded in Table 18.2.

The Influence of Pressure on the HRV

In Table 18.2, we can see that the level of variability in the heart rate actually decreases with the increase in the pressure. This means that the body interprets the increase in pressure as stress that requires a more systematic beating of the heart, that is, a more closely regulated blood flow. This reaction is reinforced as the body is further stressed by means of exercise, resulting in the variability completely disappearing. This loss of variability is certainly a reaction to stress, but is an unhealthy dynamical state that the body cannot maintain for too long a period of time.

Correlations in HRV Data Seem to be Very Large

We cannot help but note that the correlation coefficients for the interbeat intervals are very large under all the conditions of the experiment, even larger than they

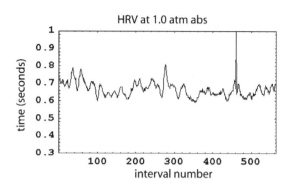

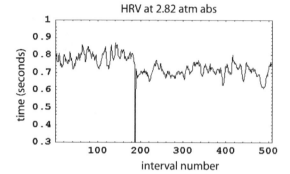

Figure 18.6. Heartbeat variability at one atmosphere and 2.82 atmospheres pressure. The heartbeat intervals are depicted in two resting situations: the upper curve is at 1 ATA and the lower curve is at 2.82 ATA. The fractal dimension of the upper curve is 1.14 and in the lower curve it is 1.09.

were for the interbreath intervals. The autocorrelation coefficient increases from 54% to 95% as an individual goes from a state of rest to that of extreme exercise, at both pressures. These large correlations are consistent with the results found for the HRV in the previous chapters, indicating that the heartbeat variability has long-time memory. In fact the qualitative behavior, that of increasing correlations with increasing levels of exercise, is the same pattern we see with the BRV in Table 18.2.

Table 18.2. Simultaneous HRV data for a variety of experimental conditions

HRV	D	H	r_1
SR	1.19 ± 0.11	0.81 ± 0.11	54%
SE1	1.05 ± 0.07	0.95 ± 0.07	87%
SE2	1.02 ± 0.05	0.98 ± 0.05	95%
DR	1.20 ± 0.10	0.80 ± 0.10	52%
DE1	1.03 ± 0.04	0.97 ± 0.04	92%
DE2	1.01 ± 0.02	0.99 ± 0.02	97%

Note. The two coefficients obtained from the least squares fitting formula for allometric scaling [Equation (16.4)] for HRV are not listed in the table. Instead, the derived fractal dimension, $D = 2 - b/2$, the Hurst coefficient $H = 2 - D$, and the autocorrelation coefficient, $r_1 = 2^{3-2D} - 1$, are listed. There were 18 individuals that participated in the experiment, so the recorded values are the average quantities $\pm$ standard deviation of the measurements.

The Fit to Allometric Scaling

We analyzed the BRV data using allometric scaling, just as we did the HRV and SRV data in the previous chapters. We calculate the mean and variance as a function of the level of aggregation of the data and plot the results on log–log graph paper. The aggregated allometric relation is then fit to the data and a mean-square fitting procedure is used to determine the parameters of the relation. A typical fit to the data is depicted in Figure 18.7. In Table 18.1, we record the fractal dimension, Hurst coefficient, and correlation coefficient resulting from this fitting procedure averaged over the 18 individuals that participated in the experiment. It is worth mentioning that the fit to the data points is quite good considering that there are only 94 data points—interbreath intervals in the case shown. The fitting curve is well away from the boundaries of uncorrelated random noise and complete regularity.

What Can We Learn from the Fit to the BRV Data?

We can see that the aggregated interbreath intervals, for a resting individual at one atmosphere pressure, falls along an allometric growth curve fairly well, which is to say that it well fit by Equation (16.7). The fractal dimension is determined by the slope of the curve, $b = 1.43$, to be $D = 1.29$, a value nearly midway between a regular curve with $D = 1.0$ and an uncorrelated random time series with $D = 1.5$. At one atmosphere, the fractal dimension of BRV decreases with light exercise, on average, and continues to decrease, on average, as the intensity of the exercise increases. The same behavior is observed at the higher pressure; in fact, there appears to be little if any effect of the increase in pressure on BRV.

The HRV Results

The aggregated heartbeat intervals under all the conditions in the above study for a single individual are qualitatively the same as those discussed in Chapter 17. We

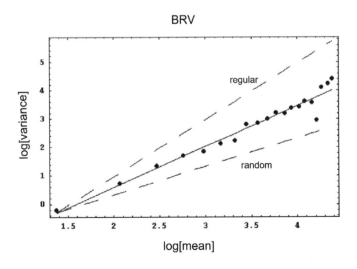

BRV

Figure 18.7. The BRV data for an individual resting at one atmosphere pressure is depicted using allometric scaling for the data obtained from the Hyperbaric Laboratory. The logarithm base 10 of the variance is plotted versus the logarithm base 10 of the mean, starting with all the data points at the lower left, to the aggregation of twenty data points at the upper right. The slope of the fitting curve is 1.43.

could, of course, display a few more graphs, but nothing in addition would be learned from them except that the HRV data fall along straight lines on log–log graph paper. The aggregated variance versus the aggregated mean satisfies an AAR for these data, just as we learned earlier. In these HRV data, the correlations are much higher than that observed in either SRV or BRV data. In fact, the fluctuations nearly stop altogether for sufficiently high levels of exercise.

Other Experimental Results

We have only used the measurements of the BRV for a single individual in this analysis. Therefore, it would not be prudent to draw too far-reaching a conclusion from these results. However, we can support our interpretation using earlier findings from other investigators, not on human interbreath intervals but those of fetal lambs. The dynamical pattern of breathing was studied in 17 fetal lambs [140]. The time series consisted of instantaneous breathing rates that appeared similar on different time scales. The clusters of faster breathing rates was interspersed with periods of relative quiescence and suggested to Szeto et al. [141] that the process of breathing was self-similar. Distribution histograms of the interbreath intervals showed log-normal behavior for low breathing rates and a transition to inverse power-law behavior for higher rates. This kind of transition is well described by the scaling argument presented here; see also West and Deering [38].

The Silence of the Lambs

Sezeto et al. interpret their data in such a way that the mechanisms that govern fetal breathing are complex and involve many processes. They point out that fetal breathing patterns are not homogeneous. The more closely the breathing patterns were examined, the more detail was revealed. Cluster of faster rates were seen within what Dawes et al. [142] called breathing episodes. When examined on even finer time scales, clusters could be found within these clusters. In addition, the pattern in the time series on the smaller time scale resembled the pattern observed on the larger time scales, thereby suggesting self-similarity to the investigators. This self-similarity is not unlike both the qualitative and quantitative features observed by the authors with regard to SRV, HRV, and BRV. In the same way the fetal breathing time series yields the typical inverse power-law spectrum, in which the slope of the spectrum increases as a function of the age of the fetus. The increase in the slope of the spectrum with fetal age suggests a reduction in complexity of the regulatory system as the fetus matures.

18.4 SUMMARY

The Interaction of Subsystems

The specific mechanisms underlying BRV are not always clear. However, it is known that the rate of breathing is regulated by a complex feedback control system. Data suggest that this control system has the following features, which are notably the same as those of the HRV control system. First, it is nonlinear. This property is manifest as nonlinear interactions that directly controls breath-to-breath changes in breath rate through the central nervous system (CNS), which determines the output level of autonomic nerve activity [130]. Second, the system has multiple input signals originating from sensors located in different parts of the body for monitoring changes in blood pressure, CO_2 levels, and tissue oxygenation. These input signals are converted into and interact with each other at the respiratory center in the CNS. Third, the respiratory system is a closed feedback and feedforward control system.

The Effect of Scaling

Chaotic variations in BRV would be consistent with the fractal characteristics of BRV revealed in the above analyses. Furthermore, the long-term correlation or memory observed in BRV time series indicates that although different regulatory mechanisms may act independently on different time scales, their effects on dynamical changes in the rate of breathing may be tied together through scaling. Thus, impairment of one individual component of BR regulation may influence other regulatory mechanisms via interdependence.

Epilogue

In this book, we have projected a picture of an acrobat dancing on a wire, moving slow and fast, to keep from falling and to control his motion and locomotion from one side of the circus tent to the other. We pointed out in the Prologue that the flowing movements and sharp hesitations of the wirewalker are all part of the spectacle. In the text, we argue that this metaphor captures a new metaphor in medicine and bioengineering on how we control the various systems within our own bodies that regulate, maintain, and guide locomotion. The wirewalker does not walk in a low measured step, nor do the physiological systems act in an orchestrated sinus rhythm; both leap and change in unpredicted and unexpected ways to perform their respective functions. The science of this combined regularity and randomness is *biodynamics* and it uses the mathematical disciplines of nonlinear dynamics, chaos theory, fractal geometry, and nonlinear data processing. We were able to circumvent much of this technical material in our discussions, but the mathematical ideas are always there in the background, forming the theoretical basis of a future holistic theory of medicine and health. This approach to biodynamics takes the initiative in reversing the historical trend towards scientific specialization in the field of medicine.

For thousands of years, heart rate, breathing rate, and steadiness of walking have been used as indicators of health and disease. The research activity in biodynamics for the past two decades establishes the limitations of these traditional health indicators and proposes new, more flexible metrics. For example, replacing average heart rate with variability in heart rate suggested changes in the operation of pacemakers. The intervals between excitation pulses, generated by modern pacemakers, are no longer constant but have a built-in randomness to more closely mimic actual heartbeats. The average breathing rate has been replaced with the variation in

Biodynamics: Why the Wirewalker Doesn't Fall. By Bruce J. West and Lori A. Griffin
ISBN 0-471-34619-5 © 2004 John Wiley & Sons, Inc.

breathing rate, suggesting changes in how we ventilate patients. Ventilators, supporting respiration in post-operative patients, have been shown in the laboratory to be significantly more effective in exchanging oxygen between air and blood when the inspiration and expiration phases are not regular but contain fractal variability. Finally, replacing the average stride interval with the variation in stride interval, suggests that the regularity in treadmill walking and other such protocols is not always desirable in physical therapy.

Consequently, we argue that the medical community should relinquish its reliance on the averages of physiological variables for understanding wellness and disease and add measures of the fluctuations to its diagnostic repertoire in bedside medicine. The fluctuations are a much more sensitive indicator of a physiological state than is an average. In this book we have discussed a number of methods for exploiting this sensitivity that have grown out of the innovative research done on the application of the principles of nonlinear dynamics systems theory, nonlinear data processing, and fractal geometry to biodynamics.

The fractal dimension and other scaling indices of time series, can be used as indicators of well being for the cardiovascular system, the respiratory system, and motor control systems. These results were anticipated, in part, 20 years ago. To paraphrase Goldberger and West [143], the notion of the chaotic (fractal) nature of healthy dynamics supports the general concept of health as an "information-rich" state. Highly periodic behavior, by contrast, seen in a variety of pathologies, reflect a loss of physiological information. As a corollary, time series analysis of EEG data may provide a new means of monitoring patients at high risk for sudden death who would be predicted to show alterations in the healthy broadband (inverse power law) inter-beat interval spectrum. Loss of sinus rhythm heart rate variability, sometimes associated with low frequency oscillations, had already been observed both in adults with severe heart failure and in the fetal distress syndrome. Loss of this broadband spectral reserve was also considered to be a marker of pathological dynamics in other diseases, including certain malignancies. The conjectured importance of chaos and fractals to healthy structure and function has been born out through the applications of abstract nonlinear models to bedside medicine in the intervening years.

PART V

MICROBIOGRAPHIES— THE SCIENTISTS WHO MADE HISTORY

Archimedes was believed to have been born in Syracuse, Sicily, which was then a Greek colony, in 287 BC. He died at the hands of a Roman soldier during the siege of Syracuse by Rome in 212 BC. He was perhaps the greatest scientist of the ancient world, being the first to systematically use mathematics in the experimental study of physical phenomena. He asked to have his gravestone inscribed with a cylinder enclosing a sphere together with the formula for the ratio of their volumes, indicating what he considered to be his greatest achievement. He contributed to both statics and hydrostatics, proving the mathematical relations for the laws of levers, and Archimedes' principle for the displacement of fluid.

Jacques (James I or Jakob) Bernoulli was born in Basel, Switzerland in 1654 and died in Basel in 1705. He was the first of the Bernoulli clan to attain a reputation in mathematics and was a pupil of Leibniz. As a young man he studied theology, astronomy, mathematics, and physics and he also traveled extensively. In 1682, he returned to Switzerland and in 1687 became professor of mathematics at the University of Basel. He wrote the second book ever published on the theory of probability, which appeared posthumously in 1713. He and his brother, Jean I, were bitter rivals, but they obtained many joint results through correspondence with Leibniz; in fact, much of what is contained in an elementary course of differential and integral calculus can be found in the correspondence of the two brothers and their mentor.

Biodynamics: Why the Wirewalker Doesn't Fall. By Bruce J. West and Lori A. Griffin
ISBN 0-471-34619-5 © 2004 John Wiley & Sons, Inc.

Jacques discovered the logarithmic spiral and was so attracted to it he requested that it be engraved on his tombstone.

Jean I (John or Johann) Bernoulli was born in Basel, Switzerland in 1667 and died in Basel in 1748. He was the bother of James I, who was 13 years his senior. Jean first studied medicine, at his father's bidding, but after he completed a PhD thesis in this area, he turned to the study of mathematics and studied with Leibniz. His ability and accomplishments were eventually recognized and he became professor of mathematics at the University of Groningen, Germany in 1695. In 1705, with the death of his brother, he was elected to fill the then vacant position at the University of Basel. Jean contributed to a broad spectrum of mathematical problems, including the calculus (he invented the variational calculus), that he made popular in the scientific community. It is often difficult to separate the mathematical results that Jean obtained from those obtained by his brother, Jacques.

Daniel Bernoulli was born in Groningen, Holland in 1700 and died in Basel, Switzerland in 1782. He was the son of Jean I. He was professor of mathematics at the Academy of Petrograd, Russia from 1725 to 1733, after which time he returned to Basel, where, like his father and uncle before him, he became a professor at the university. Daniel is credited with significant contributions to hydrodynamics, the kinetic theory of gases, and the equations determining the propagation of sound waves. Thus, when the Bernoulli name is used in physics, the reference is usually to Daniel. However, there are many allusions from the applications of mathematics, particularly probability theory, that refer to his father, uncle, brother, cousin, and nephews.

Nicolas I Bernoulli was born in Basel in 1687 and died there in 1759. He was the nephew of Jacques I and Jean I. He was a professor of mathematics at Padua from 1716 to 1719, but returned to Basel to become a professor at the university there. His formal training was in law and his first mathematical treatise concerned the legal applications of probability theory. He also wrote extensively on differential equations and geometry.

Nicolas II Bernoulli was born in Basel, Switzerland in 1695 and died in Petrograd in 1726, and was the son of Jean I. He was called to St. Petersburg by Peter the Great, after being professor of law at Bern (1723–1725) but, unfortunately, he died quite young. His major contribution is the so-called St. Petersburgh Paradox in probability theory, which is discussed in the context of fractal random processes today. Both he and his brother, Daniel, studied under their father, Jean I, along with their fellow pupil, Euler.

Jean II Bernoulli, the youngest son of Jean I, was born in Basel in 1710 and died there in 1790. Like others in his family, he initially studied law, but spent his later years as professor of mathematics at the university in Basel. His research was mainly in the area of physics.

Jean III Bernoulli was the son of Jean II, born in Basel in 1744 and died at Köpnick, near Berlin, in 1807. He also studied law, but the family business in mathematics eventually won him over. He became the director of the mathematics class at the Academy of Sciences in Berlin (they probably ran out of positions in Basel). His research was primarily in mathematics.

Ludwig Boltzmann was born in Vienna in 1844 and committed suicide at Duino, near Trieste, in 1906. He was a theoretical physicist who contributed to the development of the kinetic theory of gases, electromagnetism, and thermodynamics. His work lead to the development of statistical mechanics, based on the application of probability theory to atomic events. His expression for the entropy of a closed system in terms of the logarithm of the number of microscopic states of a system is engraved on his tombstone, and established the relation between entropy and probability theory. The proportionality constant, now known as Boltzmann's constant, 1.38×10^{-23} J/K, appears in every mathematical formulation of the statistical nature of matter. Unfortunately the experimental evidence necessary to establish many of his theoretical results did not occur until a few years after his death and the criticism his work received was thought to have contributed to his suicide.

Robert Boyle was born in Ireland in 1627 and died there in 1691. He was both a physicist and chemist, and he formulated his famous law in 1662. He was a pioneer in the use of experiment and the scientific method. He coined the word *analysis* to emphasize that the function of chemistry was to determine the composition of substances. Boyle was the first to apply the plant extract litmus as an indicator of acids and bases.

Nicolas Leonard Sadi Carnot was born in Paris, France in 1796 and died of cholera in 1831. According to the custom of the time, all his personal belongings, including his research notes, were burned. In the work that survived, he reviewed the industrial, political, and economic importance of the steam engine. In addition, he determined that the maximum amount of work that an engine can produce depends only on the temperature difference that occurs in the steam engine. This was known as Carnot's theorem and became the basis for the second law of thermodynamics, which states that heat cannot flow spontaneously from a colder to a hotter substance. Carnot's work is the foundation of the discipline of thermodynamics.

Rudolf Julius Emmanuel Clausius was born in Köslin in Pomerania, now Koszalin, Poland, in 1822 and died in Bonn in 1888. He received a PhD in physics from the University of Halle in 1848, became a Professor of Physics at the Zurich Polytechnic in 1855, and then moved to Bonn in 1869, where he held the chair of physics until his death. In 1850, Clausius introduced the concept of entropy for the first time in his mathematical formulation of the second law of thermodynamics. He argued that heat could be transformed into work and that heat was transferred from regions of high temperature to regions of low temperature. In his formulation, the entropy of the universe must increase and the universe proceeds from order to disorder.

Charles Coulomb was born in Angoulême in 1736 and died in Paris in 1806. He established the laws governing electric charge and magnetism. Although not the only researcher involved in these areas, for various reasons he became the most visible, so that the unit of electric charge is named the coulomb in his honor. He also carried out research in the area of ergonomics, which led to an understanding of the ways in which people and animals can best do work.

Democritus was a Greek philosopher about whose personal life little is known. He was probably born in Abdera around 460 BC and lived to be approximately 70 years old. He taught that all matter, both animate and inanimate, solid, liquid, or gas, is made up of tiny, indivisible particles called atoms. Atoms of liquids were smooth and easily slid over one another, whereas atoms of solids were angular, jagged, and caught onto each other to form arrangements of solids. It is tempting to attribute more to this early theory than it contains, since it was not a scientific theory but a construction of Democritus' mind alone, having no experimental content, and remained so until the time of Galileo.

Albert Einstein was born in Ulm, Germany in 1879 and died in Princeton, New Jersey in 1955. Einstein was probably the greatest physicist since Newton. He revolutionized our understanding of matter, space, and time with his theories of special and general relativity. He established the particle nature of light and deduced the photoelectric law that governs the production of electricity from light-sensitive metals. After graduating in Zurich, Switzerland in 1900, he became a clerk in the Swiss patient office (1901–1909). He worked on theoretical physics in his spare time and in 1905 published three seminal papers on Brownian motion, the photoelectric effect, and special relativity. Brownian motion concerns the description of the motion of a heavy pollen mote suspended in water in terms of the collisions of the much lighter water molecules with the pollen mote. He was able to make predictions of the movement and size of the particles, which were later verified by experiment. His work on photoelectricity introduced the notion of light quanta (photons) to explain how the surface of certain metals emit electrons when exposed to light. His special theory of relativity will not be discussed here.

Leonard Euler was born in Basel, Switzerland in 1707. His father was an amateur mathematician who studied under Jacques Bernoulli, while he (Leonard) studied under Jean I, along with Daniel and Nicolaus II. The latter two Bernoullis became Euler's life-long friends. In 1727 Euler followed Nicolaus II and Daniel Bernoulli to St. Petersburg. Euler was the most prolific scientist of his, or any other, generation. He produced 530 books and papers during his life, and the manuscripts that he left which were published after his death, bring the total to 771; subsequent research by scholars brings this impressive total to 886.

As with Newton, he created, developed, and completed whole areas of investigation. For example the entire discipline of trigonometry, as it is used today in the classroom, can be traced directly to him. He wrote books on the differential and integral calculus, mechanics (containing, for example, the Euler angles of rotation), hydraulics, ship construction, celestial mechanics, optics, and many many other subjects. He was the first person to actually write Newton's second law of motion as a partial differential equation, that is, in the familiar form $F = ma$.

He lost one eye in 1735 and the other in 1766, but even his blindness did not keep him from doing research. He continued to dictate his discoveries for the next 17 years, and by some accounts was more productive during this period than before. He died in St. Petersburg at the age of 76, while playing with his grandson, soon after doing some calculations on the orbit of the newly discovered planet, Uranus. He was engaged in writing a book on hydromechanics at the time of his death.

Michael Faraday was born in Newington, Surrey in 1791 and died in Hampton Court, Middlesex in 1867. He is considered to be the greatest experimentalist of the 18th century and was mostly self-taught. In addition to his pioneering contributions to electricity, such as the introduction of the electric field, he invented the electric motor, generator, and transformer, and discovered electromagnetic induction and the laws of electrolysis.

Richard Phillips Feynman was born in New York City in 1918 and died in Los Angeles, California in 1988. He was a physicist whose work laid the foundation for quantum electrodynamics, based on a simple and elegant system involving drawings known as Feynman diagrams. These diagrams represent the interaction between particles and how they move form one space–time point to another. His rules for the implementation of these diagrams vastly simplified the calculations in quantum field theory and contributed to our understanding of elementary particle physics.

Galileo Galilei was born in Pisa, Italy in 1564 and died in Florence, Italy in 1642. He was born on the day of Michelangelo's death and died in the year of Newton's birth. He entered the University of Pisa in 1581 to study medicine (at his father's insistence), but took up the study of geometry and eventually convinced his father that his career lay in the study of science. At the age of 25, he became professor at the University of Pisa. After two years, he vacated this position. His move was prompted by a scandal over public experiments on the fall of heavy bodies, attacks on his elders, and by offending the likes of John de Medici. Leaving Pisa, he went to a professorship of mathematics at Padua. This is where he carried out his most important scientific work, including the experiments upon which the science of statics and dynamics, or mechanics, as we know it today, is built. He developed the astronomical telescope and was the first to see sunspots. He discovered that freely falling bodies have the same constant acceleration and was the first to articulate the law of inertia. Galileo invented a thermometer, a hydrostatic balance, and a compass. His work founded the modern scientific method of deducing laws to explain the results of observations and experiment.

Johann Carl Friedrich Gauss was born at Braunschweig, Germany in 1777 and died at Göttingen in 1855. He was the son of a day laborer, but his mathematical gifts were so evident at such an early age that he was able to attend the university in Göttingen. He developed the method of least squares while at the university. He contributed to virtually every field of mathematics and became the leading mathematician of Europe. Of him it was said: "the mathematical giant who from his lofty heights embraces in one view the stars and the abysses." Today we often encounter his contributions through the law of frequency of errors and the Gaussian or normal distribution. Gauss argued, in 1833, that all units should be assembled from a few basic units or absolute units, specifically, length, mass, and time. The CGS unit for magnetic flux density is called the gauss.

Robert Hooke was born in Freshwater, Isle of Wight, in 1635 and died in London in 1703. Hooke is an unusual figure in the history of science in that although he made no major discovery himself, his experimental observations and intuition stimulated important contributions on the part of his contemporaries, including Newton. For example, he suggested in 1679 an inverse square power law for the law of gravitational attraction, which Newton used in 1687. Hooke is remembered primarily for the linear law of elasticity, coining the term *cell* in biology, and for the development of the air pump.

James Prescott Joule was born in Salford, England in 1818 and died in Sale, Cheshire, in 1889. His work established the equivalence of electrical, mechanical, and chemical energy and his experiments on the interrelation of these phenomena resulted in the first law of thermodynamics. He determined the mechanical equivalent of heat in 1843, and the international standard unit of heat, the *joule,* is named after him. He also discovered Joule's law, which defines the relation between heat and electricity. Joule had established the first law of thermodynamics—energy cannot be created nor destroyed but only transformed.

Compte Joseph Louis Lagrange was born in Turin, Italy in 1736. He spent his early years in Italy, but most of his scientific career took place in Paris, where he died in 1813. He became a professor of mathematics at the age of nineteen at the artillery school in Turin (1755), succeeded Euler as mathematical director in the Berlin Academy in 1766, and after going to Paris in 1787, became professor of mathematics at the newly founded École Normale in 1795 and then at the École Polytechnique in 1797. He managed to avoid the guillotine and, unlike other foreigners, was expressly allowed to remain in France because of his recognized mathematical genius. In the physical sciences, Lagrange is perhaps best known for his work in the variational calculus. This work enabled him to construct the equations of motion for mechanical systems using the system's energy rather than Newton's force laws.

Marquis Pierre-Simon de Laplace was born in poor circumstances at Beaumont-en-Auge, Calvados in 1740 and died the greatest scientist in Paris, France in 1827. He was an astronomer and mathematician who studied the motions of the moon and planets, and published a five-volume survey of celestial mechanics: *Traité de méchanique céleste* (1799–1825). This work contained the law of gravity as applied to the earth, explaining such phenomena as the ebb and flow of tides and the precession of the equinoxes. Laplace also wrote on probability, the calculus, differential equations, and geodesy. He was also one of the organizers of the École Polytechnique and the École Normale.

Benoit B. Mandelbrot was born in Poland in 1924 and today lives in the United States. He is the inventor of the term *fractal* that is used to describe a motif that repeats itself on ever diminishing scales. The term fractal applies to structures that have a geometrical self-similarity and to statistical processes that have a self-similarity in the random process. He also invented fractal geometry and fractal statistics. The two applications of the concept blend in the dynamical phenomenon of chaos, in which the attractors on which the dynamics unfold are fractal, leading to a time series for the trajectory that is random.

Isaac Newton was born in Lincolnshire, England in 1642 and died in Kensington, England in 1727. He was buried at Westminster Abbey, among kings. He was a sickly child who showed little promise of academic achievement. After a slow start, he began to show considerable ability and attended Trinity College in 1660. By the time he was 26, he received his MA and was recognized as the most promising mathematician and physicist in England. He had begun his work on the calculus, which he called fluxions, in the same year he received his BA, 1665, at the age of 23. The year after receiving his MA, he worked on the theory of light and began his investigations into the theory of gravity. He studied these and other phenomena using his method of fluxions.

His recognition as a scientist came early in his life. He succeeded Barrow as the Lucasian Professor of Physics at Cambridge, one of the highest academic positions in the world, at the tender age of 27. The honors kept coming for the next 60 years. He made fundamental contributions in every area of natural philosophy, today called physics, and established the modern paradigm of science. His equations of motion define the discipline of mechanics; his calculus allows us to establish and solve the equations of dynamics; his theory of universal gravitation has been called the greatest intellectual achievement of the human mind. He is considered by many to be the greatest natural scientist who ever lived.

Jules Henri Poincaré was born in France in 1854 and died there in 1912. He was a mathematician who developed the geometrical interpretation of differential equations and was a pioneer in the theory of relativity. He was the first to show that three bodies interacting by means of the law of gravity generate equations of motion that cannot, in general, be solved in closed form. This has since become known as "the three-body problem." He was able to use a combination of topological and geometrical arguments to establish that the motion of a light body in the gravitational field of two heavy bodies would be so erratic that it could not be described by an analytic function. Today, we would say that the trajectory is fractal. Poincaré developed several new branches of mathematics, including the theories of asymptotic expansions and integral invariants, and the new subject of topological dynamics.

Erwin Schrödinger was born in Austria in 1887 and died in 1961. He developed the equation that describes the wave nature of matter, and advanced the study of wave mechanics, also called quantum mechanics, to describe the behavior of electrons in atoms. In 1926, he produced a consistent mathematical explanation of the quantum theory and the structure of the atom. In 1946, he applied his vast physical insight into what the physical basis of life might be and launched the modern approach to biophysics.

Leonardo da Vinci was born near Florence, Italy in 1452 and died in Amboise, France in 1519. He was famous as a painter, sculptor, goldsmith, investigator of the circulation of blood, general scientist, architect, and writer on mechanics, optics, and perspective in art. He would have ranked high as a mathematician and scientist had not these talents been obscured by his unusual gifts in so many other areas. In physics he knew the theory of the inclined plane, determined the center of gravity of a pyramid, worked in the field of capillarity and diffraction, and studied the resistance of the air and the effect of friction.

Alessandro Volta was born in Como, Italy in 1745 and died there in 1827. He discovered how to produce electric current and constructed the first electric battery. His research in electricity was based on that of Luigi Galvani, who, in 1791, discovered that the muscles of dead frogs contract when two dissimilar metal rods are brought into contact with the muscle and each other. The unit of electromagnetic force or potential was ultimately named the *volt* in his honor.

Thomas Young was born in Milverton, Somerset in 1773 and died in London in 1829. He was a physicist and physician who established that the interference of light is due to its wave nature. Young also made important discoveries in the physiology of vision and developed the ratio of stress to strain in elasticity, called Young's modulus, to characterize materials. He was also the first to use the modern definition of kinetic energy to denote the work done by a body, that is, the product of the force over the distance it acts.

References

1. D. A. Winter, *Biomechanics and Motor Control of Human Movement,* 2nd ed., John Wiley and Sons, New York (1990).
2. J. G. Hay, *The Biomechanics of Sports Techniques,* 3rd ed., Prentice-Hall, New Jersey (1985).
3. C. P. Snow, *The Two Cultures and the Scientific Revolution,* Cambridge University Press, Cambridge (1959).
4. E. W. Montroll and B. J. West, "On an enriched collection of stochastic processes," in *Fluctuation Phenomena,* Eds. E. W. Montroll and J. L. Lebowitz, North-Holland, Amsterdam (1979); 2nd Edition (1987).
5. J. B. Bassingthwaighte, L. S. Liebovitch, and B. J. West, *Fractal Physiology,* Oxford University Press, New York (1994).
6. F. Turner, Foreword to *Chaos, Complexity and Sociology,* Eds. R. A. Eve, S. Horsfall, and M. E. Lee, Sage, Thousand Oaks (1997).
7. L. von Bertalanffy, *General Systems Theory,* G. Braziller, New York (1968).
8. B. J. West, *Fractal Physiology and Chaos in Medicine,* World Scientific, Singapore (1990).
9. B. J. West, *Physiology, Promiscuity and Prophecy at the Millennium: A Tale of Tails,* World Scientific, Singapore (1999).
10. D. S. Riggs, *The Mathematical Approach to Physiological Problems,* Williams and Wilkins, Baltimore, (1963).
11. H. E. Huntley, *Dimensional Analysis,* Dover Publications, New York (1967).
12. Galileo Galilei, *Two New Sciences,* translated by H. Crew and A. deSalvio, Dover Publications, New York (1954).
13. M. Mott-Smith, *The Concept of Energy Simply Explained,* Dover Publications, New York (1964).
14. H. Poincaré, *The Foundations of Science,* Science Press, New York (1929).
15. J. Kepler, *The New Astronomy,* Prague (1609).
16. J. Kepler, *Harmony of the Worlds,* Linz (1619).

Biodynamics: Why the Wirewalker Doesn't Fall. By Bruce J. West and Lori A. Griffin
ISBN 0-471-34619-5 © 2004 John Wiley & Sons, Inc.

17. L. A. Sena, *Units of Physical Quantities and Their Dimensions,* Mir Publishers, Moscow (1972).

18. R. K. Barnhart, *Dictionary of Science,* Houghton Mifflin, Boston (1986).

19. I. Prigogine, *Thermodynamics of Irreversibile Processes,* 2nd rev. ed., Wiley-Interscience, New York (1961); 1st ed. (1955).

20. E. Schrödinger, *What is Life?,* Cambridge University Press, New York (1995); first published in 1944.

21. M. Kleiber, "Body size and metabolism," *Hilgardia* **6**, 315–353 (1932).

22. S. Brody, *Bioenergetics and Growth,* Reinhold, New York (1945).

23. K. Schmidt-Nielsen, *Scaling, Why Is Animal Size So Important?,* Cambridge University Press, Cambridge (1984).

24. C. R. Taylor and E R. Weibel, "Design of the mammaliian respiratory system. I. Problem and strategy," *Respir. Physiol.* **44**, 1–10 (1981).

25. W. A. Calder III, *Size, Function, and Life History,* Harvard University Press, Cambridge MA (1984).

26. A. C. Brown and G. Brengelmann, "Energy Metabolism," in *Physiology and Biophysics,* Eds. T. C. Ruch and H. D. Patton, Saunders, Philadelphia (1966).

27. B. B. Mandelbrot, *Fractals, Form, Chance and Dimension,* Freeman, San Francisco (1977).

28. B. B. Mandelbrot, *The Fractal Geometry of Nature*, Freeman, San Francisco (1982).

29. *The Notebooks of Leonardo da Vinci,* compiled and edited by J. P. Richter, Dover, New York (1970).

30. C. D. Murray, "A relationship between circumference and weight and its bearing on branching angles," *J. Gen Physiol.* **10**, 725–729 (1927).

31. B. J. West, V. Bhargava, and A. L. Goldberger, "Beyond the principle of similitude; renormalization in the bronchial tree," *J. Appl. Physiol.* **60**, 1089 (1986).

32. N. MacDonald, *Trees and Networks in Biological Models,* Wiley, New York (1983).

33. K. Suwa, T. Nirva, H. Fukusawa, and Y. Saski, "Estimation of intravascular blood pressure by mathematical analysis of arterial casts," *Tokoku J. Exp. Med.* **79**, 168 (1963).

34. E. R. Weibel and D. M. Gomez, "Architecture of the Human Lung," *Science* **137**, 577–585 (1962).

35. T. A. Wilson, "Design of the bronchial tree," *Nature* **18**, 668–669 (1967).

36. D. W. Thompson, *On Growth and Form,* 2nd ed., Cambridge University Press, Cambridge (1963).

37. J. S. Huxley, *Problems of Relative Growth,* Dial Press, New York (1931).

38. B. J. West and W. Deering, "Fractal Physiology for Physicists: Levy Statistics," *Phys. Rept.* **246**, 1–100 (1994).

39. R. P. Feynman, R. B. Leighton and M. Sands, *The Feynman Lectures on Physics,* Vol. 1, Addison-Wesley, Reading, (1963).

40. R. Rosen, *Essays on Life Itself,* Columbia University Press, New York (1999).

41. T. A. McMahon, "Size and shape in biology," *Science* **179**, 1201–2014 (1973).

42. N. Rashevsky, *Mathematical Biophysics Physico-Mathematical Foundations of Biology,* Dover, New York (1938).

43. J. B. S. Haldane, *On Being the Right Size,* Edited by John Maynard Smith, Oxford University Press, New York (1991); essay of the same name first published in 1927.

44. E. R. Weibel, *Symmorphosis,* Harvard University Press, Cambridge (2000).

45. J. V. Basmajian and C. J. DeLuca, *Muscles Alive,* 5th ed., Williams & Wilkins, Baltimore (1985).

46. M. V. Volkenstein, *General Biophysics,* Vols. 1 and 2, Academic Press, New York (1983).

47. S. Vogel, *Life's Devices,* Princeton University Press, Princeton, (1988).

48. D. E. Smith, *History of Mathematics,* Vol. 1, Dover, New York (1958); first published in 1923.
49. A. Hill, *Proc. R. Soc. London, Ser. B* **126**, 136 (1938).
50. R. M. Alexander, *Exploring Biomechanics,* Scientific American Library, New York (1992).
51. E. Muybridge, *Animals in Motion,* Dover, New York (1957).
52. J. Burke, *Connections,* Little, Brown Boston (1978, 1995).
53. C. C. Chow and J. J. Collins, "Pinned polymer model of posture control," *Phys. Rev. E* **52**, 907–912 (1995).
54. W. N. Findley, J. S. Lai, and K. Onaran, *Creep and Relaxation of Nonlinear Viscoelastic Materials,* Dover Publications, New York (1976).
55. J. E. Gordon, *Structures, or Why Things Don't Fall Down,* Da Capo Press, New York, (1978).
56. I. Newton, *Principia, Book II* (1686).
57. V. G. Dethier and E. Stellar, *Animal Behavior,* Prentice-Hall, Englewood, (1961).
58. P. G. Hewitt, *Conceptual Physics,* 7th ed., HarperCollins (1993).
59. E. R. Kandel, "Small systems of neurons," in *Mind and Behavior,* Eds. R. L. Atkinson and R. C. Atkinson, Freeman, San Francisco (1979).
60. B. van der Pol and J. van der Mark, "The heartbeat considered as a relaxation oscillator and an electrical model of the heart," *Phil. Mag.* **6**, 763 (1928).
61. W. F. Ganong, *Review of Medical Physiology,* 19th ed., Simon & Shuster, New York (1999).
62. J. C. Eccles, *The Neurophysioloical Basis of Mind; The Principles of Neurophysiology,* Clarendon Press, Oxford (1953).
63. A. L. Hodgkin, A. F. Huxley, and B. Katz, *J. Physiol.* **116**, 424 (1952).
64. M. D. Mann, *The Nervous System and Behavior,* Harper & Row, Philadelphia, (1981).
65. W. J. Freeman, *Mass action in the nervous system,* Chapter 7, Academic Press, New York (1975).
66. W. J. Freeman, "Simulation of chaotic EEG patterns with a dynamic model of the olfactory system," *Biol. Cybern.* **56**, 139–150 (1987).
67. A. Babloyantz and A. Destexhe, "Low dimensional chaos in an instance of epilepsy," *Proc. Nat. Acad. Sci. U.S.A.,* **83**, 3515–17 (1986).
68. B. J. West, M. N. Novaes, and V. Kavcic, "Fractal Probability Density and EEG/ERP Time Series," in *Fractal Geometry in Biological Systems,* Eds. P. M. Iannaccone and M. Khokha, CRC Press, Boca Raton (1995).
69. H. E. Patton, "Reflex Regulation of Movement and Posture," in *Physiolgy and Biophysics,* 19th ed., Eds. T. C. Ruch and H. D. Patton, Saunders, Philadelpia (1966).
70. T. C. Ruch, "Pontobulbar Control of Posture and Orientation in Space," in *Physiolgy and Biophysics,* 19th Edition, Eds. T. C. Ruch and H. D. Patton, Saunders, Philadelpia (1966).
71. R. Brown, *Phil Mag.* **6**, 161 (1829); *Edinburgh J. Sci.* **1**, 344 (1829).
72. J. Perrin, *Brownian Movement and Molecular Reality,* Taylor and Francis, London (1910).
73. J. Ingen Housz, in *Dictionary of Scientific Biology,* Ed. C. C. Gillispie, Scribners, New York (1973).
74. Th. Svedberg, *The Existence of the Molecule,* Leipzig (1912).
75. A. Einstein, "On the movement of small particles suspended in a stationary liquid demanded by the molecular-kinetic theory of heat," (in German) *Ann. d Phys.* **17**, 549 (1905).
76. H. C. Berg, *Random Walks in Biology,* Princeton University Press, Princeton (1983).
77. H. R. Catchpole, "The Capillaries, Veins and Lymphatics," Chapter 27, in *Physiology*

and Biophysics, 19th ed., Eds. T. C. Ruch and H. D. Patton, Saunders, Philadelpia (1966).

78. A. L. Goldberger and E. Goldberger, *Clinical Electrocardiography,* 4th ed., Mosby Year Book, St. Louis (1990).

79. A. M. Scher, "Mechanical Events of the Cardiac Cycle," in *Physiology and Biophysics,* Eds. T. C. Ruch and H. D. Patton, Saunders, Philadelphia (1960).

80. J. J. W. Baker and G. E. Allen, *Matter Energy and Life,* 4th ed., Addison-Wesley, Reading (1981).

81. G. H. Bourne, *How Your Body Works,* Sigma Books, London (1949).

82. F. Ashcroft, *Life at the Extremes, The Science of Survival,* University of California Press, Berkeley (2000).

83. A. C. Burton, "Biophysical Principles of the Circulation," in *Physiology and Biophysics,* Eds. T. C. Ruch and H. D. Patton, Saunders, Philadelphia (1960).

84. H. T. A. Whiting, "Variations in Floating Ability with Age in the Male," *Res. Quart.* **34,** 84–90 (1963); "Variations in Floating Ability with Age in the Female," *Res. Quart.* **36,** 216–218 (1965).

85. H. Rouse, *Elementary Mechanics of Fluids,* Dover Publications, New York (1978), first published in 1946 by General Publishing Company, Toronto.

86. S. Vogel, *Life in moving fluids,* Princeton University Press, Princeton, (1981).

87. E. W. Montroll, "On the dynamics and evolution of some sociotechnological systems," *Bull. Am. Math. Soc., 16,* 1–46 (1987).

88. McGraw-Hill *Encyclopedia of Physics,* 2nd ed., p. 1221, McGraw-Hill (1992).

89. R. M. Alexander and A. S. Jayes, "A Dynamic Similarity Hypothesis for the Gaits of Quadrepedal Mammals," *J. Zoology, London* **201,** 577–582 (1983).

90. Rayleigh, Lord, "On the irregular flight of a tennis-ball," *Messenger of Mathematics* **7,** 14–16 (1878).

91. T. von Karman, *Aerodynamics,* McGraw-Hill, New York (1954).

92. S. J. Hall, *Basic Biomechanics,* 3rd ed., McGraw-Hill (1999).

93. J. E. Counsilman, *The Science of Swimming,* Prentice-Hall, Englewood Cliffs, (1970).

94. L. E. Alley, "An analysis of water resistance and propulsion in swimming the crawl stroke," *Res. Quart.* **23,** 269 (1952).

95. E. P. Kruchoski, "A performance analysis of drag and propusion in swimming; Three selected forms of the back crawl stroke," Ph.D. dissertation, State University of Iowa (1954).

96. S. Vogel, "Exposing Life's Limits with Dimensionless Numbers," *Physics Today,* Nov., 22–27 (1998).

97. L. R. Taylor, "Aggregation, vaiance and mean," *Nature* **189,** 732 (1961).

98. M. Conrad, "What is the use of chaos?," in *Chaos,* Ed. A. V. Holden, Princeton University Press, Princeton.

99. G. U. Yule and M. G. Kendal, *Introduction to the Theory of Statistics,* Charles Griffin and Co., London, 14th ed. (1950).

100. A. L. Goldberger, "Nonlinear dynamics for clinicians: Chaos theory, fractals, and complexity at the bedside," *The Lancet* **347,** 1312–1314 (1996).

101. J. B. Bassingthwaighte and J. H. G. M. van Beek, "Lightning and the heart: Fractal behavior in cardiac function," *Proc. IEEE* **76,** 693–699 (1988).

102. M. C. Teich, S. B. Lowen, B. M. Jost, K. Vibe-Rheymer, and C. Heneghan, "Heart Rate Variability: Measures and Models," in *Nonlinear Biomedical Signal Processing, Vol. II, Dynamic Analysis and Modeling,* Ed. M. Akay, IEEE Press, New York (2001).

103. K. Pearson, "Contributions to Mathematical Theory of Evolution," *ESP,* **40,** 445 (1989).

104. E. Ott, *Chaos in Dynamical Systems,* Cambridge University Press, New York (1993).

105. M. F. Shlesinger, B. J. West, and J. Klafter, "From Bernoulli scaling to turbulent diffusion," *Physica Scripta,* **40,** 445 (1989).

106. W. Weaver, *Lady Luck, The Theory of Probability,* Dover, New York (1963).

107. G. M. Viswanathan, V. Afanasyev, S. V. Buldyrev, E. J. Murphy, P. A. Prince, and H. E. Stanley, "Levy Flight Search Patterns of Wandering Albatrosses," *Nature* **381,** 413 (1996).

108. B. J. West, R. Zhang, A. W. Sanders, S. Miniyar, J. H. Zuckerman, and B. D. Levine, "Fractal fluctuations in cardiac time series," *Physica A* **270,** (1999).

109. R. A. Eve, S. Horsfall, and M. E. Lee, Eds., *Chaos, Complexity and Sociology,* Sage, Thousand Oaks (1997).

110. S. Thurner, S. B. Lowen, M. C. Feurstein, C. Heneghan, H. G. Feichtinger, and M. C. Teich, "Analysis, Synthesis, and Estimation of Fractal-Rate Stochastic Point Processes," *Fractals* **5,** 565–595 (1997).

111. L. R. Taylor and R. A. Taylor, "Aggregation, migration and population mechanics," *Nature* **265,** 415 (1977).

112. J. M. Hausdorff, L. Zemany, C. K. Peng, and A. L. Goldberger, "Maturation of gait dynamics: Stride-to-stride variability and its temporal organization in children," *J. Appl. Physiol.* **86,** 1040–1047 (1999).

113. W. Weber and E. Weber, *Mechanik der menschlichen Gehwerkzenge,* Göttingen, Germany (1836).

114. P. R. Cavanagh, Editor, *Biomechanics of Distance Running,* Human Kinetics Books, Champaign, (1990).

115. J. M Hausdorff, P. L. Purdon, C. -K. Peng, Z. Ladin, J. Y. Wei, and A. L. Goldberger, "Is walking a random walk? Evidence for long-range correlations in stride interval of human gait," *J. Appl. Physiol.* **78,** 349–358 (1995).

116. J. M Hausdorff, C. -K. Peng, Z. Ladin, J. Y. Wei, and A. L. Goldberger, "Fractal dynamics of human gait: stability of long-range correlations," *J. Appl. Physiol.* **80,** 1448–1457 (1996).

117. J. M Hausdorff, S. L. Mitchell, R. Firtion, C.-K. Peng, M. E. Cudkowicz, J. Y. Wei, and A. L. Goldberger, "Altered fractal dynamics of gait: Reduced stride-interval correlations with aging and Huntington's disease," *J. Appl. Physiol.* **82** (1997).

118. B. J. West and L. Griffin, "Allometric control and human gait," *Fractals* **6,** 101–108 (1998).

119. B. J. West and L. Griffin, "Allometric control, inverse power laws, and human gait," *Chaos, Solitons and Fractals* **10,** 1519–1527 (1999).

120. L. Griffin, PhD thesis, unpublished (2000).

121. B. B. Mandelbrot and J. R. Wallis, "Robustness of the rescaled range R/S in the measurement of noncyclic long run statistical dependence," *Water Resour. Res.* **5,** 967 (1969).

122. J. H. G. M. van Beek, S. A. Roger and J. B. Bassingthwaighte, "Regional myocardial flow heterogeneity explained with fractal networks," *Am J. Physiol.* *257* (*Heart Circ. Physiol.* **26**): H1670 (1989).

123. B. J. West and A. L. Goldberger, "Physiology in Fractal Dimensions," *Am. Sci.* **75,** 354–365 (1987).

124. A. L. Goldberger and B. J. West, "Fractals: A contemporary mathematical concept with applications to physiology and medicine," *Yale J. Biol. Med.* **60,** 104–119 (1987).

125. A. L. Goldberger, D. R. Rigney, and B. J. West, "Chaos and fractals in human physiology," *Scientific American* **262,** 42–49 (1990).

126. B. J. West, "Branching Laws," in *Encyclopedia of Nonlinear Science,* Edi. A. Scott, Fitzroy, Dearborn (2003).

127. A. L. Goldberger, V. Bhargava, B. J. West, and A. J. Mandell, "On a mechanism of cardiac electrical stability; the fractal hypothesis," *Biophys. J.* **48,** 525–528 (1985).

128. M. Kobayashi and T. Musha, "1/f fluctuations of heartbeat period," *IEEE Trans. Biomed. Eng.* **29**, 456 (1982).

129. C. K. Peng, J. Mietus, J. M. Hausdorff, S. Havlin, H. E. Staley, and A. L. Goldberger, "Long-range anticorrelations and non-Gaussian behavior of the heartbeat," *Phys. Rev. Lett.* **70**, 1343–1346 (1993).

130. S. M. Pikkujämsä, T. H. Mkikallio, L. B. Sourander, I. J. Rih Puukka, J. Skytt, C. K. Peng, A. L. Goldberger, and H. V. Huikuri, "Cardiac interbeat interval dynamics from childhood to senescence," *Circ.* **100**, 393–399 (1999).

131. D. L. Eckberg, "Nonlinearities of the human carotid baroreceptor-cardiac reflex," *Circ. Res.* **47**, 208–211 (1980).

132. R. E. Ganz, G. Weibels, K. H. Stacker, P. M. Faustmann, and C. W. Zimmermann, "The Lyapunov exponent of heart rate dynamics as a sensitive marker of central autonomic organization: An exemplary study of early multiple sclerosis," *Int. J. Neurosci.* **71**, 29–36 (1993).

133. V. Kariniemi and P. Ammal, "Short term variability of fetal heart during pregnancies with normal and insufficient placental function," *Am. J. Obstet. Gynecol.* **139**, 33 (1981).

134. G. A. Meyers, G. J. Martin, N. M. Magid, P. S. Barnett, J. W. Schaad, J. S. Weiss, M. Lesch, and D. J. Singer, "Power spectral analysis of heart rate variability in sudden cardiac death: comparison to other methods," *IEEE Trans. Biomed. Eng.* **33**, 1149–1156 (1986).

135. G. C. Casolo, E. Balli, T. Taddei et al. "Decreased spontaneous heart rate variability in congestive heart failure," *Am J. Cardiol.* **64**, 1162–1167 (1989).

136. J. Hilderbrandt and A. C. Young, "Anatomy and physics of respiration," in *Physiolgy and Biophysics,* 19th ed., Eds. T. C. Ruch and H. D. Patton, Saunders, Philadelpia (1966).

137. F. Roher, "Flow resistance in human air passages and the effect of irregular branching of the bronchial system on the respiratory process in various regions of the lungs," *Pflugers Arch* **162**, 255–299, Repr. 1975: *Translations in Respiratory Physiology,* Ed. J. B. West, Stroudsburg, Dowden, Huchinson and Ross.

138. T. R. Nelson and D. K. Manchester, "Morphological modeling using fractal geometries," *IRRR Trans. Med. Imag.* **7**, 439 (1988).

139. B. J. West, "Physiology in fractal dimension: error tolerance," *Ann. Biomed. Eng.* **18**, 135–149 (1990).

140. L. A. Griffin, H. Mummery, R. E. Moon, and B. J. West, "Fractal Nature of Respiration and Heartbeat," submitted to *J. Appl. Physiol.*

141. H. H. Szeto, P. Y. Cheng, J. A. Decena, Y. Cheng, D. Wu, and G. Dwyer, "Fractal properties of fetal breathing dynamics," *Am. J. Physiol.* **262** (*Regulatory Integrative Comp. Physiol.* **32**) R141–R147 (1992).

142. G. S. Dawes, H. E. Gox, M. B. Leduc, E. C. Liggins, and R. T. Richards, "Respiratory movement and rapid eye movements in the fetal lamb," *J. Physiol. Land.* **220**, 119–143 (1972).

143. A. L. Goldberger and B. J. West, "Chaos in Physiology: Health or Disease?", in *Chaos in Biological Systems,* Eds. H. Degan, A. V. Holden, and L. F. Olsen, Plenum (1987).

Index

Biodynamics: Why the Wirewalker Doesn't Fall. By Bruce J. West and Lori A. Griffin
ISBN 0-471-34619-5 © 2004 John Wiley & Sons, Inc.